Understanding NEC® Calculations

Third Edition

Mike Holt

an International Thomson Publishing company 

Albany • Bonn • Boston • Cincinnati • Detroit • London • Madrid
Melbourne • Mexico City • New York • Pacific Grove • Paris • San Francisco
Singapore • Tokyo • Toronto • Washington

NOTICE TO THE READER

Publisher does not warrant or guarantee any of the products described herein or perform any independent analysis in connection with any of the product information contained herein. Publisher does not assume, and expressly disclaims, any obligation to obtain and include information other than that provided to it by the manufacturer.

The reader is expressly warned to consider and adopt all safety precautions that might be indicated by the activities herein and to avoid all potential hazards. By following the instructions contained herein, the reader willingly assumes all risks in connections with such instructions.

The publisher makes no representation or warranties of any kind, including but not limited to, the warranties of fitness for particular purpose or merchantability, nor are any such representations implied with respect to the material set forth herein, and the publisher takes no responsibility with respect to such material. The publisher shall not be liable for any special, consequential, or exemplary damages resulting, in whole or part, from the readers' use of, or reliance upon, this material.

Cover Design: Courtesy of Brucie Rosch

Delmar Staff
Publisher: Alar Elken
Acquisitions Editor: Mark Huth
Project Editor: Barbara Diaz
Production Coordinator: Toni Hansen
Art/Design Coordinator: Rachel Baker
Editorial Assistant: Dawn Daugherty
Production Manager: Mary Ellen Black
Marketing Manager: Mona Caron

Design Layout and Typesetting: Paul Wright

Graphic Illustrations: Mike Culbreath

For more information contact:

Delmar Publishers
3 Columbia Circle, Box 15015
Albany, New York 12212-5015

International Thomson Publishing Europe
Berkshire House
168 - 173 High Holborn
London WC1V 7AA
United Kingdom

Nelson ITP, Australia
102 Dodds Street
South Melbourne
Victoria, 3205 Australia

Nelson Canada
1120 Birchmount Road
Scarborough, Ontario
M1K 5G4, Canada

International Thomson Publishing France
Tour Maine-Montparnasse
33 Avenue du Maine
75755 Paris Cedex 15, France

International Thomson Editores
Seneca 53
Colonia Polanco
11560 Mexico D. F. Mexico

International Thomson Publishing GmbH
Königswinterer Strasβe 418
53227 Bonn
Germany

International Thomson Publishing Asia
60 Albert Street
#15 - 01 Albert Complex
Singapore 189969

International Thomson Publishing Japan
Hirakawa-cho Kyowa Building, 3F
2-2-1 Hirakawa-cho, Chiyoda-ku
Tokyo 102, Japan

ITE Spain/Paraninfo
Calle Magallanes, 25
28015-Madrid, Espana

1 2 3 4 5 6 7 8 9 10 XXX 03 02 01 00 99

Library of Congress Cataloging-in-Publication Data
Holt, Charles Michael.
Understanding NEC calculations / Charles Michael Holt, Sr. — 3rd ed.
p. cm.
ISBN 0-7668-1130-1 (alk. paper)
1. Electric engineering—United States—Insurance requirements.
2. Electric engineering—Mathematics. I. Title.
TK260.H65 1999
621.319'24'076—dc21 99-20861
CIP

I dedicate this book to God, who has given me the ability to write it.

Delmar Publishers Is Your Electrical Book Source!

Whether you're a beginning student or a master electrician, Delmar Publishers has the right book for you. Our complete selection of proven best-sellers and all-new titles is designed to bring you the most up-to-date, technically-accurate information available.

NATIONAL ELECTRICAL CODE

National Electrical Code® 1999/NFPA
Revised every three years, the *National Electric Code®* is the basis of all U.S. electrical codes.
Order # 0-8776-5432-8
Loose-leaf version in binder
Order # 0-8776-5433-6

National Electrical Code® Handbook 1999/NFPA
This essential resource pulls together all the extra facts, figures, and explanations you need to interpret the 1999 *NEC®*. It includes the entire text of the Code, plus expert commentary, real-world examples, diagrams, and illustrations that clarify requirements.
Order # 0-8776-5437-9

Illustrated Changes in the 1999 National Electrical Code®/O'Riley
This book provides an abundantly-illustrated and easy-to-understand analysis of the changes made to the 1999 *NEC®*.
Order # 0-7668-0763-0

Understanding the National Electrical Code®, 3E/Holt
This book gives users at every level the ability to understand what the *NEC®* requires, and simplifies this sometimes intimidating and confusing code.
Order # 0-7668-0350-3

Illustrated Guide to the National Electrical Code®/Miller
Highly-detailed illustrations offer insight into Code requirements, and are further enhanced through clearly-written, concise blocks of text that can be read very quickly and understood with ease. Organized by classes of occupancy.
Order # 0-7668-0529-8

Interpreting the National Electrical Code®, 5E/Surbrook
This updated resource provides a process for understanding and applying the *National Electrical Code®* to electrical contracting, plan development, and review.
Order # 0-7668-0187-X

Electrical Grounding, 5E/O'Riley
Electrical Grounding is a highly illustrated, systematic approach for understanding grounding principles and their application to the 1999 *NEC®*.
Order # 0-7668-0486-0

ELECTRICAL WIRING

Electrical Raceways and Other Wiring Methods, 3E/Loyd
The most authoritive resource on metallic and nonmetallic raceways, provides users with a concise, easy-to-understand guide to the specific design criteria and wiring methods and materials required by the 1999 *NEC*®.
Order # 0-7668-0266-3

Electrical Wiring Residential, 13E/Mullin
Now in full color! Users can learn all aspects of residential wiring and how to apply them to the wiring of a typical house from this, the most widely-used residential wiring book in the country.
Softcover Order # 0-8273-8607-9
Hardcover Order # 0-8273-8610-9

House Wiring with the NEC®/Mullin
The focus of this new book is the applications of the *NEC*® to house wiring.
Order # 0-8273-8350-9

Electrical Wiring Commercial, 10E/Mullin and Smith
Users can learn commercial wiring in accordance with the *NEC*® from this comprehensive guide to applying the newly revised 1999 *NEC*®.
Order # 0-7668-0179-9

Electrical Wiring Industrial, 10E/Smith and Herman
This practical resource has users work their way through an entire industrial building—wiring the branch-circuits, feeders, service entrances, and many of the electrical appliances and sub-systems found in commercial buildings.
Order # 0-7668-0193-4

Cables and Wiring, 2E/AVO
This concise, easy-to-use book is your single-source guide to electrical cables—it's a "must-have" reference for journeyman electricians, contractors, inspectors, and designers.
Order # 0-7668-0270-1

ELECTRICAL MACHINES AND CONTROLS

Industrial Motor Control, 4E/Herman and Alerich
This newly-revised and expanded book, now in full color, provides easy-to-follow instructions and essential information for controlling industrial motors. Also available are a new lab manual and an interactive CD-ROM.
Order # 0-8273-8640-0

Electric Motor Control, 6E/Alerich and Herman
Fully updated in this new sixth edition, this book has been a long-standing leader in the area of electric motor controls.
Order # 0-8273-8456-4

Introduction to Programmable Logic Controllers/Dunning
This book offers an introduction to Programmable Logic Controllers.
Order # 0-8273-7866-1

Technician's Guide to Programmable Controllers, 3E/Cox
Uses a plain, easy-to-understand approach and covers the basics of programmable controllers.
Order # 0-8273-6238-2

Programmable Controller Circuits/Bertrand
This book is a project manual designed to provide practical laboratory experience for one studying industrial controls.
Order # 0-8273-7066-0

Electronic Variable Speed Drives/Brumbach
Aimed squarely at maintenance and troubleshooting, *Electronic Variable Speed Drives* is the only book devoted exclusively to this topic.
Order # 0-8273-6937-9

Electrical Controls for Machines, 5E/Rexford
State-of-the-art process and machine control devices, circuits, and systems for all types of industries are explained in detail in this comprehensive resource.
Order # 0-8273-7644-8

Electrical Transformers and Rotating Machines/Herman
This new book is an excellent resource for electrical students and professionals in the electrical trade.
Order # 0-7668-0579-4

Delmar's Standard Guide to Transformers/Herman
Delmar's Standard Guide to Transformers was developed from the best-seller *Standard Textbook of Electricity* with expanded transformer coverage not found in any other book.
Order # 0-8273-7209-4

DATA AND VOICE COMMUNICATION CABLING AND FIBER OPTICS

Complete Guide to Fiber Optic Cable System Installation/Pearson
This book offers comprehensive, unbiased, state-of-the-art information and procedures for installing fiber optic cable systems.
Order # 0-8273-7318-X

Fiber Optics Technician's Manual/Hayes
Here's an indispensible tool for all technicians and electricians who need to learn about optimal fiber optic design and installation as well as the latest troubleshooting tips and techniques.
Order # 0-8273-7426-7

A Guide for Telecommunications Cable Splicing/Highhouse
A "how-to" guide for splicing all types of telecommunications cables.
Order # 0-8273-8066-6

Premises Cabling/Sterling
This reference is ideal for electricians, electrical contractors, and inspectors needing specific information on the principles of structured wiring systems.
Order # 0-8273-7244-2

ELECTRICAL THEORY

Delmar's Standard Textbook of Electricity, 2E/Herman
This exciting full-color book is the most comprehensive book on DC/AC circuits and machines for those learning the electrical trades.
Order # 0-8273-8550-1

Industrial Electricity, 6E/Nadon, Gelmine, and Brumbach
This revised, illustrated book offers broad coverage of the basics of electrical theory and industrial applications. It is perfect for those who wish to be industrial maintenance technicians.
Order # 0-7668-0101-2

EXAM PREPARATION

Electrician's Exam Preparation, Holt
This comprehensive exam prep guide includes all of the topics on both the master and journeyman electrician competency exams.
Order # 0-7668-0376-7

REFERENCE

ELECTRICAL REFERENCE SERIES

This series of technical reference books is written by experts and designed to provide the electrician, electrical contractor, industrial maintenance technician, and other electrical workers with a source of reference information about virtually all of the electrical topics that they encounter.

Electrician's Technical Reference—Motor Controls/Carpenter
Electrician's Technical Reference—Motor Controls is a source of comprehensive information on understanding the controls that start, stop, and regulate the speed of motors.
Order # 0-8273-8514-5

Electrician's Technical Reference—Motors/Carpenter
Electrician's Technical Reference—Motors builds an understanding of the operation, theory, and applications of motors.
Order # 0-8273-8513-7

Electrician's Technical Reference—Theory and Calculations/Herman
Electrician's Technical Reference—Theory and Calculations provides detailed examples of problem-solving for different kinds of DC and AC circuits.
Order # 0-8273-7885-8

Electrician's Technical Reference—Transformers/Herman
Electrician's Technical Reference—Transformers focuses on the theoretical and practical aspects of single-phase and 3-phase transformers and transformer connections.
Order # 0-8273-8496-3

Electrician's Technical Reference—Hazardous Locations/Loyd
Electrician's Technical Reference—Hazardous Locations cover electrical wiring methods and basic electrical design considerations for hazardous locations.
Order # 0-8273-8380-0

Electrician's Technical Reference—Wiring Methods/Loyd
Electrician's Technical Reference—Wiring Methods covers electrical wiring methods and basic electrical design considerations for all locations, and shows how to provide efficient, safe, and economical applications of various types of available wiring methods.
Order # 0-8273-8379-7

Electrician's Technical Reference—Industrial Electronics/Herman
Electrician's Technical Reference—Industrial Electronics covers components most used in heavy industry, such as silicon control rectifiers, triacs, and more. It also includes examples of common rectifiers and phase-shifting circuits.
Order # 0-7668-0347-3

RELATED TITLES

Common Sense Conduit Bending and Cable Tray Techniques/Simpson
Now geared especially for students, this manual remains the only complete treatment of the topic in the electrical field.
Order # 0-8273-7110-1

Practical Problems in Mathematics for Electricians, 5E/Herman
This book details the mathematics principles needed by electricians.
Order # 0-8273-6708-2

Electrical Estimating/Holt
This book provides a comprehensive look at how to estimate electrical wiring for residential and commercial buildings with extensive discussion of manual versus computer-assisted estimating.
Order # 0-8273-8100-X

Electrical Studies for Trades/Herman
Based on *Delmar's Standard Textbook of Electricity,* this new book provides non-electrical trades students with the basic information they need to understand electrical systems.
Order # 0-8273-7845-9

Preface

INTRODUCTION

The writing style of this book is informal and relaxed, and the book contains clear graphics and examples that apply to the electrical exam.

To get the most out of this book, you should answer the two hundred questions at the end of each unit. The eight hundred questions contained in this book are typical calculation questions from electrician exams across the country. If you have difficulty with a question, skip it and go back to it later. You will find that the Instructor's Guide contains detailed explanations for each question.

HOW TO USE THIS BOOK

Each unit of this book contains objectives, explanations with graphics, examples, steps for calculations, formulas, and practice calculation questions. As you read this book, review the author's comments, graphics, and examples with your 1999 Code book.

Note: This book contains many cross references to other related Code rules. Please take the time to review the cross references.

As you progress through this book, you will find some formulas, rules, or comments that you don't understand. Don't get frustrated. Highlight the section in the book that you are having a problem with. Discuss it with your boss, inspector, co-worker etc.; maybe they'll have some additional feedback. If necessary just skip the difficult points. Once you have completed the book, review those highlighted sections again and see if you understand.

Some words are italicized to bring them to your attention. Be sure that you understand the terms before you continue with each unit.

DIFFERENT INTERPRETATIONS

Electricians, contractors, some inspectors, and others love arguing Code interpretations and discussing Code requirements. As a matter of fact, discussing the NEC® and its application is a great way to increase your knowledge of the Code and its intended use. The best way to discuss Code requirements with others is by referring to a specific Section in the Code, rather than by talking in vague generalities.

I have taken great care in researching the Code rules in this book, but I'm not perfect. If you feel that I have made an error, please let me know by contacting me at Delmar Publishers by mail addressed to Electrical Editor, Delmar Publishers, 3 Columbia Circle, Box 15015, Albany, NY 12212. You can also send e-mail to info@delmar.com.

ABOUT THE AUTHOR

I have worked my way up through the electrical trade from an apprentice electrician to a master electrician and electrical inspector. I did not complete high school due to circumstances and dropped out after completing 11th grade. Realizing that success depends on one's education, I immediately attained my GED, and ten years later I attended the University of Miami's Graduate School for a Master's in Business Administration (MBA).

I am nationally recognized as one of America's most knowledgeable electrical trainers and have touched the lives of thousands of electricians, inspectors, contractors, and engineers.

I reside in Central Florida, am the father of seven children, and have many outside interests and activities. I am a National Barefoot Waterskiing Champion, have set five barefoot waterski records and am currently ranked No. 2. I enjoy white water rafting, racquetball, playing my guitar, and spending time with my family. My commitment to God has helped me develop a lifestyle that balances family, career, and self.

ACKNOWLEDGMENTS

I would like to say thank you to all the people in my life who believed in me, even to those who didn't. There are many people who played a role in the development and production of this book. I will start with Mike Culbreath (Master Electrician), who helped me transform my words and visions into lifelike graphics. I could not have produced such a fine book without his help.

Next, Paul Wright of Digital Design Group, for the electronic production and typesetting. A very special thanks goes to Brooke Stauffer from NECA, Ray Cotter from North Tech Education Center, Elzy R. Williams, P.E., Adjunct Professor at John Tyler Community College, and Craig H. Matthews, P.E., from Austin Brockenbrough and Associates, L.L.P, for their incredible job of proofing this textbook for technical correctness.

To my beautiful wife, Linda, and my seven children, Belynda, Melissa, Autumn, Steven, Michael, Meghan, and Brittney, thank you for loving me.

Also thanks to all those who helped me in the electrical industry, *Electrical Construction* and *Maintenance* magazines for my first "big break". To Joe McPartland, my "mentor," who was there to help and encourage me. Joe, I'll never forget to help others as you helped me.

In addition, I would like to thank Joe Salimando, the former publisher of *Electrical Contractor* and Dan Walters of the National Electrical Contractors Association (NECA) for my second "big break", and for putting up with all of my crazy ideas.

I would also like to thank James Stallcup, Dick Lloyd, Mark Ode, D.J. Clements, Joe Ross, John Calloggero, Tony Selvestri, and Marvin Weiss for being special people in my life.

The final personal thank you goes to Sarina, my long-time friend and office manager. Thank you for covering the office for me while I spend so much time writing books, doing seminars, and producing videos and software. Your love and concern for me has helped me through many difficult times.

Delmar Publishers as well as the author would also like to thank those individuals who reviewed the manuscript and offered invaluable suggestions and feedback. Their assistance is greatly appreciated.

Robert Blakely
Mississippi Gulf Coast Community College

Ray Cotter, Electrical Instructor
North Tech Education Center Palm Beach, Florida

John P. Cox
Brevard Community College

Tim DeDonder
Kaw Area Vocational-Technical School

William Elarton
Los Angeles Trade and Technical College

David Gehlauf
Tri-County Vocational Technical

Larry Killebrew
Mid-America Vocational-Technical School

Craig H. Matthews, P.E.
Austin Brockenbrough and Associates, L.L.

John Mills, Master Electrician, Instructor
Dade County, Florida

John Penley
Albert Lea Technical College

Larry C. Phillips
Minneapolis Technical College

Charles Plimpton
New Hampshire Technical College

Rodney Stanley
Morehead State University

Brooke Stauffer, NECA
National Electrical Contractors Association

Kurt A. Stout, Electrical Inspector
Plantation, Florida

Elzy R. Williams, P.E.
Adjunct Professor at John Tyler Community College

GETTING STARTED

THE EMOTIONAL ASPECT OF LEARNING

To learn effectively, you must develop an attitude that learning is a process that will help you grow both personally and professionally. The learning process has an emotional as well as an intellectual component that we must recognize. To understand what affects our learning, consider the following:

Positive Image. Many feel disturbed by the expectation of being treated like children, and we often feel threatened by the learning experience.

Uniqueness. Each of us will understand the subject matter from different perspectives, and we all have some unique learning problems and needs.

Resistance to Change. People tend to resist change and information that appear to threaten their comfort level of knowledge. However, we often support new ideas that support our existing beliefs.

Dependence and Independence. The dependent person is afraid of disapproval and often will not participate in class discussion and will tend to wrestle alone. The independent person spends too much time asserting differences and too little time trying to understand others' views.

Fear. Most of us feel insecure and afraid of learning until we understand the process. We fear that our performance will not match the standard set by us or by others.

Egocentricity. Our ego tendency is to prove someone is wrong, feeling a victorious surge of pride. Learning together without a win/lose attitude can be an exhilarating experience.

Emotion. It is difficult to discard our cherished ideas in the face of contrary facts when overpowered by the logic of others.

HOW TO GET THE BEST GRADE ON YOUR EXAM

Studies have concluded that for students to get their best grades, they must learn to get the most from their natural abilities. It's not how long you study or how high your IQ is: it's what you do and how you study that count the most. To get your best grade, you must make a decision to do your best and follow as many of the following techniques as possible.

Reality. These instructions are a basic guide to help you get the maximum grade. It is unreasonable to think that all of the instructions can be followed to the letter all of the time. Day-to-day events and unexpected situations must be taken into consideration.

Support. You need encouragement in your studies, and support from your loved ones and employer. To properly prepare for your exam, you need to study ten to fifteen hours per week for about three to six months.

Communication with Your Family. Good communication with your family members is very important. Studying every night and on weekends may cause tension. Try to get their support, cooperation, and encouragement during this trying time. Let them know the benefits. Be sure to plan some special time with them during this preparation period; don't go overboard and leave them alone too long.

Stress. Stress can really take the wind out of you. It takes practice, but get into the habit of relaxing before you begin your studies. Stretch; do a few sit-ups and push-ups; take a 20-minute walk or a few slow, deep breaths. Close your eyes for a couple of minutes; deliberately relax the muscle groups that are associated with tension, such as the shoulders, back, neck, and jaw.

Attitude. Maintaining a positive attitude is important. It helps keep you going and helps keep you from getting discouraged.

Training. Preparing for the exam is the same as training for any event. Get plenty of rest and avoid intoxicating drugs, including alcohol. Stretch or exercise each day for at least 10 minutes. Eat light meals such as pasta, chicken, fish, vegetables, fruit, etc. Try to avoid heavy foods, such as red meats, butter, and other high-fat foods. They slow you down and make you tired and sleepy.

Eye Care. It is very important to have your eyes checked! Human beings were not designed to do constant seeing less than arm's length away. Our eyes were designed for survival and for spotting food and enemies at a distance. Your eyes will be under tremendous stress because of prolonged, near-vision reading, which can result in headaches, fatigue, nausea, squinting, or eyes that burn, ache, water, or tire easily.

Be sure to tell the eye doctor that you are studying to pass an exam (bring this book and the Code Book) and that you expect to do a tremendous amount of reading and writing. Reading glasses can reduce eye discomfort.

Reducing Eye Strain. Be sure to look up occasionally from near tasks to focus on distant objects. Your work area should be three times brighter than the rest of the room. Don't read under a single lamp in a dark room. Try to eliminate glare. Mixing fluorescent and incandescent lighting can be helpful.

Sit up straight with your chest up and shoulders back so that both eyes are an equal distance from what you are viewing.

Getting Organized. Our lives are so busy that simply making time for homework is almost impossible. You can't waste time looking for a pencil or missing paper. Keep everything you need together. Maintain folders, one for notes, one for exams and answer keys, and one for miscellaneous items.

It is very important that you have a private study area available at all times. Keep your materials there. The dining room table is not a good spot.

Time Management. Time management and planning are very important. There simply are not enough hours in the day to get everything done. Make a schedule that allows time for work, rest, study, meals, family, and recreation. Establish a schedule that is consistent from day to day.

Have a calendar and immediately plan your exam preparation schedule. Try to follow the same routine each week, and try not to become overtired. Learn to pace yourself to accomplish as much as you can without the need for cramming.

Learn How to Read. Review the book's contents and graphics. This will help you develop a sense of the material.

Clean Up Your Act. Keep all of your papers neat, clean, and organized. Now is not the time to be sloppy. If you are not neat, now is an excellent time to begin.

Speak Up in Class. If you are in a classroom setting, the most important part of the learning process is class participation. If you don't understand the instructor's point, ask for clarification. Don't try to get attention by asking questions you already know the answer to.

Study with a Friend. Studying with a friend can make learning more enjoyable. You can push and encourage each other. You are more likely to study if someone else is depending on you.

Students who study together perform above average because they try different approaches and explain their solutions to each other. Those who study alone spend most of their time reading and rereading the text and trying the same approach time after time even though it is unsuccessful.

Study Anywhere/Anytime. To make the most of your limited time, always keep a copy of the book(s) with you. Any time you get a minute free, study. Continue to study any chance you get. You can study at the supply house while waiting for your material; you can study during your coffee break, or even while you are at the doctor's office. Become creative!

You need to find your best study time. For some it could be late at night when the house is quiet. For others, it's the first thing in the morning before things get going.

Set Priorities. Once you begin your study, stop all phone calls, TV shows, radio, snacks, and other interruptions. You can always take care of it later.

HOW TO TAKE THE EXAM

Being prepared for an exam means more than just knowing electrical concepts, the Code, and the calculations. Have you felt prepared for an exam, then choked when actually taking it? Many good and knowledgeable electricians couldn't pass their exam because they did not know how to take an exam.

Taking exams is a learned process that takes practice and involves strategies. The following suggestions are designed to help you learn these methods.

Relax. This is easier said than done, but it is one of the most important factors in passing your exam. Stress and tension cause us to choke or forget. Everyone has had experiences where they got tense and couldn't think straight. The first step is to become aware of the tension, and the second step is to make a deliberate effort to relax. Make sure you're comfortable; remove clothes if you are hot, or put on a jacket if you are cold.

There are many ways to relax, and you have to find a method that works for you. Two of the easiest methods that work very well for many people follow:

Breathing Technique: Take a few slow, deep breaths every few minutes. Do not confuse this with hyperventilation, which is abnormally fast breathing.

Single-Muscle Relaxation: When we are tense or stressful, many of us do things like clench our jaws, squint our eyes, or tense our shoulders without even being aware of it. If you find a muscle group that does this, deliberately relax that one group. The rest of the muscles will automatically relax. Try to repeat this every few minutes, and it will help you stay more relaxed during the exam.

Have the Proper Supplies. First of all, make sure you have everything you need several days before the exam. The night before the exam is not the time to be out buying pencils, calculators, and batteries. The night before the exam, you should have a checklist (prepared in advance) of everything you could possibly need.

The following is a sample checklist to get you started:

- Six sharpened #2H pencils or two mechanical pens with extra #2H leads. The type with the larger lead is better for filling in the answer key.
- Two calculators. Most examining boards require quiet, paperless calculators. Solar calculators are great, but there may not be enough light to operate them.
- Spare batteries. Take two sets of extra batteries. It's very unlikely you'll need them, but it is possible.
- Extra glasses, if you use them.
- A watch for timing questions.
- All your reference materials, even the ones not on the list. Let the proctors tell you which ones are not permitted.
- A Thermos of something you like to drink. Coffee is excellent.
- Some fruit, nuts, candy, aspirin, analgesic, etc.
- Know where the exam is going to take place and how long it takes to get there. Arrive at least 30 minutes early.

Note. It is also a good idea to pack a lunch rather than go out. It can give you a little time to review the material for the afternoon portion of the exam, and it reduces the chance of coming back late.

Understand the Question. To answer a question correctly, you must first understand the question. One word in a question can totally change the meaning of it. Carefully read every word of every question. Underlining key words in the question will help you focus.

Skip the Difficult Questions. Contrary to popular belief, you do not have to answer one question before going on to the next one. The irony is that the question you get stuck on is one that you'll probably get wrong. This will result in your not having enough time to answer the easy questions. You will get all stressed-out, and a chain reaction will start. More people fail their exams for this reason than for any other.

The following strategy should be used to avoid getting into this situation.

- **First Pass:** Answer the questions you know. Give yourself about thirty seconds for each question. If you can't find the answer in your reference book within the thirty seconds, go on to the next question. Chances are that you'll come across the answers while looking up another question. The total time for the first pass should be 25 percent of the exam time.
- **Second Pass:** This pass is done the same way as the first pass except that you allow a little more time for each question, about sixty seconds. If you still can't find the answer, go on to the next one. Don't get stuck. Total time for the second pass should be about 30 percent of the exam time.
- **Third Pass:** See how much time is left and subtract thirty minutes. Spend the remaining time equally on each question. If you still haven't answered the question, it's time to make an educated guess. Never leave a question unanswered.
- **Fourth pass:** Use the last thirty minutes of the exam to transfer your answers from the exam booklet to the answer key. Read each question and verify that you selected the correct answer on the test book. Transfer the answers carefully to the answer key. With the remaining time, see if you can find the answer to those questions you guessed at.

Guessing. When time is running out and you still have questions remaining, GUESS! Never leave a question unanswered.

You can improve your chances of getting a question correct by the process of elimination. When one of the choices is "none of these," or "none of the above," it is usually not the correct answer. This improves your chances from one-out-of-four (25 percent), to one-out-of-three (33 percent). Guess "All of these" or "All of the Above," and don't select the high or low number.

How do you pick one of the remaining answers? Some people toss a coin, others will count up how many of the answers were As, Bs, Cs, and Ds and use the one with the most tallied as the basis for their guess.

CHECKING YOUR WORK

The first thing to check (and you should be watching out for this during the whole exam) is to make sure you mark the answer in the correct spot. People have failed the exam by $^1/_2$ of a point. When they reviewed their exam, they found they correctly answered several questions on the test booklet but marked the wrong spot on the exam answer sheet. They knew the answer was "(b) False" but marked "(d)" in error.

Another thing to be very careful of is marking the answer for, let's say, question 7, in the spot reserved for question 8.

CHANGING ANSWERS

When re-reading the question and checking the answers during the fourth pass, resist the urge to change an answer. In most cases, your first choice is best. If you aren't sure, stick with the first choice. Only change answers if you are sure you made a mistake. Multiple choice exams are graded electronically, so be sure to thoroughly erase any answer that you changed. Also erase any stray pencil marks from the answer sheet.

ROUNDING OFF

You should always round your answers to the same number of places as the exam's answers. Numbers below "5" are rounded down, while numbers "5" and above are rounded up.

Example: If an exam has multiple choice of:

(a) 2.2 (b) 2.1
(c) 2.3 (d) none of these

And your calculation comes out to 2.16, do not choose the answer "(d) none of these." The correct answer is (a) 2.2, because the answers in this case are rounded off to the tenth.

Example: It could be rounded to tens, such as:

(a) 50 (b) 60
(c) 70 (d) none of these.

For this group, an answer such as 67 would be "(c) 70," while an answer of 63 would be "(b) 60." The general rule is to check the question's choice of answers, then round off your answer to match it.

SUMMARY

- Make sure everything is ready and packed the night before the exam.
- Don't try to cram the night before the exam if you don't know the material by then.
- Have a good breakfast. Get the Thermos and energy snacks ready.
- Take all of your reference books. Let the proctors tell you what you can't use.
- Know where the exam is to be held, and be there early.
- Bring ID and your confirmation papers, if there are any, from the license board.
- Review the NEC® while you wait for your exam to begin.
- Try to stay relaxed.
- Determine the time per question for each pass, and don't forget to save thirty minutes for transferring your answers to the answer key.
- Remember, in the first pass, answer only the easy questions. In the second pass, spend a little more time per question, but don't get stuck. In the third pass, use the remainder of the time minus thirty minutes. In the

fourth pass, check your work and transfer the answers to the answer key.

THINGS TO BE CAREFUL OF

- Don't get stuck on any one question.
- Read each question carefully.
- Be sure you mark the answer in the correct spot on the answer sheet.
- Don't get flustered or extremely tense.

EXAMINATION COPIES

To request examination copies of this book, call or write to:

Delmar Publishers
3 Columbia Circle
P.O. Box 15015
Albany, NY 12212-5015
Phone: 1-800-347-7707 • 1-518-464-3500 •
Fax: 1-518-464-0301

CHAPTER 1
Electrical Theory Calculations

Scope of Chapter 1

Unit 1

Electrician's Math and Basic Electrical Formulas

OBJECTIVES

After reading this unit, the student should be able to briefly explain the following concepts:

Part A – Electrician's Math
- Fractions
- Kilo
- Knowing your answer
- Multiplier
- Parentheses
- Percent increase
- Percentage R
- Reciprocals
- Square root
- Squaring
- Transposing formulas

Part B – Basic Electrical Formulas
- Conductance and resistance
- Electric meters
- Electrical circuit
- Electrical circuit values
- Ohm's law
- PIE circle formula
- Power changes with the square of the voltage
- Power source
- Power wheel

After reading this unit, the student should be able to briefly explain the following terms:

Part A – Electrician's Math
- Fractions
- Kilo
- Multiplier
- Parentheses
- Percentage
- Ratio
- Reciprocals
- Rounding off
- Square root
- Squaring a number
- Transposing

Part B – Basic Electrical Formulas
- A – Ampere
- Alternating current
- Ammeter
- Ampere
- Armature
- Clamp-on ammeters
- Conductance
- Conductors
- Current
- Direct current
- Directly proportional
- E – Electromotive Force
- Electric meters
- Electromagnetic
- Electromagnetic field
- Electron pressure
- Helically wound
- Intensity
- Inversely proportional
- Megohmeter
- Ohmmeter
- Ohms
- P – Power
- Perpendicular
- Polarity
- Polarized
- Power
- Power source
- Resistance
- Shunt bar
- Shunt meter
- Solenoid
- V – Voltage
- Voltmeter
- W – Watt

PART A – ELECTRICIAN'S MATH

1–1 FRACTIONS

Fractions represent parts of numbers. To change a fraction to a decimal form, divide the numerator (top number of the fraction) by the denominator (bottom number of the fraction).

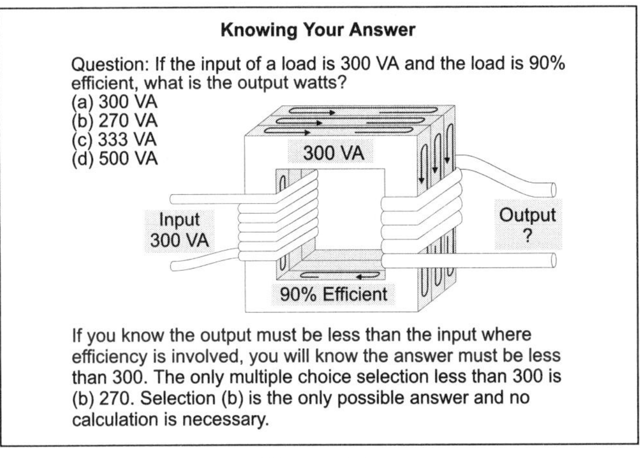

Figure 1–1
Knowing Your Answer

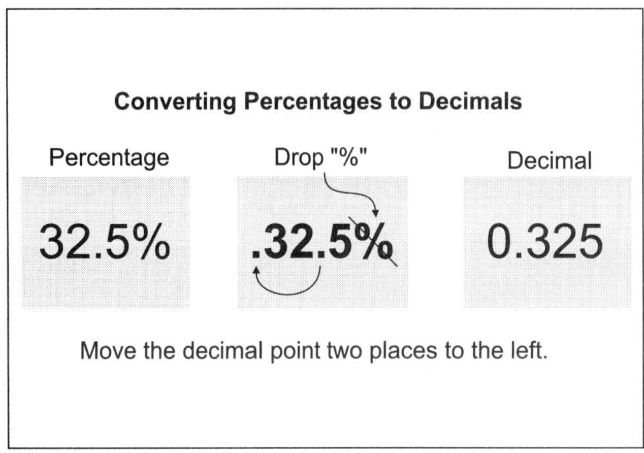

Figure 1–2
Converting Percentages to Decimals

❑ **Fractions to Decimal**

Convert the following fractions to a decimal.

$^1/_6$ = one divided by six = 0.167

$^5/_4$ = five divided by four = 1.25

$^7/_2$ = seven divided by two = 3.5

1–2 KILO

The letter k is the abbreviation of *kilo,* which represents 1,000 in the metric system of measurement.

❑ **Kilo**

What is the wattage for an 8kW rated range?

(a) 8 watts (b) 8,000 watts (c) 4,000 watts (d) none of these

• Answer: (b) 8,000 watts

Wattage = kW × 1,000. In this case 8 kW × 1,000 = 8,000 watts.

❑ **Kilo**

What is the kVA rating of a 300 VA load?

(a) 300 kVA (b) 3,000 kVA (c) 30 kVA (d) 0.3 kVA

• Answer: (d) 0.3 kVA, kVA = VA divided by 1,000.

In this case 300 VA/1,000 = 0.3 kVA

Note: The use of k is not limited to kW or kVA; it is also used to express wire size, such as 250 kcmil.

1–3 KNOWING YOUR ANSWER

When working with mathematical calculations, you should know if the answer is greater than or less than the values given.

❑ **Knowing Your Answer**

If the input of a load is 300 watts and the load is 90 percent efficient, what is the output watts?

Note. Because efficiency is always less than 100 percent, the output is always less than the input (Figure 1–1).

(a) 300 VA (b) 270 VA (c) 333 VA (d) 500 VA

• Answer: (b) 270 VA

Since the use of efficiency in the question implied that the output had to be less than the input, the answer must be less than 300 watts. The only choice that is less than 300 watts is (b) 270 VA.

1–4 MULTIPLIER

Often a number is required to be increased or decreased by a percentage. When a percentage or fraction is used as a multiplier, follow these steps:

Step 1: ➛ Convert the multiplier to a decimal form, then

Step 2: ➛ Multiply the number by the decimal value from Step 1.

❑ **Increase by 125 Percent**

An overcurrent protection device (breaker or fuse) must be sized no less than 125 percent of the continuous load. If the load is 80 ampere, the overcurrent protection device would have to be sized no less than _____ ampere.

(a) 80 ampere (b) 100 ampere (c) 125 ampere (d) none of these

• Answer: (b) 100 ampere

Step 1: ➛ Convert 125% to a decimal: 1.25

Step 2: ➛ Multiply the load rating by the multiplier: 80 ampere × 1.25 = 100 ampere

❑ **Limit to 80 Percent**

The maximum continuous load on an overcurrent protection device is limited to 80 percent of the device rating. If the device is rated 50 ampere, what is the maximum continuous load?

(a) 80 ampere (b) 125 ampere (c) 50 ampere (d) 40 ampere

• Answer: (d) 40 ampere

Step 1: ➛ Convert 80% to a decimal: 0.8

Step 2: ➛ Multiply the overcurrent protection device rating by the multiplier: 50 ampere × 0.8 = 40 ampere

1–5 PARENTHESES

Whenever numbers are in *(parentheses),* we must complete the mathematical function within the parentheses before proceeding with the rest of the problem.

❑ **Parenthesis**

What is the voltage drop of two No. 14 conductors carrying 16 ampere for a distance of 100 feet? Use the following example for the answer:

$$VD = \frac{2 \times K \times I \times D}{CM}, = \frac{(2 \text{ wires} \times 12.9 \text{ ohm} \times 16 \text{ ampere} \times 100 \text{ feet})}{4{,}110 \text{ circular mils}}$$

(a) 3 volts (b) 3.6 volts (c) 10.04 volts (d) none of these

• Answer: (c) 10.04 volts

Step 1: ➛ Calculate the value of the parentheses first: (2 wires × 12. 9 ohm × 16 ampere × 100 feet) = 41,280.

Step 2: ➛ Divide the top number by the bottom number: 41,280/4,110 = 10.4 volts dropped.

1–6 PERCENTAGES

A *percentage (%)* is a ratio of two numbers. When changing a percent to a decimal or whole number, simply move the decimal point two places to the left (Figure 1–2).

❑ **Percentage**

32.5% = 0.325 100% = 1.00 125% = 1.25 300% = 3.00

1–7 PERCENT INCREASE

Increasing a number by a specific percentage is accomplished by:

Step 1: ➛ Converting the percentage to a decimal.

Step 2: ➛ Determining the multiplier: Add one to the decimal value from Step 1.

Step 3: ➛ Multiply the number by the multiplier from Step 2.

❑ **Percent Increase**

Increase the whole number 45 by 35%.

(a) 61   (b) 74   (c) 83   (d) 104

• Answer: (a) 61

Step 1: ➛ Convert 35% to 0.35.

Step 2: ➛ Multiplier: Add one to the decimal value from Step 1: 1 + 0.35 = 1.35.

Step 3: ➛ Multiply the number 45 by the multiplier: 45 × 1.35 = 60.75.

Step 4: ➛ Round 60.75 up to 61.

1–8 PERCENTAGE RECIPROCALS

A reciprocal is a whole number converted into a fraction with the number *one* as the numerator (top number). This fraction is then converted to a decimal.

Step 1: ➛ Convert the number to a decimal.

Step 2: ➛ Divide the number into one.

❑ **Reciprocal**

What is the reciprocal of 80 percent?

(a) 0.80 percent   (b) 100 percent   (c) 125 percent   (d) none of these

• Answer: (c) 1.25 or 125 percent

Step 1: ➛ Convert 80% to a decimal: 80% = 0.8.

Step 2: ➛ Divide 0.80 into one: $^1/_{0.80}$ = 1.25, which is the same as 125%.

❑ **Reciprocal**

A continuous load requires an overcurrent protection device sized no smaller than 125 percent of the load [*Section 210–20(a)*]. What is the maximum continuous load permitted on a 100-ampere overcurrent protection device?

(a) 100 ampere   (b) 125 ampere   (c) 80 ampere   (d) none of these

• Answer: (c) 80 ampere

Step 1: ➛ Convert 125% to a decimal: 125% = 1.25

Step 2: ➛ Divide 1.25 into one: $^1/_{1.25}$ = 0.8 or 80%.

If the overcurrent device is sized no less than 125 percent of the load, the load is limited to 80 percent of the overcurrent protection device rating (reciprocal). Therefore, the maximum load is limited to:
100 ampere × 0.8 = ampere.

1–9 ROUNDING

Numbers below 5 are rounded down, while numbers 5 and above are rounded up.

Note: Rounding to three significant figures should be sufficient for most calculations, such as:

0.1245 = 0.125
1.674 = 1.67
21.94 = 21.9
367.28 = 367

Rounding for Exams

You should always round your answer to the same magnitude as the answers. Do not choose "none of these" in an exam until you have checked all the answers. If after rounding your answer to the exam format there is no answer, then you should choose "none of these."

❑ **Rounding**

The sum of 12, 17, 28, and 40 is equal to _____?

(a) 80 (b) 90 (c) 100 (d) none of these

• Answer: (c) 100

The answer is actually 97, but there is no 97 as a choice. Do not choose "none of these" in an exam until you have checked how the choices are rounded off. The choices in this case are all rounded off to the nearest ten.

1–10 SQUARING

Squaring a number is multiplying a number by itself, such as: $23^2 = 23 \times 23 = 529$.

❑ **Squaring**

What is the power consumed, in watts, of a No. 12 conductor that is 200 feet long and has a resistance of 0.4 ohm? The current flowing in the circuit is 16 ampere. Formula: $I^2 \times R$

(a) 50 watts (b) 150 watts (c) 100 watts (d) 200 watts

• Answer: (c) 100 watts

Step 1: ➵ $P = I^2 \times R$, I = 16 ampere, R = 0.4 ohm.

Step 2: ➵ P = 16 ampere $^2 \times$ 0.4 ohm.

Step 3: ➵ P = 102.4 watts; answers are rounded to 50s.

❑ **Squaring**

What is the area, in square inches, of a 1-inch trade size raceway whose actual internal diameter is 1.049 inches? Formula: Area = $\pi \times r^2$, $\pi = 3.14$, r = radius ($1/2$ the diameter).

(a) 1 square inch (b) 0.86 square inch (c) 0.34 square inch (d) 0.5 square inch

• Answer: (b) 0.86 square inch

Step 1: ➵ Raceway area = $\pi \times r^2$, = $3.14 \times (1/2 \times 1.049)^2$, = 3.14×0.5245^2.

Step 2: ➵ Raceway area = $3.14 \times (0.5245 \times 0.5245)$, = 3.14×0.2751 = 0.86 square inch.

1–11 SQUARE ROOT

The *square root* of a number is the opposite of squaring a number. For all practical purposes you must use a calculator with a square root key to determine the square root of a number. For an electrician's exam, the only square root number you need to know is the square root of $\sqrt{3}$ is 1.732. To multiply, divide, add, or subtract a number by a square root value, determine the square root value first, then perform the math function.

Step 1: ➵ Enter the number in the calculator.

Step 2: ➵ Press the $\sqrt{\_}$ key of the calculator.

❑ **Square Root Sample**

What is the $\sqrt{3}$?

(a) 1.55 (b) 1.73 (c) 1.96 (d) none of these

• Answer: (b) 1.73

Step 1: ➵ Enter the number 3 in a calculator.

Step 2: ➵ Press the $\sqrt{\_}$ key = 1.732.

❑ **Square Root**

36,000 watts/(208 volts $\times \sqrt{3}$) is equal to _____ ampere?

(a) 120 (b) 208 (c) 360 (d) 100

• Answer: (d) 100

Step 1: ➵ Determine the $\sqrt{3}$ = 1.732.

Step 2: ➵ Multiply 208 volts $\times$ 1.732 = 360 volts.

Step 3: ➵ 36,000 watts/360 volts = 100 ampere.

❑ **Square Root**

The phase voltage is equal to $\frac{208 \text{ volts}}{\sqrt{3}}$ _____?

(a) 120 volts (b) 208 volts (c) 360 volts (d) none of these

• Answer: (a) 120 volts

Step 1: ➵ Determine the $\sqrt{3} = 1.732$.

Step 2: ➵ Divide 208 volts by 1.732 = 120 volts.

1–12 TRANSPOSING FORMULAS

Transposing is an algebraic function used to rearrange formulas. Example: the formula $I = P/E$ can be transposed to $E = P/I$ or $P = E \times I$ (Figure 1–3):

❑ **Transpose**

Transpose the formula $CM = \frac{(2 \times K \times I \times D)}{VD}$

to find the voltage drop of the circuit (Figure 1–4).

(a) VD = CM

(b) VD = (2 × K × I × D)

(c) VD = (2 × K × I D)/CM

(d) none of these

• Answer: (c)

$VD = \frac{(2 \times K \times I \times D)}{CM}$

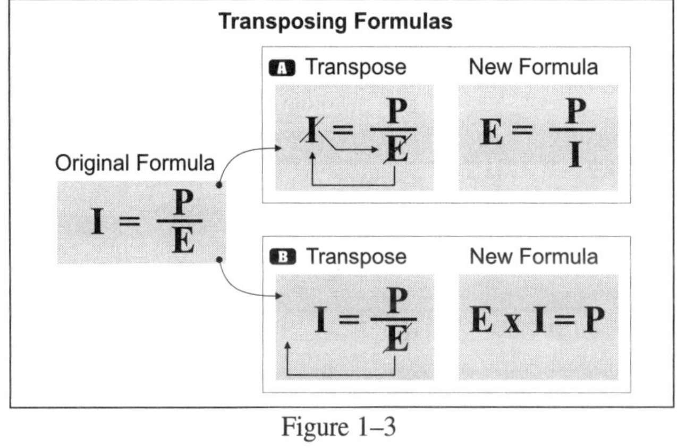

Figure 1–3
Transposing Formulas

Note: Most people have no idea of how to transpose formulas. I only add this to the book to refresh the memory of those who do understand. If you don't understand, it's okay; don't worry about it.

PART B – BASIC ELECTRICAL FORMULAS

1–13 ELECTRICAL CIRCUIT

An electric circuit consists of power source, conductors, and load. For current to travel in the circuit, there must be a complete path from one terminal of the power supply, through the conductors and the load and back to the other terminal of the power supply (Figure 1–5).

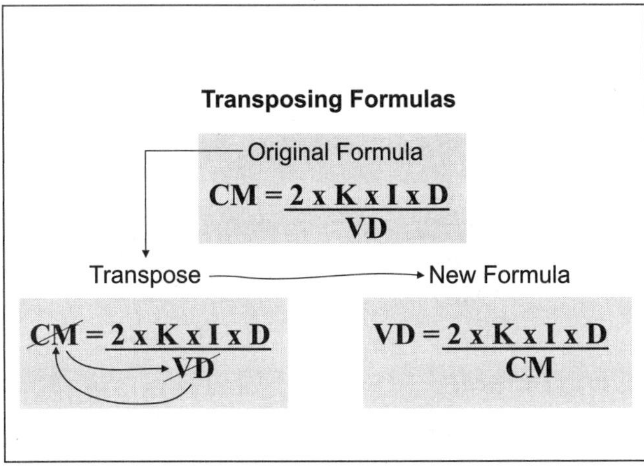

Figure 1–4
Transposing Formulas

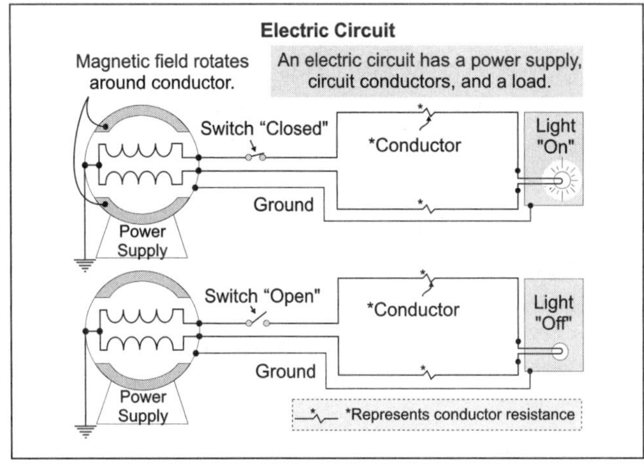

Figure 1–5
Electric Circuit

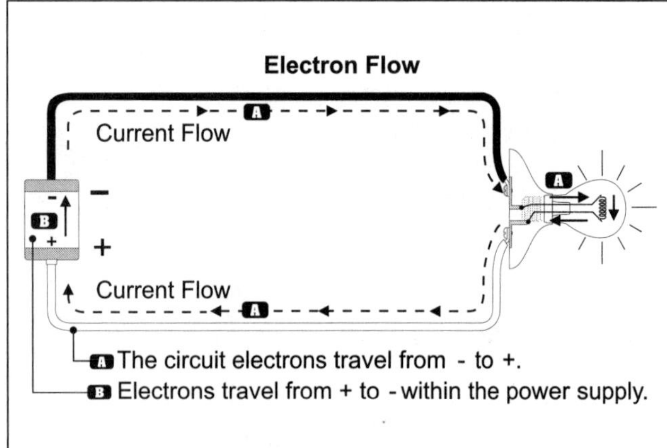
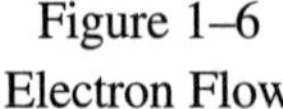

Figure 1–6
Electron Flow

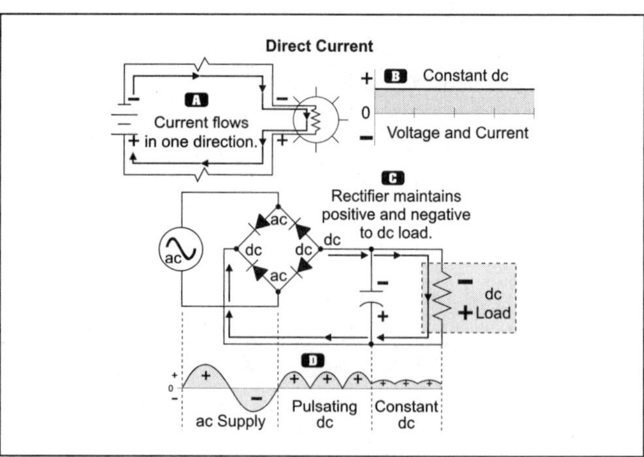

Figure 1–7
Direct Current

1–14 ELECTRON FLOW

Inside a direct current power source (such as a battery) the electrons travel from the positive terminal to the negative terminal; however, outside of the power source, electrons travel from the negative terminal to the positive terminal (Figure 1–6).

1–15 POWER SOURCE

In any completed circuit, it takes a force to push the electrons through the power source, conductor, and load. The two types of electric current are *direct current* and *alternating current.*

Direct Current

The polarity from direct current power sources never changes. That is, the current always flows out of the negative terminal of the power source in the same direction. When the power supply is a *battery,* the polarity and the voltage magnitude remain the same (Figure 1–7).

Alternating Current

Alternating current power sources produce a voltage and current that has a constant change in polarity and magnitude at a constant frequency. Alternating current flow is produced by a *generator* or an *alternator* (Figure 1–8).

1–16 CONDUCTANCE AND RESISTANCE

Conductance

Conductance is the property of metal that permits current to flow. The best conductors, in order of their conductivity, are: silver, copper, gold, and aluminum. Although silver is a better conductor of electricity than copper, copper is used most widely because it is less expensive (Figure 1–9, Part A).

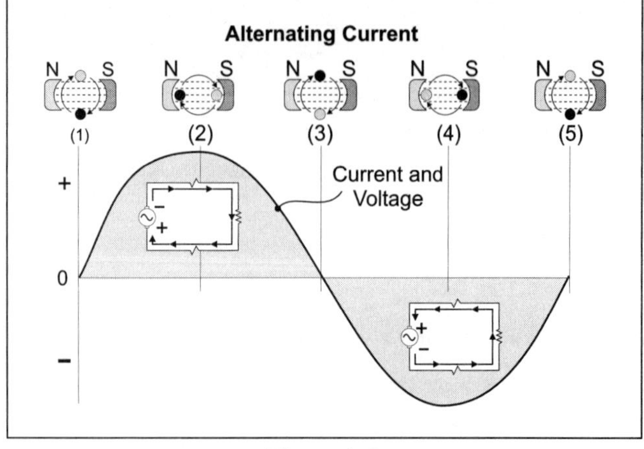

Figure 1–8
Alternating Current

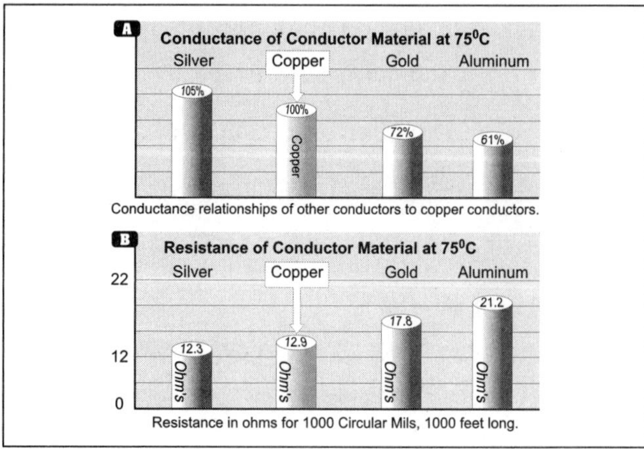

Figure 1–9
Conductance and Resistance

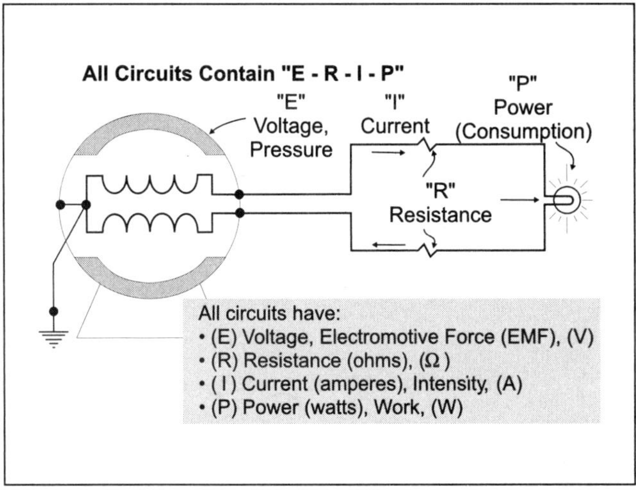

Figure 1–10
Electrical Circuit Values

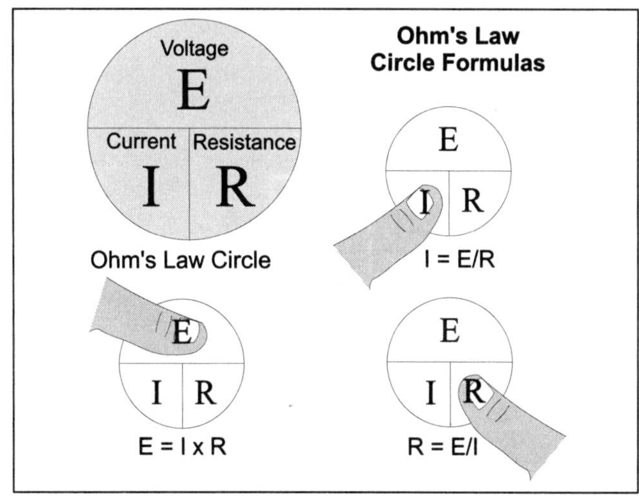

Figure 1–11
Ohm's Law Circle Formulas

Resistance

Resistance is the opposite of conductance. It is the property that opposes the flow of electric current. The resistance of a conductor is measured in ohms according to a standard length of 1,000 feet (Figure 1–9, Part B). This value is listed in the *National Electrical Code®, Chapter 9, Table 8* for direct current circuits and in *Chapter 9, Table 9* for alternating current circuits.

1–17 ELECTRICAL CIRCUIT VALUES

In an electrical circuit there are four circuit values that must be understood. They are *voltage, resistance, current,* and *power* (Figure 1–10).

Voltage

Electron pressure is called *electromotive force (E* or *EMF),* measured by the unit *volt (V),* and is abbreviated by the letters *E, EMF* or *V.* Voltage is also a term used to described the difference of potential between any two points.

Resistance

The friction opposition to the flow of electrons is called *resistance (R),* and the unit of measurement is the *ohm.* Every component of an electric circuit contains resistance, including the power supply.

Current

Free electrons moving in the same direction in a conductor produce an electrical *current* sometimes called *intensity (I).* The rate at which electrons move is measured by the unit called *ampere (A).*

Power

The rate of work that can be produced by the movement of electrons is called *power (P),* and the unit is the *watt (W).*

Note: A 100 watt lamp consumes 100 watts of power per hour.

1–18 OHM's LAW, I = E/R

Ohm's Law, $I = {}^{E}/_{R}$, demonstrates the relationship between current, voltage, and resistance in a direct current, or in an alternating current circuit that supplies only resistive loads (Figure 1–11). Other derived formulas include:

I = E/R **E = I × R** **R = E/I**

Ohm's Law states that:

Current is directly proportional to voltage. If the voltage is increased by a given percentage, current increases by that same percentage. If the voltage is decreased by a given percentage, current decreases by the same percentage (Figure 1–12, Part A).

Current is inversely proportional to resistance. An increase in resistance results in a decrease in current. A decrease in resistance results in an increase in current (Figure 1–12, Part B).

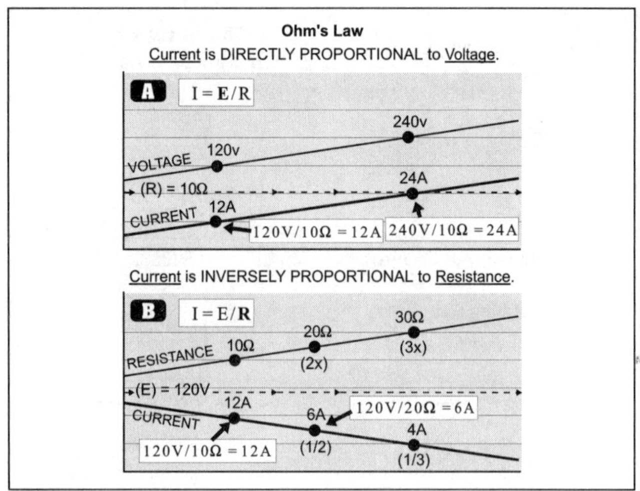

Figure 1–12
Part A – Current Proportional to Voltage
Part B – Current Inversely Proportional to Resistance

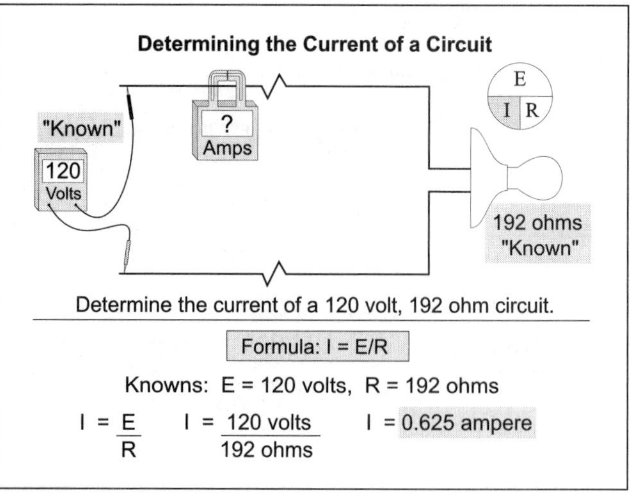

Figure 1–13
Determining the Current of a Circuit

Opposition to Current Flow

In a direct current (dc) circuit, the physical resistance of the conductor opposes the flow of electrons. In an alternating current (ac) circuit, three factors oppose current flow. They are *conductor resistance, inductive reactance*, and *capacitive reactance.* The opposition to current flow, due to a combination of resistance and reactance, is called *impedance,* measured in ohms, and abbreviated with the letter Z. Impedance will be covered later, so for now assume that all circuits have very little or no reactance.

❑ Ampere

A 120 volt power source supplies a lamp with a resistance of 192 ohm. What is the current flow of the circuit (Figure 1–13)?

(a) 0.6 ampere (b) 0.5 ampere (c) 2.5 ampere (d) 1.3 ampere

• Answer: (a) 0.6 ampere

Step 1: ➛ *What is the question?* What is the current (I)?

Step 2: ➛ *What do you know?*
E = 120 volts, R = 192 ohm.

Step 3: ➛ The formula is: I = E/R

Step 4: ➛ I = 120 volts/192 ohm = 0.625 ampere.

❑ Voltage

What is the voltage drop of two No. 12 conductors that supply a 16 ampere load located 50 feet from the power supply? The total resistance of both conductors is 0.2 ohm (Figure 1–14).

(a) 16 volts (b) 32 volts
(c) 1.6 volts (d) 3.2 volts

• Answer: (d) 3.2 volts

Step 1: ➛ *What is the question?* What is voltage (E) drop?

Step 2: ➛ *What do you know about the conductors?*
I = 16 ampere, R = 0.2 ohm.

Step 3: ➛ The formula is: E = I × R.

Step 4: ➛ E = 16 ampere × 0.2 ohm = 3.2 volts.

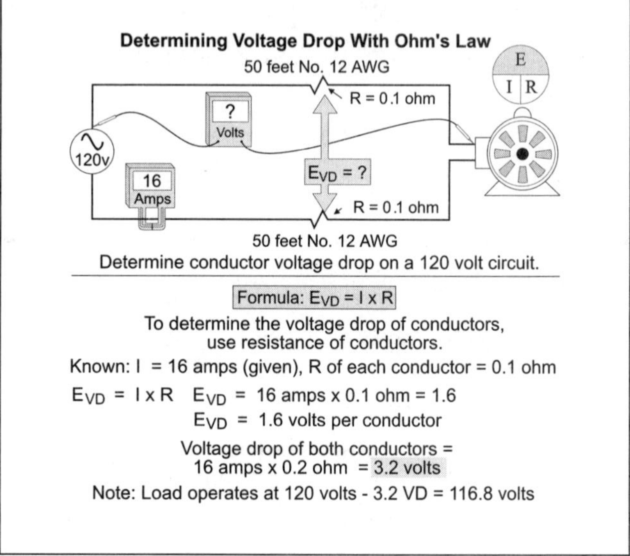

Figure 1–14
Voltage Example

❑ **Resistance**

What is the resistance of the circuit conductors when the conductor voltage drop is 3 volts and the current flowing through the conductors is 100 ampere (Figure 1–15)?

(a) 0.03 ohm (b) 0.2 ohm (c) 3 ohm (d) 30 ohm

• Answer: (a) 0.03 ohm

Step 1: ➵ *What is the question?* What is the resistance (R)?

Step 2: ➵ *What do you know about the conductors?* E = 3 volts dropped, I = 100 ampere.

Step 3: ➵ The formula is: R = E/I.

Step 4: ➵ R = 3 volts/100 ampere, = 0.03 ohm.

1–19 PIE CIRCLE FORMULA

The PIE circle formula shows the relationships between power, current, and voltage (Figure 1–16).

P = E × I **I = P/E** **E = P/I**

❑ **Power**

What is the power loss, in watts, for two conductors that carry 12 ampere and have a voltage drop of 3.6 volts (Figure 1–17)?

(a) 4.3 watts (b) 43 watts (c) 432 watts (d) none of these

• Answer: (b) 43 watts

Step 1: ➵ *What is the question?* It is: What is the power (P)?

Step 2: ➵ *What do you know?* E = 3.6 volts dropped, I = 12 ampere.

Step 3: ➵ The formula is: P = E × I.

Step 4: ➵ The answer is: P = 3.6 volts × 12 ampere = 43.2 watts per hour.

❑ **Current**

What is the current flow, in ampere, in the circuit conductors that supply a 7.5 kW heat strip rated 240 volts when connected to a 240 volt power supply (Figure 1–18)?

(a) 25 ampere (b) 31 ampere (c) 39 ampere (d) none of these

• Answer: (b) 31 ampere

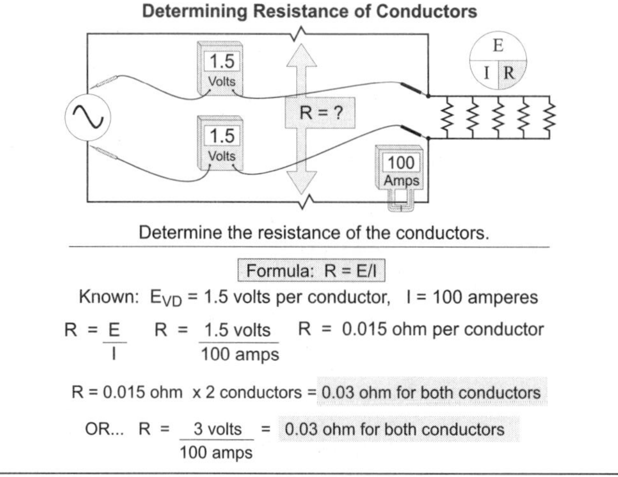

Figure 1–15
Resistance Example

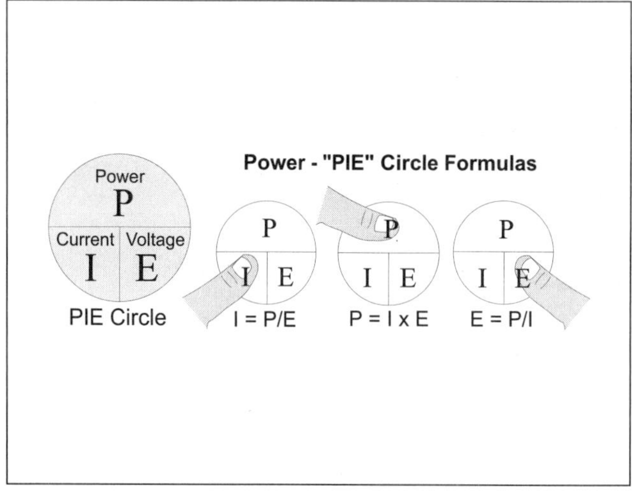

Figure 1–16
Pie Circle Formulas

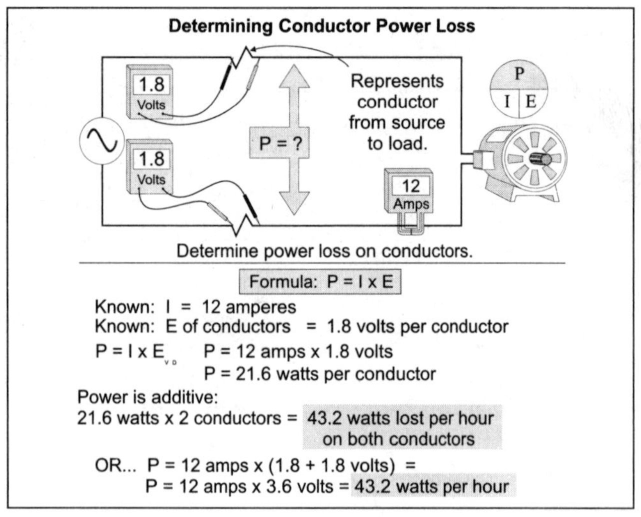

Figure 1–17
Determining Conductor Power Loss

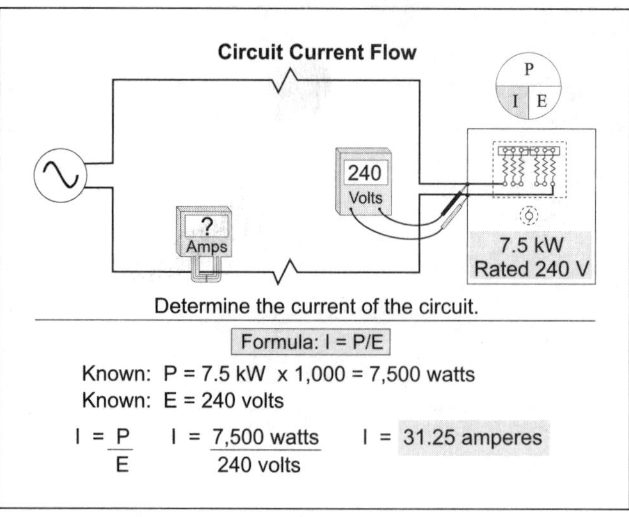

Figure 1–18
Circuit Current Flow

Step 1: ➸ *What is the question? What is the current* (I)?

Step 2: ➸ *What do you know?* P = 7,500 watts, E = 240 volts

Step 3: ➸ The formula is: I = P/E.

Step 4: ➸ I = 7,500 watts/240 volts = 31.25 ampere

1–20 FORMULA WHEEL

The formula wheel combines Ohm's Law and PIE formulas. The formula wheel is divided up into four sections with three formulas in each section (Figure 1–19).

❑ **Resistance**

What is the resistance of a 75-watt light bulb rated 120 volts (Figure 1–20)?

(a) 100 ohm (b) 192 ohm (c) 225 ohm (d) 417 ohm

• Answer: (b) 192 ohm

$R = E^2/P$ (Formula 1 of 12), E = 120 volt rating, P = 75 watt rating, R = 120 volts^2/75 watts = 192 ohm

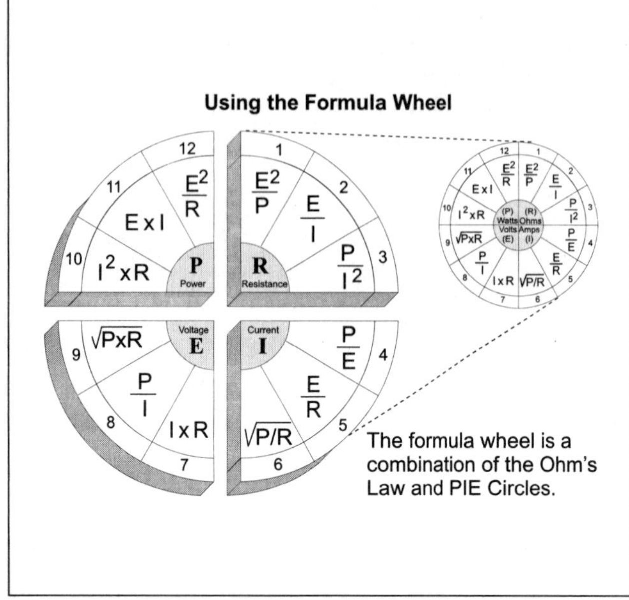

Figure 1–19
Using the Formula Wheel

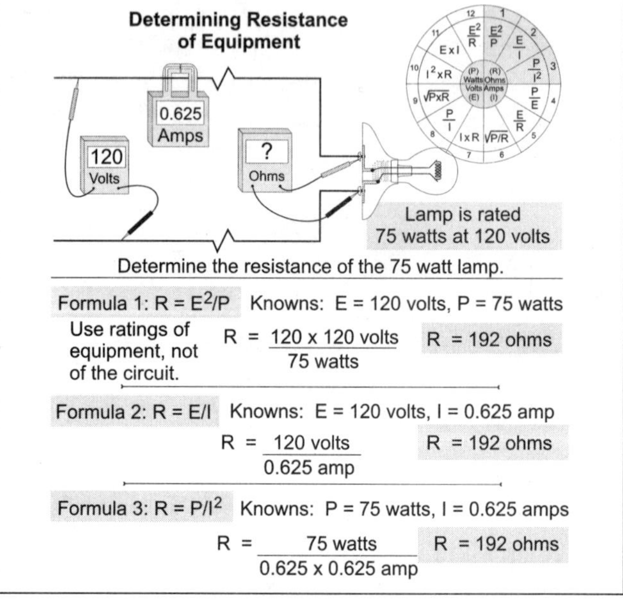

Figure 1–20
Determining Resistance of Equipment

❑ Current

What is the current flow of a 10 kW heat strip connected to a 230 volt (single-phase) power supply (Figure 1–21)?

(a) 13 ampere (b) 26 ampere
(c) 43 ampere (d) 52 ampere

• Answer: (c) 43 ampere

I = P/E (Formula 4 of 12)

P = 10,000 watts, E = 230 volts

I = 10,000 watts/230 volts = 43 ampere

Note: Always assume single-phase, unless three-phase is specified in the question.

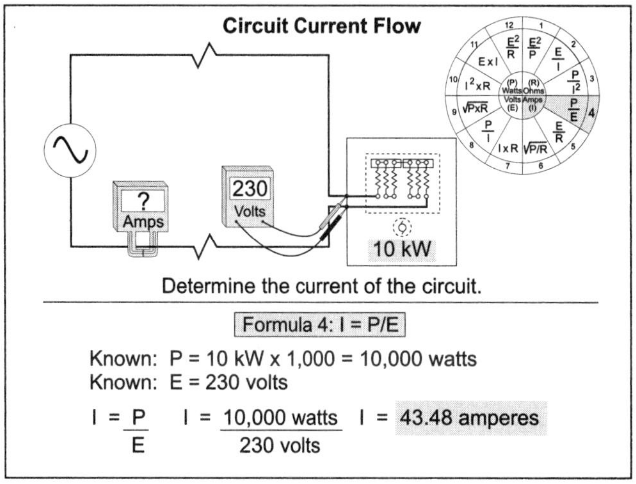

Circuit Current Flow

? Amps
230 Volts
10 kW

Determine the current of the circuit.

Formula 4: I = P/E

Known: P = 10 kW x 1,000 = 10,000 watts
Known: E = 230 volts

$I = \frac{P}{E}$ $I = \frac{10{,}000 \text{ watts}}{230 \text{ volts}}$ I = 43.48 amperes

Figure 1–21
Circuit Current Flow

❑ Voltage

What is the voltage drop of 200 feet of No. 12 conductor that carries 16 ampere (Figure 1–22)? The resistance of No. 12 copper conductor is 2 ohm per 1,000 feet.

(a) 1.6 volts dropped (b) 2.9 volts dropped (c) 3.2 volts dropped (d) 6.4 volts dropped

• Answer: (d) 6.4 volts dropped

E = I × R (Formula 7 of 12)

I = 16 ampere, R = 2 ohm/1,000 = 0.002 ohm per foot × 200 feet = 0.4 ohm

E = 16 ampere × 0.4 ohm = 6.4 volts dropped

❑ Power

The total resistance of two No. 12 copper conductors, 75 feet long, is 0.3 ohm (0.15 ohm for each conductor). The current of the circuit is 16 ampere. What is the power loss of the conductors in watts per hour (Figure 1–23)?

(a) 19 watts (b) 77 watts (c) 172.8 watts (d) none of these

• Answer: (b) 77 watts

$P = I^2R$ (Formula 10 of 12)

I = 16 ampere, R = 0.3 ohm

P = 16 ampere^2 × 0.3 ohm = 76.8 watts per hour

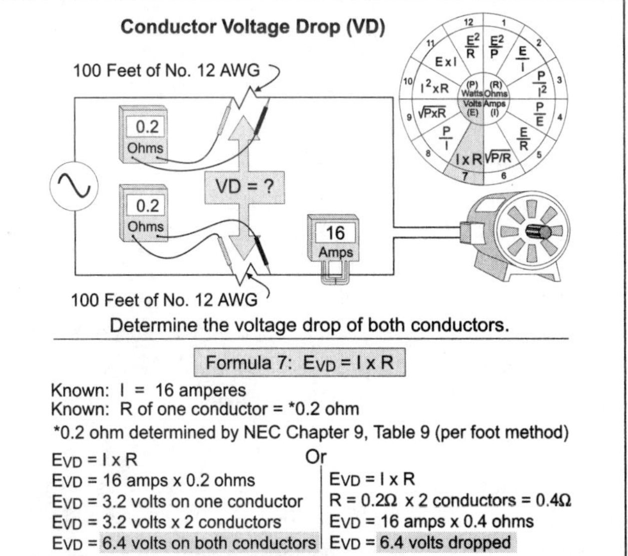

Figure 1–22
Conductor Voltage Drop (VD)

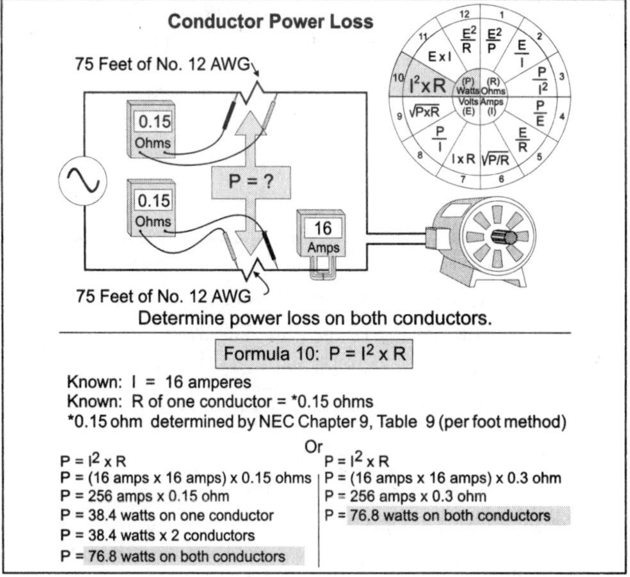

Figure 1–23
Conductor Power Loss

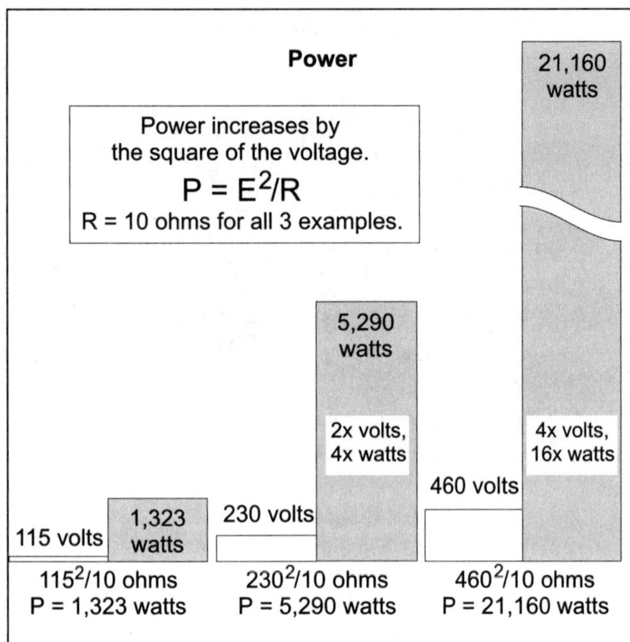

Figure 1–24
Power Changes with the Square of the Voltage

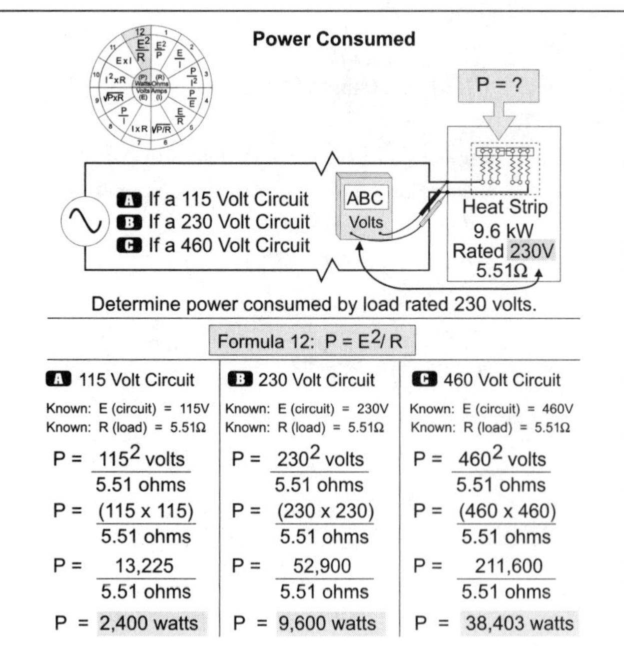

Figure 1–25
Power Changes with the Square of the Voltage

1–21 POWER CHANGES WITH THE SQUARE OF THE VOLTAGE

The power consumed by a resistor is affected by the voltage applied. Power is proportional to the square of the voltage and directly proportional to the resistance (Figure 1–24).

$P = E^2/R$

❑ Power Changes with the Square of the Voltage

What is the power consumed of a 9.6 kW heat strip rated 230 volts connected to 115-, 230-, and 460-volt power supplies? Note: The resistance of the heat strip is 5.51 ohm (Figure 1–25).

Step 1: ➼ *What is the question?* What is the power (P) consumed?

Step 2: ➼ *What do you know about the heat strip?*
E = 115 volts, 230 volts and 460 volts, R = 5.51 ohm

Step 3: ➼ The formula to determine power is $P = E^2/R$.
P at 115 volts = 115 volts²/5.51 ohm, P = 2,400 watts
P at 230 volts = 230 volts²/5.51 ohm, P = 9,600 watts (2 times volts = 4 times power)
P at 460 volts = 460 volts²/5.51 ohm, P = 38,403 watts (4 times volts = 16 times power)

1–22 ELECTRIC METERS

Basic electrical meters use a helically wound coil of conductor called a *solenoid* to produce a strong electromagnetic field to attract an iron bar inside the coil. The iron bar that moves inside the coil is called an *armature.* When a meter has positive (+) and negative (–) shown for the meter leads, the meter is said to be *polarized.* The negative (–) lead must be connected to the negative terminal of the power source, and the positive (+) lead must be connected to the positive terminal.

Ammeter

An ammeter is a meter that has a *helically* (spirally) wound coil, and it uses the circuit energy to measure direct current. As current flows through the meter's coil, the coil's electromagnetic field draws in the iron bar (armature). The greater the current flow through the meter's coil, the greater the electromagnetic field and the further the armature is drawn into the coil. Ammeters are connected in series with the circuit and are used to measure only direct current (Figure 1–26).

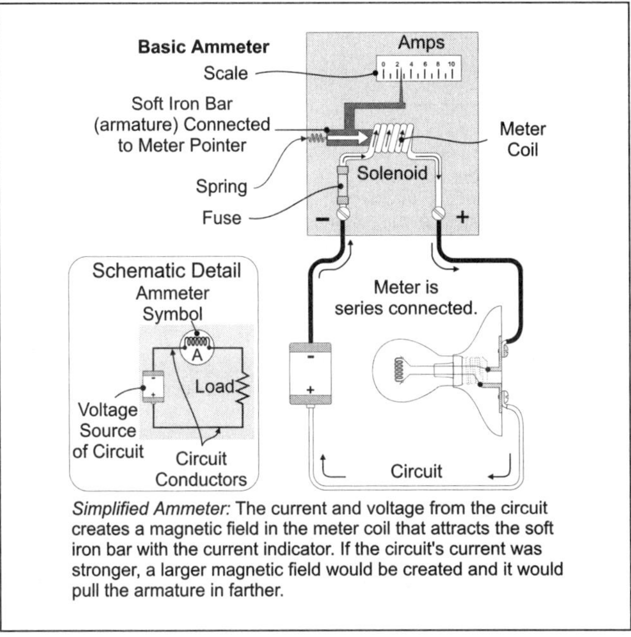

Simplified Ammeter: The current and voltage from the circuit creates a magnetic field in the meter coil that attracts the soft iron bar with the current indicator. If the circuit's current was stronger, a larger magnetic field would be created and it would pull the armature in farther.

Figure 1–26
Basic Ammeter

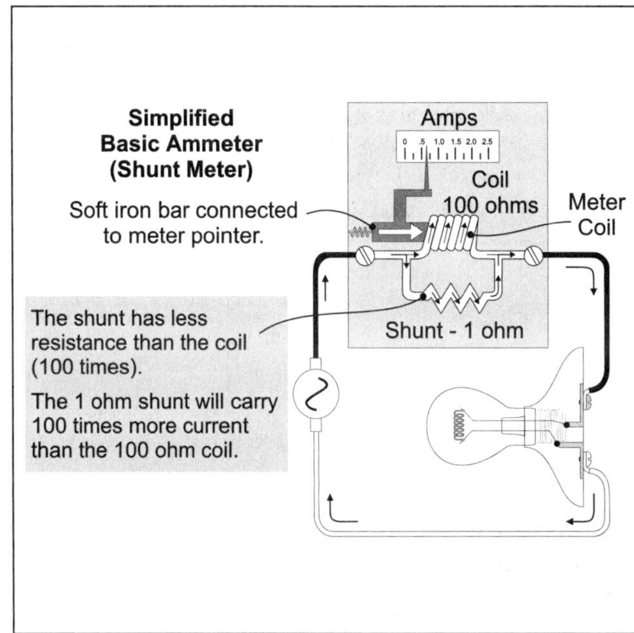

Figure 1–27
Simplified Basic Ammeter (Shunt Meter)

Ammeters are connected in *series* with the power supply and the load. If the ammeter is accidentally connected in *parallel* to the power supply, the current flow through the meter will be extremely high. The excessive high current through the meter will destroy the meter due to excessive heat.

If the ammeter is not connected to the proper *polarity* when measuring direct current, the meter's needle will quickly move in the reverse direction and possibly damage the meter's calibration.

Ammeters that measure currents larger than 10 milliampere often contain a device called a *shunt* which is placed in parallel with the meter coil. This permits the current flow to divide between the meter's coil and the shunt bar. The current through the meter coil depends on the resistance of the shunt bar.

❑ **Shunt Ammeter Current**

What is the current flow through the meter if the shunt bar is 1 ohm and the coil is 100 ohm (Figure 1–27)?

(a) the same as the shunt
(b) 1/10 the shunt amperage
(c) 1/100 the shunt amperage
(d) 1/1,000 the shunt amperage

• Answer: (c) 1/100 the shunt amperage

Since the shunt is 100 times less resistant than the meter's coil, the shunt bar will carry 100 times more current than the meter's coil, or the meter's coil will carry 1/100 the shunt amperage.

Clamp-on Ammeter

Clamp-on ammeters are used to measure alternating current. They are connected *perpendicularly* (at a 90° angle) around the conductor without breaking the circuit. A clamp-on ammeter indirectly utilizes the circuit energy by induction of the electromagnetic field. The clamp-on ammeter is actually a transformer (sometimes called a *current transformer*). The primary winding is the phase conductor, and the secondary winding is the meter's coil (Figure 1–28).

The electromagnetic field around the phase conductor expands and collapses, which causes electrons to flow in the meter's circuit. As current flows through the meter's coil, the electromagnetic field of the meter draws in the armature. Since the phase conductor serves as the primary (one turn), the current to be measured must be high enough to produce an electromagnetic field that is strong enough to cause the meter to operate.

Ohmmeter and Megohmeter

Ohmmeters are used to measure the resistance of a circuit or component and can be used to locate open circuits or shorts. An ohmmeter has an armature, a coil, and its own power supply, generally a battery. Ohmmeters are always connected to de-

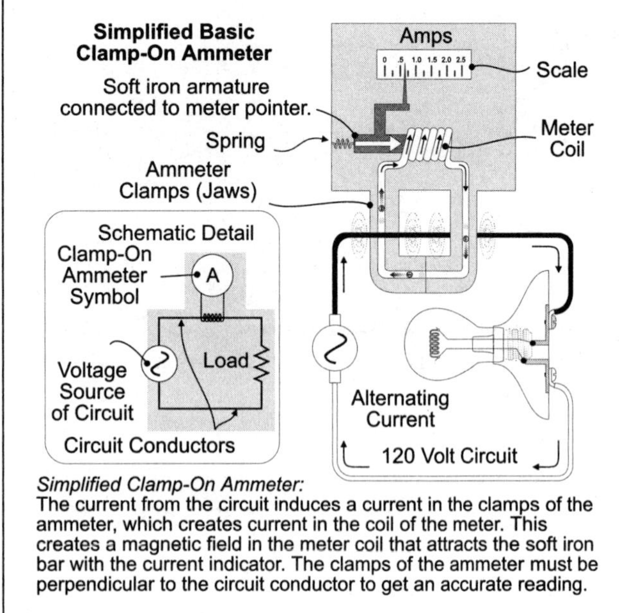

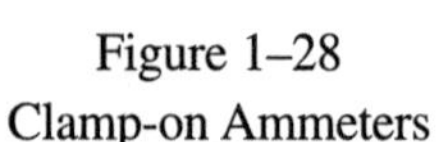
Figure 1–28
Clamp-on Ammeters

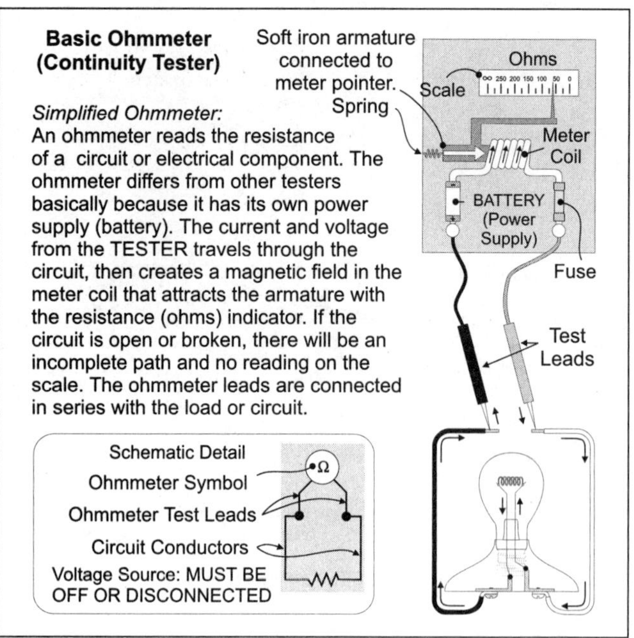

Figure 1–29
Ohm Meter

energized circuits, and polarity need not be observed. When an ohmmeter is used, current flows through the meter's coil, causing an electromagnetic field around the coil and drawing in the armature. The greater the current flow, as a result of lower resistance ($I = E/R$), the greater the magnetic field and the further the armature is drawn into the coil. A short circuit is indicated by a reading of zero, and an open circuit is indicated by infinity (∞) (Figure 1–29).

A *megohmmeter* or *megohmer* is an instrument designed to measure very high resistances such as those found in cable insulation between motor or transformer windings.

Voltmeter

Voltmeters are used to measure both dc and ac voltage. A voltmeter contains a *resistor* in *series* with the coil and utilizes the circuit energy for its operation. The purpose of the resistor in the meter is to reduce the current flow through the meter. As current flows through the meter's helically wound coil, the combined electromagnetic field of the coil draws in the iron bar. The greater the circuit voltage, the greater the current flow through the meter's coil ($I = {}^{E}/_{R}$). The greater the current flow through the meter's coil, the greater the electromagnetic field and the further the armature is drawn into the coil (Figure 1–30).

Polarity must be observed when connecting voltmeters to direct current circuits. If the meter is not connected to the proper polarity, the meter's needle will quickly move in the reverse direction and damage the meter's calibration. Polarity is not required when connecting voltmeters to alternating current circuits.

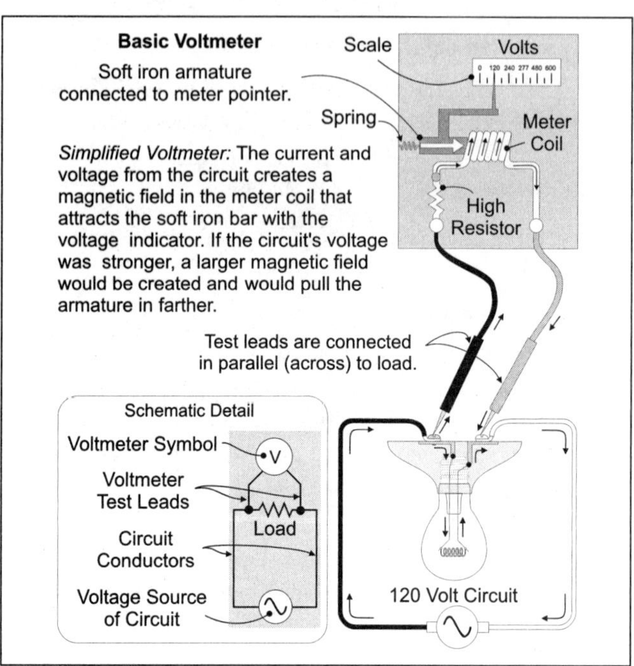

Figure 1–30
Volt Meter

Unit 1 – Electrician's Math and Basic Electrical Formulas

Part A – Electrician's Math (• Indicates that 75% or less of exam takers get the question correct)

Note: Always assume copper conductors for all answers, unless aluminum is specified in the question.

1–1 Fractions

1. The decimal equivalent for the fraction $^1/_2$ is _____ .
(a) 0.5 (b) 5 (c) 2 (d) 0.2

2. The decimal equivalent for the fraction $^4/_{18}$ is _____ .
(a) 4.5 (b) 1.5 (c) 2.5 (d) 0.2

1–2 Kilo

3. What is the kW of a 75-watt load?
(a) 75 kW (b) 7.5 kW (c) 0.75 kW (d) 0.075 kW

1–3 Knowing Your Answer

4. • The output of a transformer is 100 VA. The transformer efficiency is 90 percent. What is the transformer input power?
Note: Because efficiency is always less than 100%, the input is always greater than the output.
(a) 90 watts (b) 110 watts (c) 100 watts (d) 125 watts

1–4 Multiplier

5. The method of increasing a number by multiplying it by another number is call the _____ .
(a) percentage (b) decimal (c) fraction (d) multiplier

6. An overcurrent protection device (breaker or fuse) is required to be sized no less than 115 percent of the load. If the load is 20 ampere, the overcurrent protection device would have to be sized at no less than _____ .
(a) 20 ampere (b) 23 ampere (c) 17 ampere (d) 30 ampere

7. The maximum continuous load on an overcurrent protection device is limited to 80 percent of the device rating. If the device is rated 90 ampere, the maximum continuous load is _____ ampere.
(a) 72 (b) 90 (c) 110 (d) 125

8. A 50 ampere rated wire is required to be adjusted for temperature. If the correct multiplier is 0.80, which of the following statements is/are correct?
(a) the answer will be less than 50 ampere
(b) 80 percent of the ampacity (50) can be used
(c) the formula is 50 ampere × 0.8
(d) all the above

1–5 Parentheses

9. What is the distance of two No. 14 conductors carrying 16 ampere with a voltage drop of 10 volts?
Formula: D = (4,100 circular mils × 10 volts dropped)/(2 wires × 12.9 ampere × 16 ohm)
(a) 50 feet (b) 75 feet (c) 100 feet (d) 150 feet

10. What is the current in ampere of a three-phase, 18 kW, 208-volt load? Formula: $I = \frac{W}{E \times \sqrt{3}}$
(a) 25 ampere (b) 50 ampere

(c) 100 ampere (d) 150 ampere

1–6 Percentages

11. When changing a percent value to a decimal or whole number, simply move the decimal point two places to the _____ .
(a) right (b) left (c) depends (d) none of these

12. The decimal equivalent for 75 percent is _____ .
(a) 0.075 (b) 0.75 (c) 7.5 (d) 75

13. The decimal equivalent for 225 percent is_____ .
(a) 225 (b) 22.5 (c) 2.25 (d) 0.225

14. The decimal equivalent for 300 percent is _____ .
(a) 0.03 (b) 0.3 (c) 3 (d) 30.0

1–7 Percent Increase

15. The feeder demand load for an 8 kW load increased by 20 percent is _____ kVA.
(a) 8 (b) 9.6 (c) 6.4 (d) 10

1–8 Percentage Reciprocals

16. What is the reciprocal of 125 percent?
(a) 0.8 (b) 100 percent (c) 125 percent (d) none of these

17. A continuous load requires an overcurrent protection device sized no smaller than 125 percent of the load. What is the maximum continuous load permitted on a 100-ampere overcurrent protection device?
(a) 100 ampere (b) 150 ampere (c) 80 ampere (d) 110 ampere

1–9 Rounding Off

18. • The sum of 5, 7, 8, and 9 is approximately _____ .
(a) 20 (b) 25 (c) 30 (d) 35

1–10 Squaring

19. What is the power consumed in watts of a No. 12 conductor that is 100 feet long and has a resistance of (R) 0.2 ohm, the current (I) in the circuit is 16 ampere? Formula: Power = I^2R.
(a) 75 watts (b) 50 watts (c) 100 watts (d) 200 watts

20. • What is the area in square inches of a 2-inch raceway?

Formula: Area = πr^2 π, $\pi = 3.14$, r = radius ($^1/_2$ the diameter)
(a) 1 square inch (b) 2 square inches (c) 3 square inches (d) 4 square inches

21. The numeric equivalent of 4^2 is _____ .
(a) 2 (b) 8 (c) 16 (d) 32

22. The numeric equivalent of 12^2 is _____ .
(a) 3.46 (b) 24 (c) 144 (d) 1,728

1–11 Square Root

23. What is the square root of 1,000 ($\sqrt{1000}$)?
(a) 3 (b) 32 (c) 100 (d) 500

24. The square root of $\sqrt{3}$ is _____ .
(a) 1.732 (b) 9 (c) 729 (d) 1.5

1–12 Transposing Formulas

25. • Transform the formula $CM = \frac{(2 \times K \times I \times D)}{VD}$ to find the distance of the circuit.

(a) $D = VD \times CM$
(b) $D = \frac{(2 \times K \times I \times C \times M)}{VD}$
(c) $D = (2 \times K \times I) \times CM$
(d) $D = \frac{CM \times VD}{2 \times K \times I}$

26. If $I = {}^{P}/_{E}$, which of the following statements contain the correct transposed formula?
(a) $P = {}^{E}/_{I}$ (b) $P = {}^{I}/_{E}$ (c) $P = I \times E$ (d) $P = I^2E$

Part B – Basic Electrical Formulas

1–13 Electrical Circuit

27. An electric circuit consists of the _____ .
(a) power source (b) conductors (c) load (d) all of these

1–14 Electron Flow

28. Inside the power source, electrons travel from the positive terminal to the negative terminal.
(a) True (b) False

1–15 Power Source

29. The polarity of a(n) _____ current power source(s) never change. One terminal is always negative, and the other is always positive. _____ current flows out of the negative terminal of the power source at the same polarity.
(a) Static (b) Direct (c) Alternating (d) all of the above

30. _____ current power sources produce a voltage that has a constant, equal change in polarity and magnitude in both directions.
(a) Static (b) Direct (c) Alternating (d) all of the above

1–16 Conductance and Resistance

31. Conductance is the property of metal that permits current to flow. The best conductors, in order of conductivity, are: _____ .
(a) gold, silver, copper, aluminum
(b) copper, gold, copper, aluminum
(c) gold, copper, silver, aluminum
(d) silver, copper, gold, aluminum

1–17 Electrical Circuit Values

32. The _____ is the pressure required to force one ampere of electrons through a one ohm resistor.
(a) ohm (b) watt (c) volt (d) ampere

33. All conductors have resistance that opposes the flow of electrons. Some materials have more resistance than others. _____ has the lowest resistance, and _____ is more resistant than copper.
(a) Silver, gold (b) Gold, aluminum (c) Gold, silver (d) None of these

34. Resistance is represented by the letter *R*, and it is expressed in _____ .
(a) volts (b) impedance (c) capacitance (d) ohms

35. The opposition to the flow of current can be thought of as restricting the flow of electrons in the circuit. Every component of an electric circuit contains resistance, except the power supplies such as the generator or transformer.
(a) True (b) False

36. In electrical systems, the volume of electrons that moves through a conductor is called the _____ of the circuit.
(a) intensity (b) voltage (c) power (d) resistance

37. The rate of work that can be produced by the movement of electrons is called _____ .
(a) voltage (b) current (c) power (d) none of these

1–18 Ohm's Law (I = E/R)

38. The Ohm's Law formula demonstrates that current is _____ proportional to the voltage and _____ proportional to the resistance.
(a) indirectly, inversely (b) inversely, directly (c) inversely, indirectly (d) directly, inversely

39. In an alternating current circuit, which factors oppose current flow?
(a) resistance (b) capacitance reactance (c) induction reactance (d) all of these

40. The opposition to current flow due in an alternating current circuit is called _____ and is often represented by the letter Z.
(a) resistance (b) capacitance (c) induction (d) impedance

41. • What is the voltage drop of two No. 12 conductors supplying a 16-ampere load located 100 feet from the power supply? Formula: $E_{VD} = I \times R$, I = 16 ampere, R = 0.4 ohms (200 feet of No. 12 copper wire)
(a) 6.4 volts (b) 12.8 volts (c) 1.6 volts (d) 3.2 volts

42. What is the resistance of the circuit conductors when the conductor voltage drop is 7.2 volts and the current flow is 50 ampere?
(a) 0.14 ohm (b) 0.3 ohm (c) 3 ohm (d) 14 ohm

1–19 Pie Circle Formula

43. What is the power loss in watts for a conductor that carries 24 ampere and has a voltage drop of 7.2 volts?
(a) 173 watts (b) 350 watts (c) 700 watts (d) 2,400 watts

44. What is the current flow of a 10 kW heat strip rated 240 volts, single-phase?
(a) 35 ampere (b) 38 ampere (c) 42 ampere (d) 60 ampere

1–20 Formula Wheel

45. • The formulas listed in the formula wheel apply to _____ .
(a) direct current circuits only
(b) alternating current circuits with unity power factor
(c) a and b
(d) none of these

46. When working any formula, the key to getting the correct answer is following these four simple steps:
Step 1: ➻ Know what the question is asking.
Step 2: ➻ Determine the knowns of the circuit or resistor.
Step 3: ➻ Select the formula.
Step 4: ➻ Work out the formula calculation.
(a) True (b) False

47. The total resistance of two No. 12 copper conductors 150 feet long is 0.6 ohm and the current of the circuit is 16 ampere. What is the power loss of the conductors in watts per hour?
(a) 50 watts per hour (b) 150 watts per hour (c) 300 watts per hour (d) 600 watts per hour

48. • What is the conductor power loss in watts for a 120-volt circuit that has a 3 percent voltage drop and carries a current flow of 12 ampere? The load operates 24 hours per day, 365 days each year.
(a) 43 watts (b) 86 watts (c) 172 watts (d) 722 watts

49. • What does it cost per year (24 hours per day, 365 day per year at 8.6 cents per kW) for the power loss of a conductor? The No. 12 copper conductor resistance is 0.3 ohm and the current flow is 12 ampere.
(a) $33.55 (b) $13.10 (c) $130.50 (d) $140.21

1–21 Power Changes with the Square of the Voltage

50. • What is the power consumed of a 10 kW heat strip rated 230 volts connected to a 115-volt circuit?
(a) 10 kW (b) 2.5 kW (c) 5 kW (d) 20 kW

1–22 Electrical Meters

51. A(n) _____ is connected in series with the load.
(a) watt-hour meter (b) voltmeter (c) power meter (d) ammeter

52. • Ammeters are used to measure _____, are connected in series with the circuit, and are said to shunt the circuit.
(a) current (b) power (c) voltage (d) all of these

53. Clamp-on ammeters have one coil connected _____ around the circuit conductor.
(a) series (b) parallel (c) series-parallel (d) at right angles (perpendicular)

54. An ohmmeter has _____ connected in series with the resistor. As current flows through the meter coil, the magnetic field around the coil draws in the soft iron bar. The greater the current flow through the circuit, the greater the magnetic field and the further the armature is drawn into the coil.
(a) coil and resistor (b) coil and power supply (c) two coils (d) none of these

☆ Challenge Questions

1–2 Kilo

55. • kVA is equal to _____ .
(a) 100 VA (b) 1,000 volts (c) 1,000 watts (d) 1,000 VA

1–16 Conductance and Resistance

56. • _____ is not an insulator.
(a) Bakelite (b) Oil (c) Air (d) Salt water

1–17 Electrical Circuit Values

57. • _____ is not the force that moves electrons.
(a) EMF (b) Voltage (c) Potential (d) Current

58. Conductor resistance varies with _____ .
(a) material (b) voltage (c) current (d) power

1–18 Ohm's Law ($I = E/R$)

59. • If the contact resistance of a connection increases and the current remains the same, the voltage drop across the connection _____ .
(a) will increase (b) will decrease (c) will remain the same (d) cannot be determined

60. • To double the current of a circuit when the voltage remains constant, the resistance (R) must be _____ .
(a) doubled (b) reduced by half (c) increased (d) none of these

61. • An ohmmeter is being used to test a relay coil. The equipment instructions indicate that the resistance of the coil should be between 30 and 33 ohm. The ohmmeter indicates that the actual resistance is less than 22 ohm. This reading would most likely indicate_____ .
(a) the coil is okay (b) an open coil (c) a shorted coil (d) a meter problem

1–20 Formula Wheel

62. • To calculate the power consumed by a resistive appliance, one needs to know _____ .
(a) voltage and current (b) current and resistance (c) voltage and resistance (d) any of these

63. • The number of watts of heat given off by a resistor is expressed by the formula $I^2 \times R$. If 10 volts is applied to a 5 ohm resistor, then _____ of heat will be given off.
(a) 500 watts (b) 250 watts (c) 50 watts (d) 20 watts

64. • Power loss in a circuit because of heat can be determined by the formula _____ .
(a) $P = R \times I$ (b) $P = I \times R$ (c) $P = I^2 \times R$ (d) none of these

65. • If current remains the same and resistance increases, the circuit will consume _____ power.
(a) more (b) less

66. When a lamp that is rated 500 watts at 115 volts is connected to a 120-volt power supply, the current of the circuit will be _____ . Tip: Does power remain the same when voltage is changed?
(a) 3.8 ampere (b) 4.5 ampere (c) 2.7 ampere (d) 5.5 ampere

1–21 Power Changes with the Square of the Voltage

67. A 120-volt rated toaster will produce _____ heat when supplied by 115 volts.
(a) more (b) less (c) the same (d) none of these

68. • When a resistive load is operated at a voltage 10 percent higher than the nameplate rating of the appliance, the appliance will _____ .
(a) have a longer life (b) draw a lower current (c) use more power (d) none of these

69. • A 1,500-watt heater rated 230 volts is connected to a 208-volt supply. The power consumed for this load is _____ watts. Tip: When the voltage is reduced, will the power be greater or less?
(a) 1,625 (b) 1,750 (c) 1,850 (d) 1,225

70. • The total resistance of a circuit is 12 ohm; the load is 10 ohm, and the wire 2 ohm. If the current of the circuit is 3 ampere, then the power consumed by the circuit conductors is _____ .
(a) 28 watts (b) 18 watts (c) 90 watts (d) 75 watts

1–22 Electric Meters

71. • The best instrument for detecting an electric current is a(n) _____ .
(a) ohmmeter (b) voltmeter (c) ammeter (d) wattmeter

72. The polarity of a circuit being tested must be observed when connecting an ohmmeter to _____ .
(a) an alternating current circuit
(b) a direct current circuit
(c) any circuit
(d) polarity does not matter because the circuit is not energized

73. • When the test leads of an ohmmeter are shorted together, the meter will read _____ on the scale.
(a) zero ohm (b) 1,000 (c) infinity (d) all of these

74. • A short circuit is indicated by a reading of _____ when tested with an ohmmeter.
(a) zero (b) ohm (c) infinity (d) R

75. Voltmeters are used to measure _____ .
(a) voltages to ground (b) voltage differences (c) AC voltages only (d) DC voltages only

76. Voltmeters must be connected in _____ with the circuit component being tested.
(a) series (b) parallel (c) series-parallel (d) multiwire

77. To measure the voltage across a load, you would connect a(n) _____ .
(a) voltmeter across the load (b) ammeter across the load
(c) voltmeter in series with the load (d) ammeter in series with the load

78. A voltmeter is connected in _____ to the load.
(a) series (b) parallel (c) series-parallel (d) none of these

79. • In the course of normal operation, the least effective instrument in indicating that a generator may overheat because it is overloaded is a(n) _____ .
(a) ammeter (b) voltmeter (c) wattmeter (d) none of these

80. • A direct current voltmeter (not a digital meter) can be used to measure _____ .
(a) power (b) frequency (c) polarity (d) power factor

81. • Polarity must be observed when connecting an analog voltmeter to _____ current circuit.
(a) an alternating (b) direct (c) any (d) polarity does not matter

82. The minimum number of wattmeters necessary to measure the power in the load of a balanced 3-phase, 4-wire system is _____ .
(a) 1 (b) 2 (c) 3 (d) 4

Unit 2

Electrical Circuits

OBJECTIVES

After reading this unit, the student should be able to briefly explain the following concepts:

- Series circuit calculations
- Parallel circuit calculations
- Series-parallel circuit
- Calculations
- Multiwire circuit calculations
- Neutral current calculations
- Dangers of multiwire circuits

After reading this unit, the student should be able to briefly explain the following terms:

- Amp-hour
- Equal resistor method
- Kirchoff's law
- Multiwire circuits
- Neutral conductor
- Nonlinear loads
- Parallel circuits
- Phases
- Pigtailing
- Product of the sum method
- Reciprocal method
- Series circuit
- Series-parallel circuit
- Unbalanced current
- Ungrounded conductors (hot wires)

PART A – SERIES CIRCUITS

INTRODUCTION TO SERIES CIRCUITS

A series circuit is a circuit in which the current leaves the voltage source and flows through every electrical device with the same intensity before it returns to the voltage source. If any part of a series circuit is opened, the current stops flowing in the entire circuit (Figure 2–1).

For most practical purposes, series (closed-loop) circuits are not used for building wiring, but they are important for the operation of many control and signal circuits (Figure 2–2). Motor control circuit stop switches are generally wired in series with the starter's coil and the line conductors. Dual-rated motors, such as 460/230 volts, have their windings connected in series when supplied by the higher voltage, and in parallel when supplied by the lower voltage.

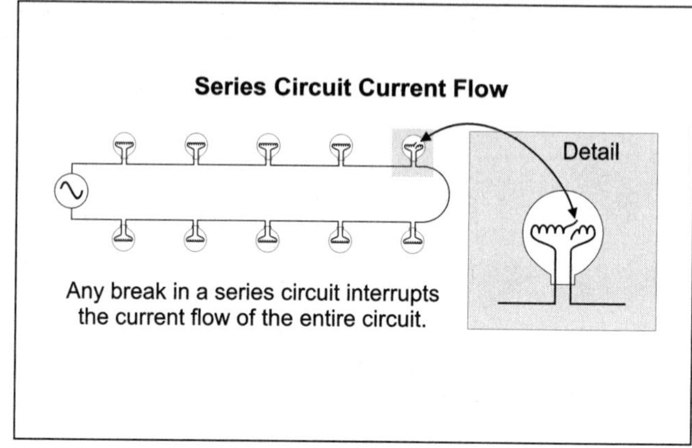

Figure 2–1
Series Circuit Current Flow

2–1 UNDERSTANDING SERIES CALCULATIONS

It is important to understand the relationship between resistance, current, voltage, and power of series circuits (Figure 2–3).

Calculating Resistance Total

In a series circuit, the total resistance of the circuit is equal to the sum of all the series resistors' resistance, according to the formula:

$$R_T = R_1 + R_2 + R_3 + R_4$$

❑ **Total Resistance**

What is the total resistance of the loads (Figure 2–4)?

(a) 2.5 ohm (b) 5.5 ohm
(c) 7.5 ohm (d) 10 ohm

• Answer: (c) 7.5 ohm

R_1 – Power Supply	0.05 ohm
R_2 – Conductor No. 1	0.15 ohm
R_3 – Appliance	7.15 ohm
R_4 – Conductor No. 2	0.15 ohm
Total Resistance:	7.50 ohm

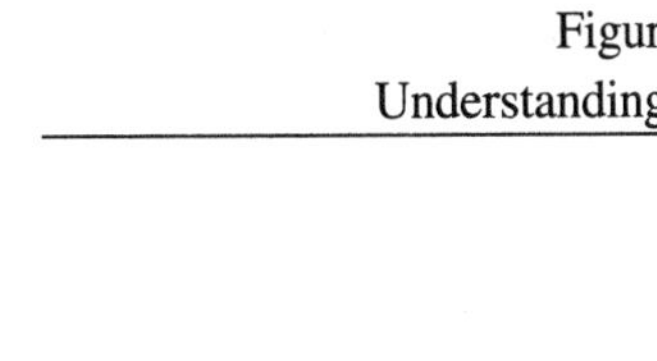

Figure 2–3
Understanding I, E, R, and P

Calculating Voltage Drop

The result of current flowing through a resistor is a reduction in voltage across the resistor, which is called voltage drop. In a closed loop circuit, the sum of the voltage drops of all the loads is equal to the voltage source (Figure 2–5). This is known as Kirchoff's First Law of Voltage: The voltage drop (VD) of each resistor can be determined by the formula:

$$E_{VD} = I \times R$$

I = Current of the circuit

R = Resistance of the resistor

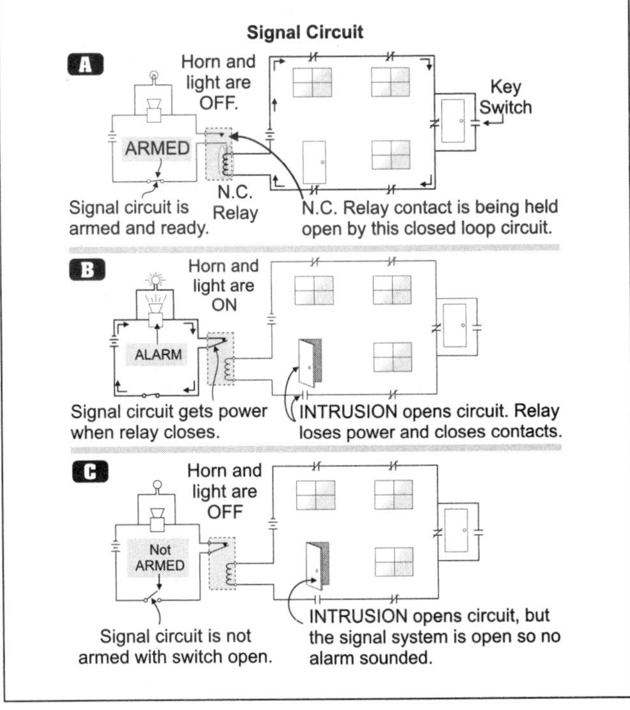

Figure 2–2
Signal Circuit

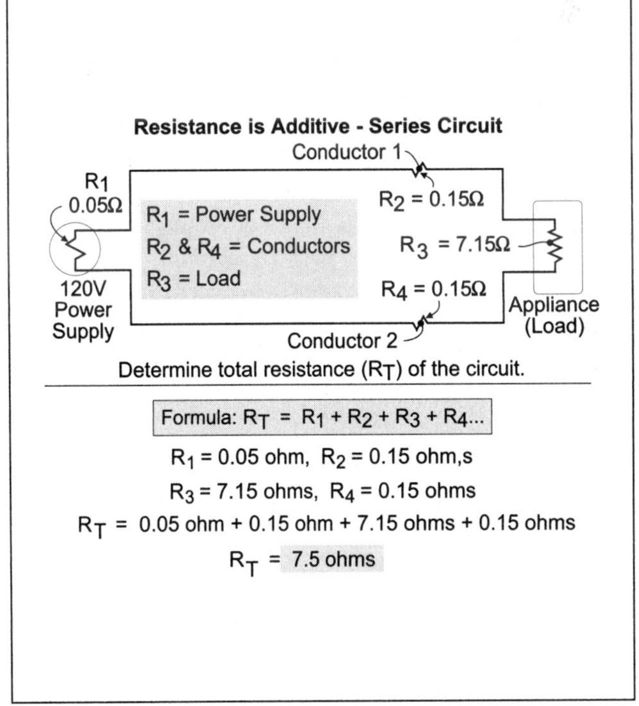

Figure 2–4
Resistance Is Additive

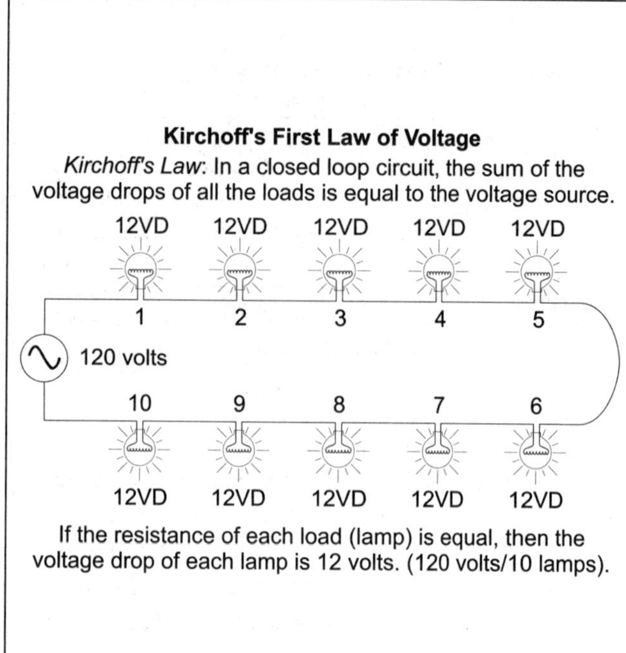

Figure 2–5
Kirchoff's First Law of Voltage

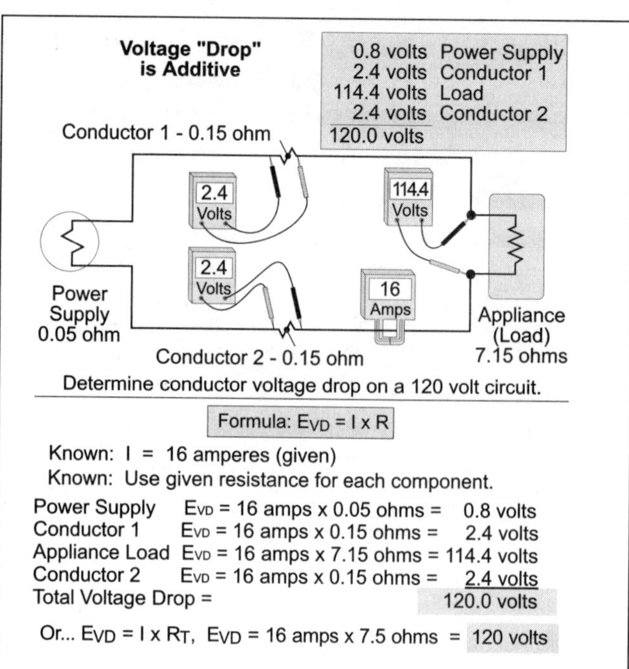

Figure 2–6
Voltage Drop Is Additive

❑ Voltage Drop

What is the voltage drop across each resistor (Figure 2–6) using the formula VD = I × R?

• Answer: VD = I × R

VD = I × R

VD_{R1} – Power Supply	16 ampere × 0.05 ohm	=	0.8 volts dropped
VD_{R2} – Conductor No. 1	16 ampere × 0.15 ohm	=	2.4 volts dropped
VD_{R3} – Appliance	16 ampere × 7.15 ohm	=	114.4 volts dropped
VD_{R4} – Conductor No. 2	16 ampere × 0.15 ohm	=	2.4 volts dropped
Total Voltage Drop	16 ampere × 7.50 ohm	=	120.0 volts dropped

Note: Due to rounding, the sum of the voltage drops may be slightly different than the voltage source.

Voltage of Series Connected Power Supplies

When *power supplies* are connected in series, the voltage of each power supply will add together (sum), providing that all the polarities are connected properly.

❑ Series Connected Power Supplies

What is the total voltage output of four 1.5-volt batteries connected in series (Figure 2–7)?

(a) 1.5 volts (b) 3.0 volts
(c) 4.5 volts (d) 6.0 volts

• Answer: (d) 6 volts

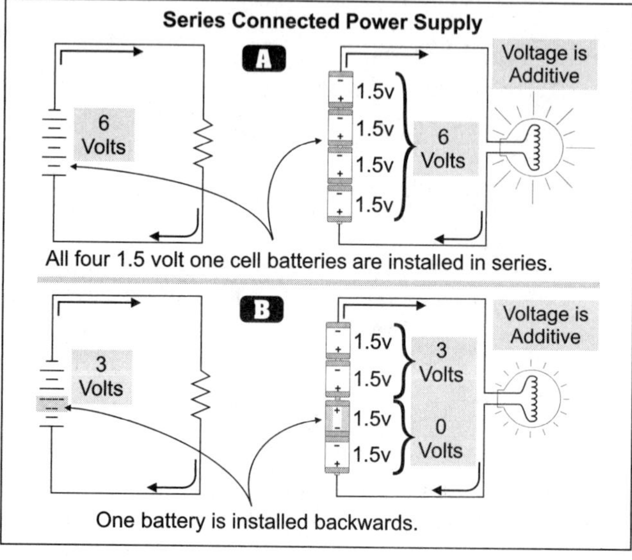

Figure 2–7
Series Connected Power Supply

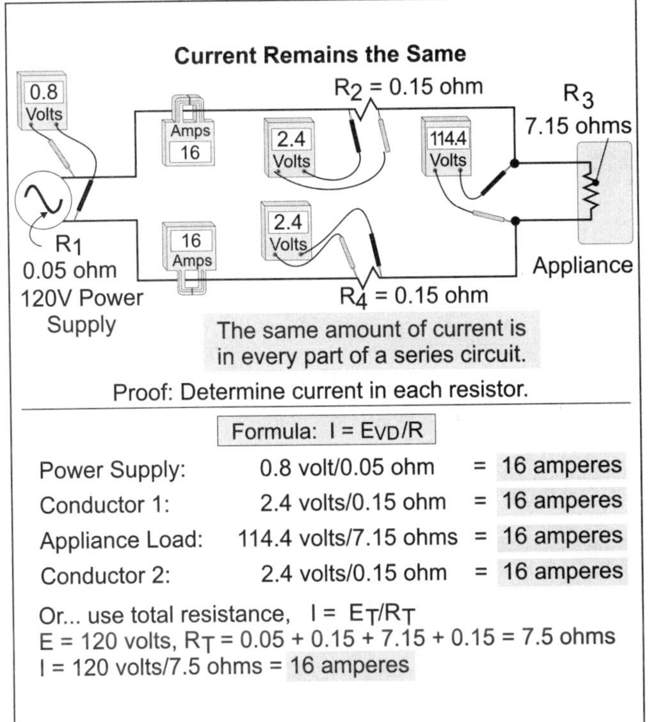

Figure 2–8
Current Remains the Same

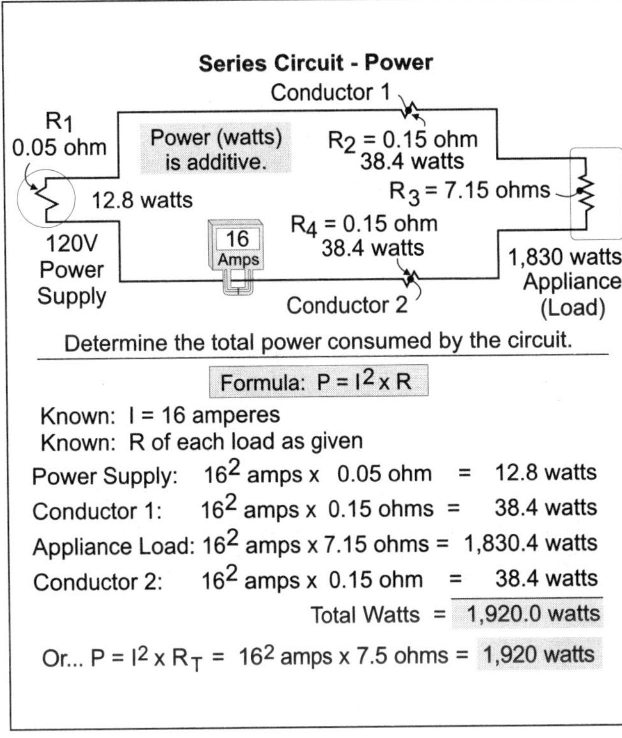

Figure 2–9
Series Circuit – Power Is Additive

Current of Resistor

In a series circuit, the current throughout the circuit is constant and does not change. The current through each resistor of the circuit can be determined by the formula:

$I = E/R$

E = Voltage drop of the load or circuit

R = Resistance of the load or circuit

Note: If the resistance is not given, you can determine the resistance of a resistor (if you know the nameplate voltage and power rating of the load) by the formula: $R = E^2/P$, E = Nameplate voltage rating (squared), P = Nameplate power rating

❑ Current of Circuit

What is the current flow through the series circuit (Figure 2–8)? Note: Current remains the same.

(a) 4 ampere (b) 8 ampere
(c) 12 ampere (d) 16 ampere

• Answer: (d) 16 ampere

$I = E/R$

I_{R1} – Power Supply	0.8 volt drop/0.05 ohm	= 16 ampere
I_{R2} – Conductor No. 1	2.4 volts dropped /0.15 ohm	= 16 ampere
I_{R3} – Appliance	114.4 volts dropped/7.15 ohm	= 16 ampere
I_{R4} – Conductor No. 2	2.4 volts dropped/0.15 ohm	= 16 ampere
Circuit Current	120 volts/7.5 ohm	= 16 ampere

Power of Resistor or Circuit

The power consumed in a series circuit is equal to the sum of the power of all of the resistors in the series circuit. The resistor with the highest resistance will consume the most power, and the resistor with the smallest resistance will consume the least power. You can calculate the power consumed (watts) of each resistor or of the circuit by the formula:

$P = I^2R$, I^2 = Current of the circuit (squared), R = Resistance of circuit or resistor

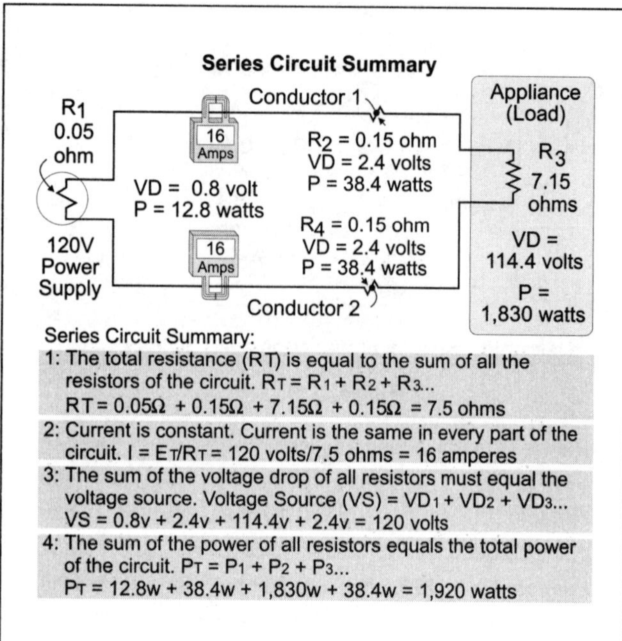

Figure 2–10
Series Circuit Summary

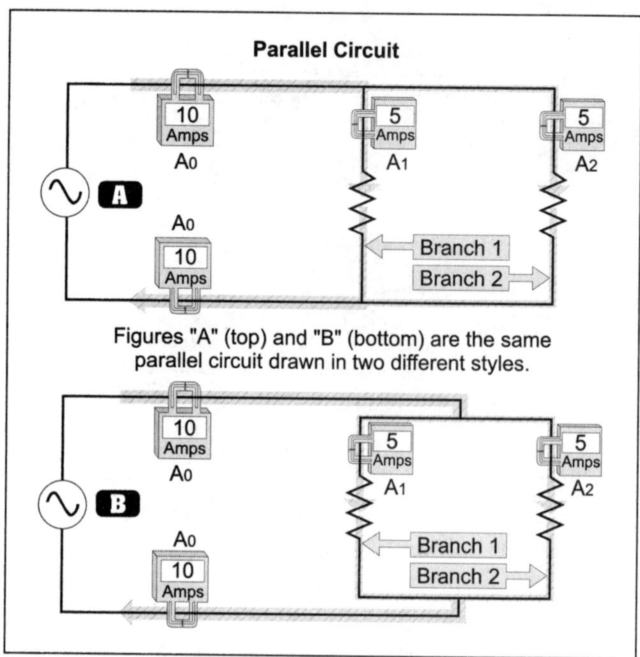

Figure 2–11
Parallel Circuit

❑ Power

What is the power consumed of each resistor (Figure 2–9)?

- Answer: $P = I^2R$

$P = I^2R$

P_{R1} – Power Supply	16 ampere^2 × 0.05 ohm = 12.8 watt
P_{R2} – Conductor No. 1	16 ampere^2 × 0.15 ohm = 38.4 watt
P_{R3} – Appliance	16 ampere^2 × 7.15 ohm = 1,830 watt
P_{R4} – Conductor No. 2	16 ampere^2 × 0.15 ohm = 38 watt

2–2 SERIES CIRCUIT SUMMARY

(Figure 2–10) Series Circuit Summary.

Note 1: The total resistance of the series circuit is equal to the sum of all the resistors of the circuit.

Note 2: Current is constant.

Note 3: The sum of the voltage drop of all resistors must equal the voltage source.

Note 4: The sum of the power of all resistors equals the total power of the circuit.

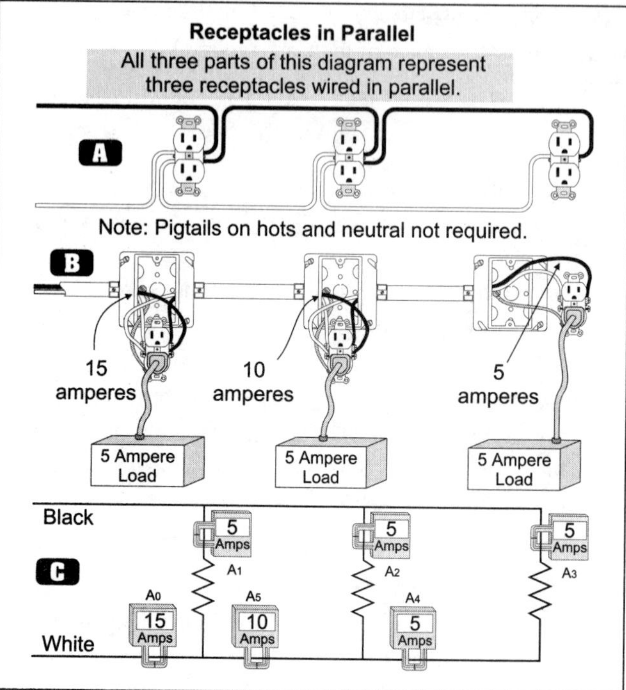

Figure 2–12
Receptacles in Parallel

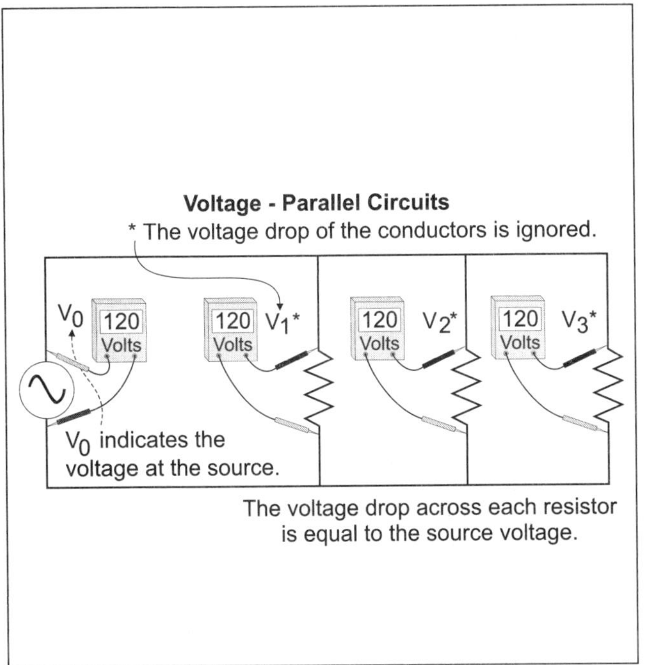

Figure 2–13
Voltage–Parallel Circuits

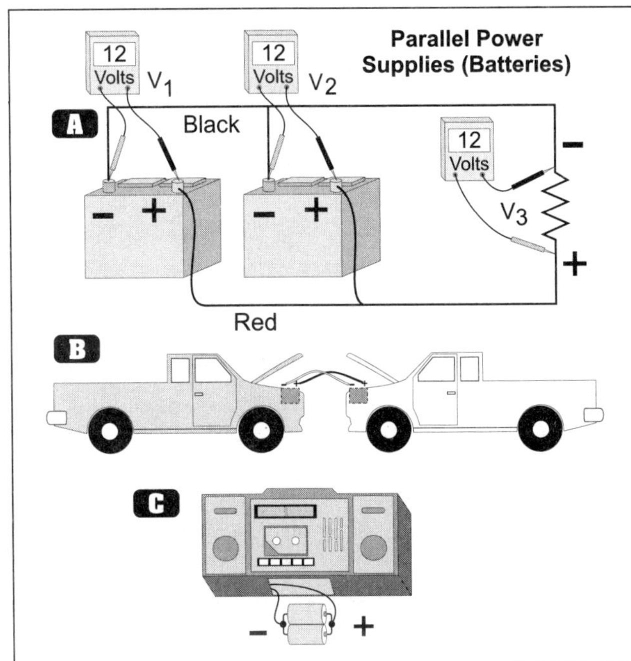

Figure 2–14
Parallel Power Supplies

PART B – PARALLEL CIRCUITS

INTRODUCTION TO PARALLEL CIRCUITS

A parallel circuit is a circuit in which current leaves the voltage source, branches through different parts of the circuit, and then returns to the voltage source (Figure 2–11). Parallel is a term used to describe a method of connecting electrical components so that the current can flow through two or more different branches of the circuit.

2–3 PRACTICAL USES OF PARALLEL CIRCUITS

Parallel circuits are commonly used for building wiring (Figure 2–12); in addition, parallel (open-loop) circuits are often used for fire alarm systems and the internal wiring of many types of electrical equipment, such as motors and transformers. Dual-rated motors, such as 460/230 volts, have their windings connected in parallel when supplied by the lower voltage, and in series when supplied by the higher voltage.

2–4 UNDERSTANDING PARALLEL CALCULATIONS

It is important to understand the relationship between voltage, current, power, and resistance of a parallel circuit.

Voltage of Each Branch

The voltage drop across loads connected in parallel is equal to the voltage that supplies each parallel branch (Figure 2–13).

Power Supplies Connected in Parallel

When power supplies are connected in parallel, voltage remains the same but the current, or amp-hour, capacity is increased. When connecting batteries in parallel, always connect batteries of the same voltage with the proper polarity.

❑ **Parallel Connected Power Supplies**

If two 12-volt batteries are connected in parallel, what is the output voltage (Figure 2–14)?

(a) 3 volts (b) 6 volts (c) 12 volts (d) 24 volts

- Answer: (c) 12 volts. The voltage remains the same when power supplies are connected in parallel.

Note: Two 12-volt batteries connected in parallel will result in an output voltage of 12 volts, but the amp-hour capacity will be increased, resulting in longer service.

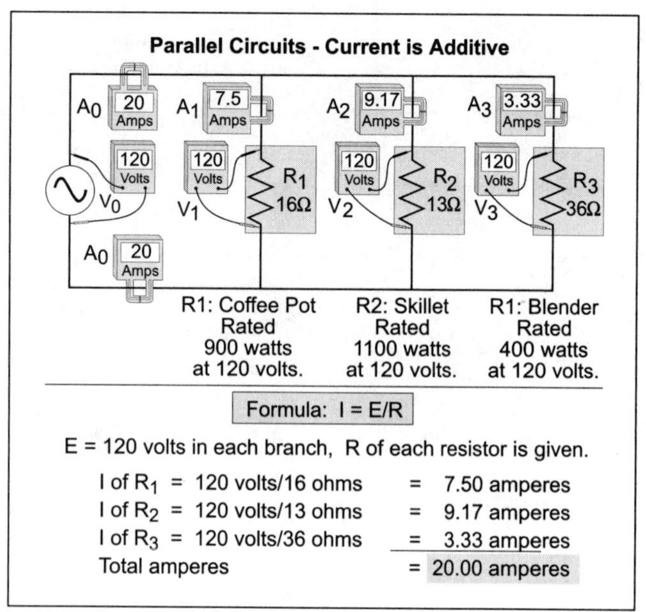

Figure 2–15
Parallel Circuits – Current Is Additive

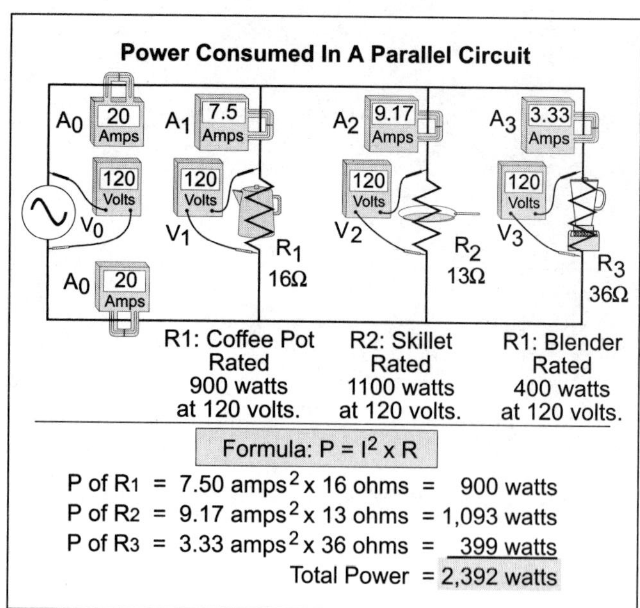

Figure 2–16
Power Consumed in a Parallel Circuit

Current through Each Branch

The current from the power supply is equal to the sum of the branch circuit currents. The current in each branch depends on the branch voltage and branch resistance and can be calculated by the formula:

$I = E/R$

E = Voltage of Branch

R = Resistance of Branch

❑ Current through Each Branch

What is the current of each appliance (Figure 2–15)?

• Answer: $I = E/R$

$I = E/R$

I_{R1} – Coffee Pot	120 volts/16 ohm	=	7.50 ampere
I_{R2} – Skillet	120 volts/13 ohm	=	9.17 ampere
I_{R3} – Blender	120 volts/36 ohm	=	3.33 ampere
Total Current (7.5 ampere + 9.17 ampere + 3.33 ampere)			20.00 ampere

Power Consumed of Each Branch

The total power consumed of any circuit is equal to the sum of the branch powers. Each branch power depends on the branch current and resistance. The power can be found by the formula:

$P = I^2R$

I = Current of Each Branch

R = Resistance of Each Branch

❑ Power of Each Branch

What is the power consumed of each appliance (Figure 2–16)?

• Answer: $P = I^2R$

$P = I^2R$

I_{R1} – Coffee Pot	$7.5\ ampere^2 \times 16\ ohm$	=	900 watts
I_{R2} – Skillet	$9.17\ ampere^2 \times 13\ ohm$	=	1,093 watts
I_{R3} – Blender	$3.33\ ampere^2 \times 36\ ohm$	=	399 watts
Total Power (900 watts + 1,093 watts + 399 watts)			2,392 watts

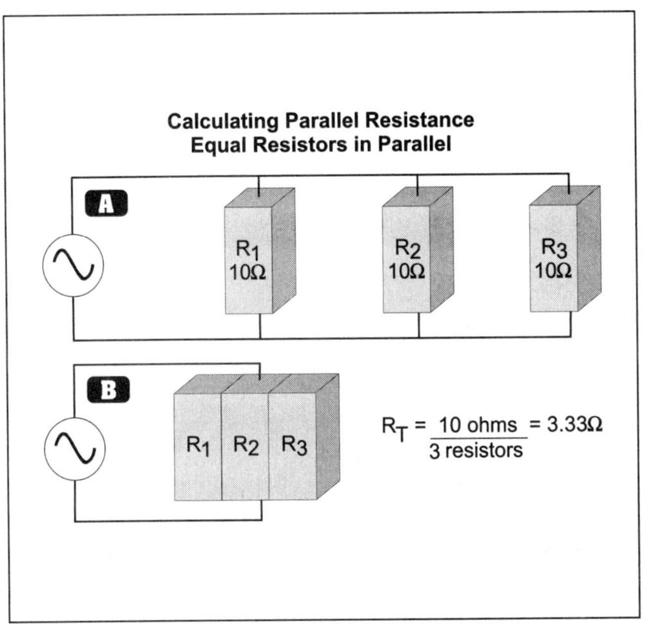

Figure 2–17
Calculating Parallel Resistance–Equal Resistors in Parallel

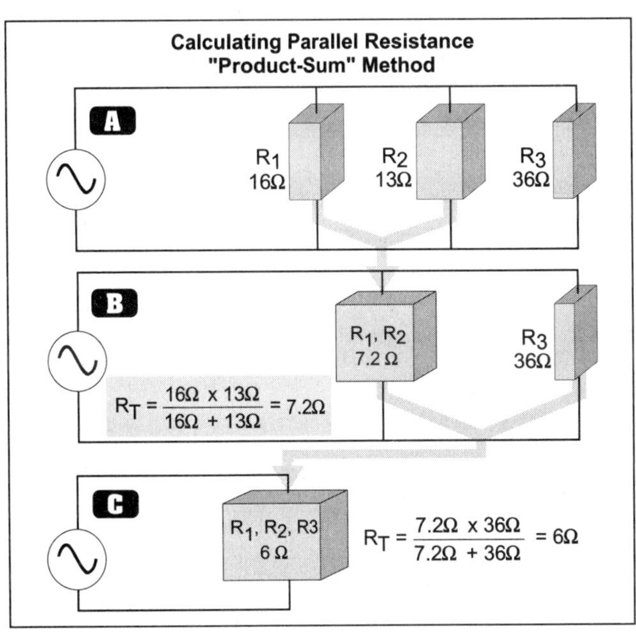

Figure 2–18
Calculating Parallel Resistance "Product–Sum" Method

2–5 PARALLEL CIRCUIT RESISTANCE CALCULATIONS

In a parallel circuit, the total circuit resistance is always less than the smallest resistor and can be determined by one of three methods:

Equal Resistor Method

The equal resistor method can be used when all the resistors of the parallel circuit have the same resistance. Simply divide the resistance of one resistor by the number of resistors in parallel.

$$R_T = R/N$$

R = Resistance of One Resistor, N = Number of Resistors

❑ **Equal Resistors Method**

The total resistance of three 10 ohm resistors is _____ (Figure 2–17).

(a) 10 ohm (b) 20 ohm (c) 30 ohm (d) none of these

- Answer: (d) none of these

$$R_T = \frac{\text{Resistance of One Resistor}}{\text{Number of Resistors}}, = \frac{10 \text{ ohms}}{3 \text{ resistors}}, = 3.33 \text{ ohm}$$

The product over the sum method

The product over the sum method can be used to calculate the resistance of two resistors.

$$R_T = \frac{R_1 \times R_2 \text{ (Product)}}{R_1 + R_2 \text{ (Sum)}}$$

The term *product* means the answer of numbers that are multiplied together. The term *sum* is the answer to numbers that are added together. The product of the sum method can be used for more than two resistors, but only two can be calculated at a time.

❑ **Product of the Sum Method**

What is the total resistance of a 16 ohm coffee pot and a 13 ohm skillet connected in parallel (Figure 2–18)?

(a) 16 ohm (b) 13.09 ohm (c) 29.09 ohm (d) 7.2 ohm

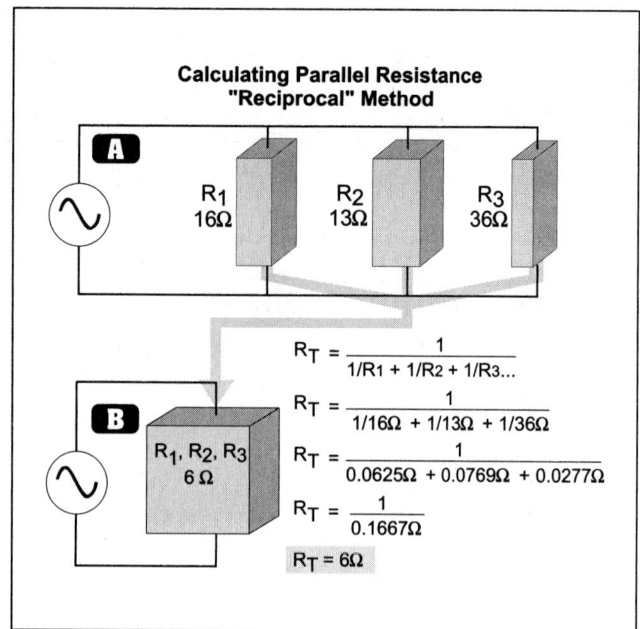

Figure 2–19
Calculating Parallel Resistance "Reciprocal" Method

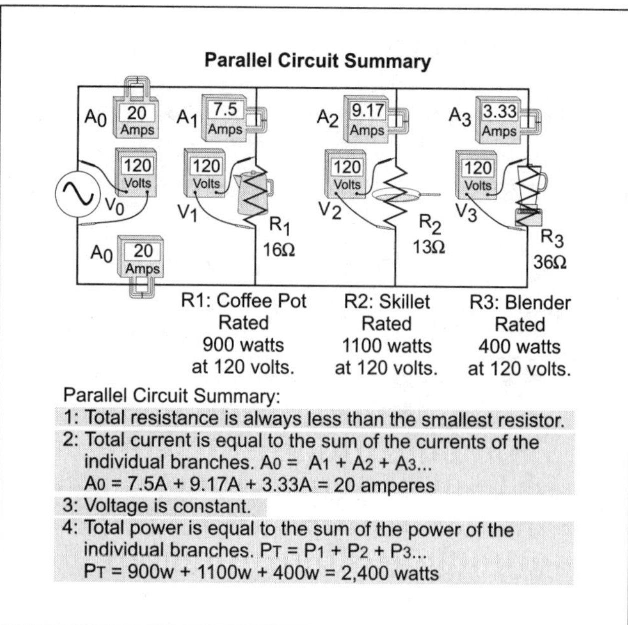

Figure 2–20
Parallel Circuit Summary

• Answer: (d) 7.2 ohm

The total resistance of a parallel circuit is always less than the smallest resistor (13 ohm).

$$R_T = \frac{R_1 \times R_2}{R_1 + R_2} = \frac{16 \text{ ohms} \times 13 \text{ ohms}}{16 \text{ ohms} + 13 \text{ ohms}} = 7.2 \text{ ohm}$$

Reciprocal Method

The advantage of the reciprocal method is that this formula can be used for an unlimited number of parallel resistors.

$$R_T = \frac{1}{1/R_1 + 1/R_2 + 1/R_3 \ldots}$$

❑ **Reciprocal Method**

What is the resistance total of a 16 ohm, 13 ohm, and 36 ohm resistor connected in parallel (Figure 2–19)?

(a) 13 ohm (b) 16 ohm (c) 36 ohm (d) 6 ohm

• Answer: (d) 6 ohm

$$R_T = \frac{1}{1/16 \text{ ohm} + 1/13 \text{ ohm} + 1/36 \text{ ohm}} \quad R_T = \frac{1}{0.0625 \text{ ohm} + 0.0769 \text{ ohm} + 0.0278 \text{ ohm}}$$

$$R_T = \frac{1}{0.1672 \text{ ohm}} \quad R_T = 6 \text{ ohm}$$

2–6 PARALLEL CIRCUIT SUMMARY

Note 1: The total resistance of a parallel circuit is always less than the smallest resistor (Figure 2–20).

Note 2: Current total of a parallel circuit is equal to the sum of the currents of the individual branches.

Note 3: Voltage is constant.

Note 4: Power total is equal to the sum of the power in all the individual branches.

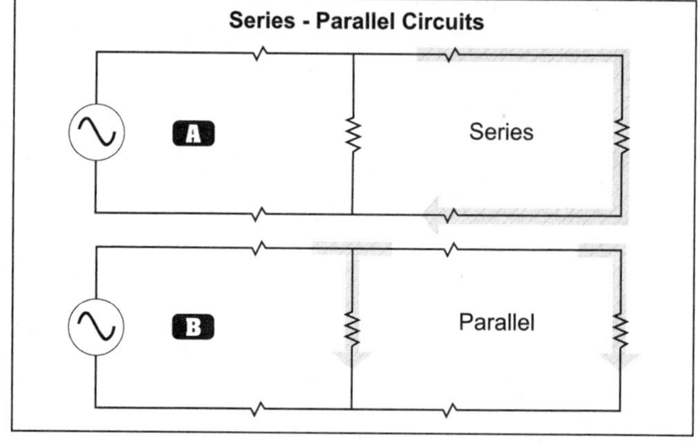

Figure 2–21
Series – Parallel Circuits

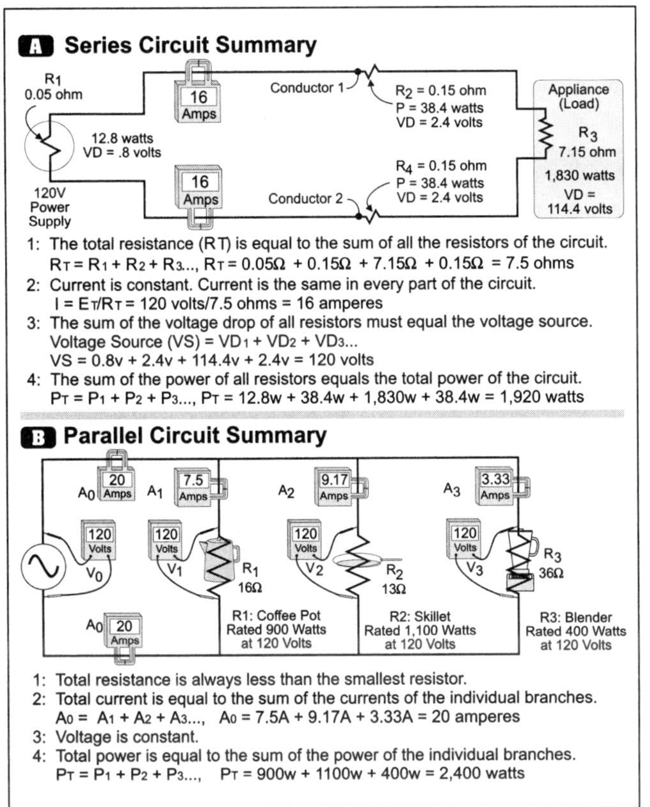

Figure 2–22
Series Circuit Summary

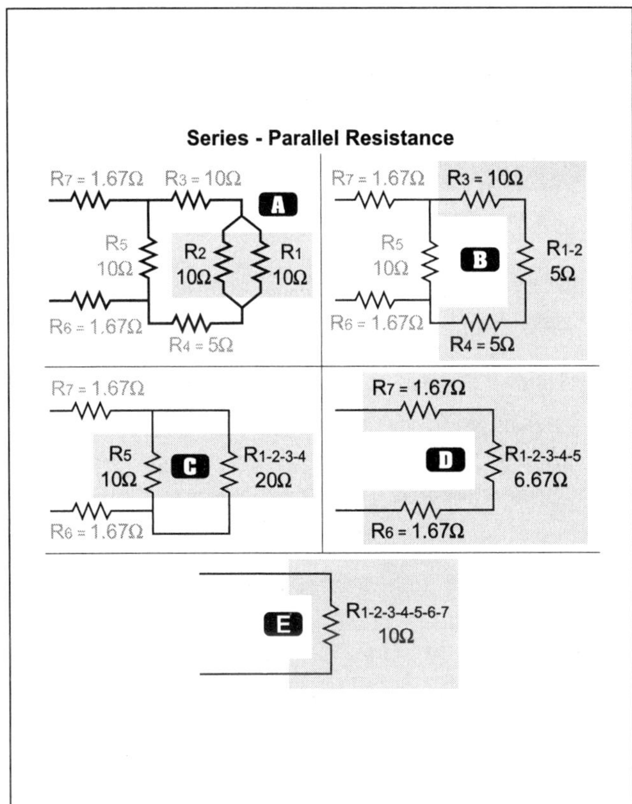

Figure 2–23
Series-Parallel Resistance

PART C – SERIES-PARALLEL CIRCUITS

INTRODUCTION TO SERIES-PARALLEL CIRCUITS

A series-parallel circuit is a circuit that contains some resistors in series and some resistors in parallel to each other (Figure 2–21).

2–7 REVIEW OF SERIES AND PARALLEL CIRCUITS

For a better understanding of series-parallel circuits, let's review the rules for series and parallel circuits. That portion of the circuit that contains resistors in series must comply with the rules of series circuits, and that portion of the circuit that is connected in parallel must comply with the rules of parallel circuits (Figure 2–22).

Series Circuit Rules:

Note 1: Resistance is additive.

Note 2: Current is constant.

Note 3: Voltage is additive.

Note 4: Power is additive.

Parallel Circuit Rules:

Note 1: Resistance is less than the smallest resistor.

Note 2: Current is additive.

Note 3: Voltage is constant.

Note 4: Power is additive.

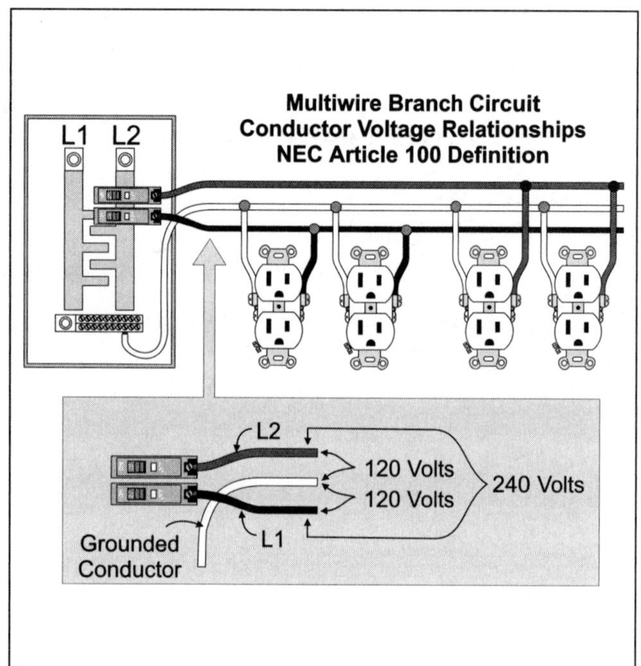

Figure 2–24
Multiwire Branch Circuit

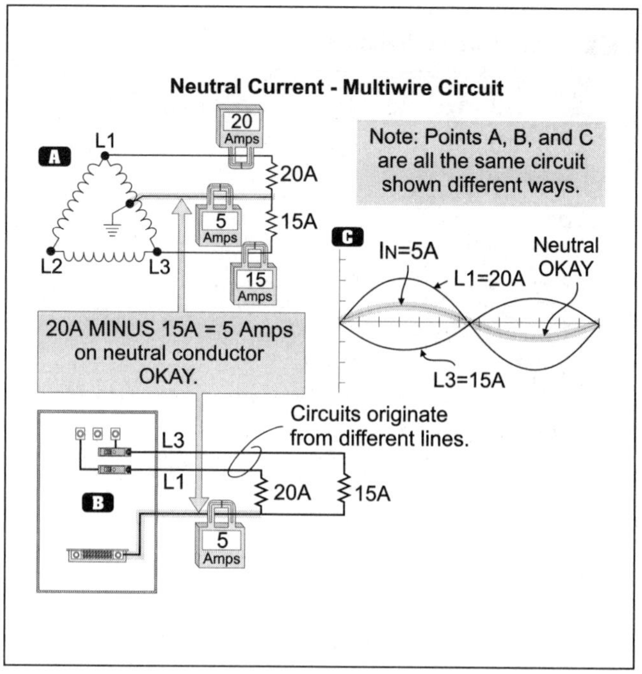

Figure 2–25
Neutral Current in Multiwire Branch Circuit

2–8 SERIES-PARALLEL CIRCUIT RESISTANCE CALCULATIONS

When determining the resistance total of a series-parallel circuit, it is best to redraw the circuit so you can see the series components and the parallel branches. Determine the resistance of the series components first or the parallel components depending on the circuit, then determine the resistance total of all the branches. Keep breaking the circuit down until you have determined the total effective resistance of the circuit as one resistor.

❑ **Series-Parallel Circuit Resistance**

What is the resistance total of the circuit (Figure 2–23, Part A)?

(a) 2 ohms (b) 5 ohms (c) 7 ohms (d) 10 ohms

- Answer: (d) 10 ohms

Step 1: ➼ Determine the equal parallel resistors: R_1 (10 ohm) and R_2 (10 ohm) (Figure 2–23, Part A):

$R_T = R/N$, R = Resistance of one resistor, N = Number of resistors

R_T = 10 ohm/2 resistors, R_T = 5 ohm

Step 2: ➼ Redraw the circuit (Figure 2–23, Part B)

Determine the series resistance of $R_{1,2}$ (5 ohm), plus R_3 (10 ohm), plus R_4 (5 ohm).

Series resistance total is equal to $R_{1,2} + R_3 + R_4$

R_T = 5 ohm + 10 ohm + 5 ohm, R_T = 20 ohm

Step 3: ➼ Redraw the circuit (Figure 2–23, Part C)

Determine the parallel resistance of R_5, plus the resistance of $R_{1,2,3,4,}$. Remember the resistance total of a parallel circuit is always less than the smallest parallel branch (10 ohm). Since we have only two parallel branches, the resistance can be determined by the product of the sum method.

$$R_T = \frac{R_5 \times R_{1,2,3,4} \text{ (Product)}}{R_1 + R_2 \text{ (Sum)}} = \frac{(10 \text{ ohms} \times 20 \text{ ohms})}{(10 \text{ ohms} + 20 \text{ ohms})} = 6.67 \text{ ohm}$$

Step 4: ➼ Redraw the circuit (Figure 2–23, Part D)

Determine the series resistance total of R_7, plus R_6, plus $R_{1,2,3,4,5}$

Series resistance total is equal to $R_7 + R_6 + R_{1,2,3,4,5}$

R_T = 1.67 ohm + 1.67 ohm + 6.67 ohm, R_T = 10 ohm

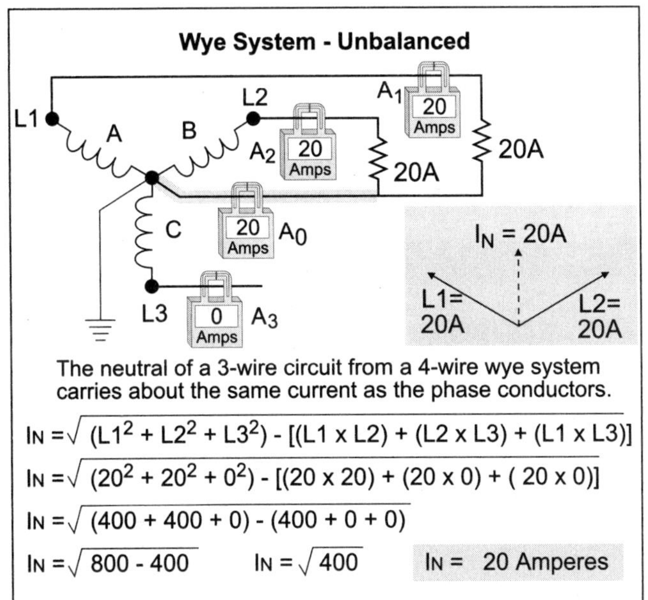

Figure 2–26
Neutral Current – Unbalanced Wye Circuit

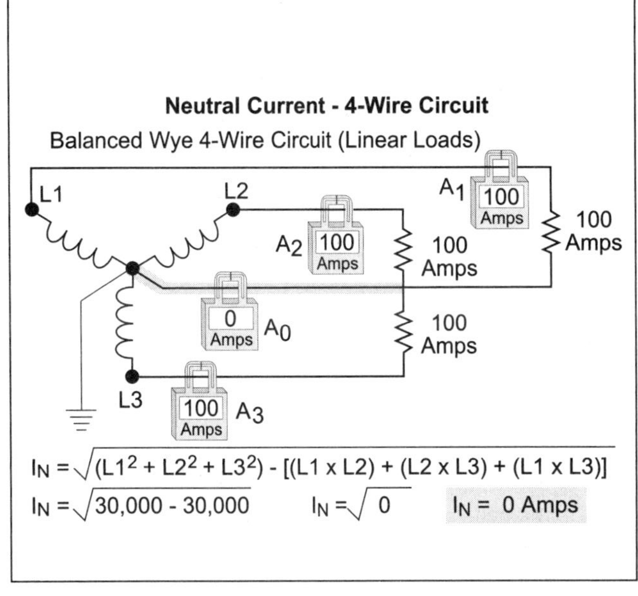

Figure 2–27
Neutral Current – Balanced Wye 4-Wire Circuit

PART D – MULTIWIRE BRANCH CIRCUITS

INTRODUCTION TO MULTIWIRE BRANCH CIRCUITS

A multiwire branch circuit consists of two or more ungrounded conductors (hot or phase conductors) having a potential difference between them and an equal difference of potential between each hot wire and grounded (neutral) conductor (Figure 2–24).

Note: The *National Electrical Code®* contains specific requirements on multiwire branch circuits. See Article 100 definition of multiwire branch circuit, Section 210-4 branch circuit requirements, and Section 300-13(b) requirements for pigtailing neutral conductors.

2–9 NEUTRAL CURRENT CALCULATIONS

When current flows in the *neutral* conductor of a multiwire branch circuit, the current is called *unbalanced current.* This current can be determined according to the following:

120/240-volt, 3-Wire Circuits

In a 120/240-volt, 3-wire circuit consisting of two hot wires and a grounded (neutral) conductor, the third wire will carry no current when the circuit is balanced. However, the neutral (or grounded) conductor will carry unbalanced current when the circuit is not balanced. The neutral current can be calculated as:

$\mathbf{I_{Neutral} = N_{Line\ 1} - N_{Line\ 2}}$

$N_{Line\ 1}$ = Line 1 Neutral Current

$N_{Line\ 2}$ = Line 2 Neutral Current

❑ **120/240-volt, 3-Wire Neutral Current**

What is the neutral current if current is: Line 1 = 20 ampere and Line 2 = 15 ampere (Figure 2–25)?

(a) 0 ampere (b) 5 ampere (c) 10 ampere (d) 35 ampere

• Answer: (b) 5 ampere

$I_{Neutral} = N_{Line\ 1} - N_{Line\ 2}$

$I_{Neutral}$ = 20 ampere less 15 ampere

$I_{Neutral}$ = 5 ampere

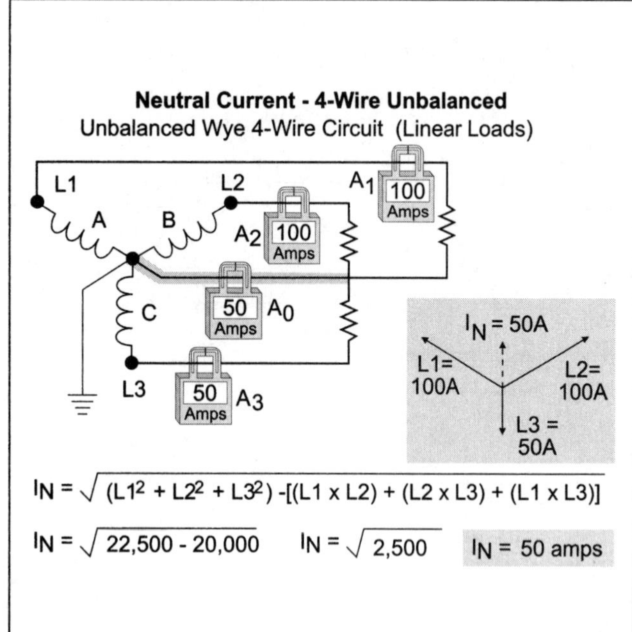

Figure 2–28
Neutral Current – 4-Wire Wye – Unbalanced

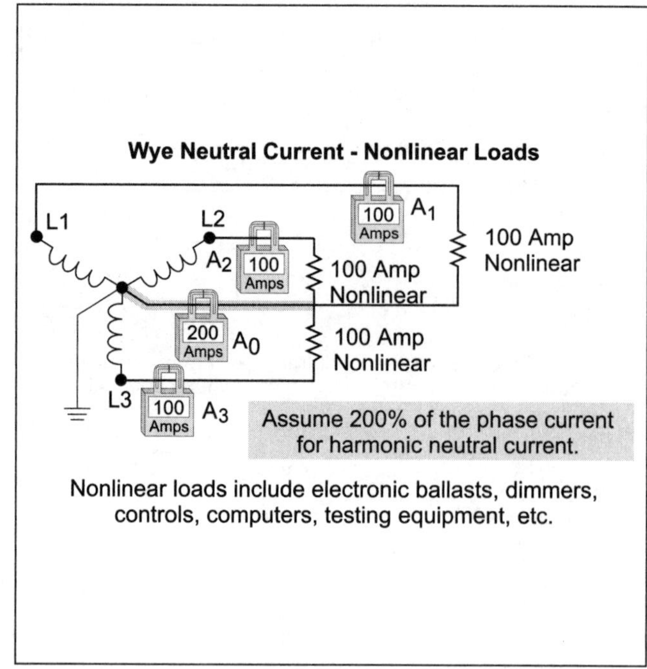

Figure 2–29
Neutral Current – Nonlinear Loads

Wye 3-Wire Neutral Current

A 3-wire, 208Y/120- or 480Y/277-volt circuit consisting of two phases and a neutral always carries unbalanced current. The current on the grounded (neutral) conductor is determined by the formula:

$$I_{Neutral} = \sqrt{(N_{Line\ 1}{}^2 + N_{Line\ 2}{}^2) - (N_{Line\ 1} \times N_{Line\ 1})}$$

❑ Three-wire Wye Circuit Neutral Current

What is the neutral current for a 20 ampere, 3-wire circuit (two hots and a neutral)? Power is supplied from a 208Y/120-volt feeder (Figure 2–26).

(a) 40 ampere (b) 20 ampere (c) 60 ampere (d) 0 ampere

• Answer: (b) 20 ampere

$I_{Neutral} = \sqrt{(N_{Line\ 1}{}^2 + N_{Line\ 2}{}^2) - (N_{Line\ 1} \times N_{Line\ 1})}$

$I_{Neutral} = \sqrt{(20\text{ ampere}^2 + 20\text{ ampere}^2) - (20\text{ ampere} \times 29\text{ ampere})} = \sqrt{400}$

$I_{Neutral} = 20$ amperes

❑ Wye 4-wire Circuit Neutral Current

A 4-wire wye, 120/208Y- or 277/480Y-volt circuit will carry no current when the circuit is balanced, but will carry unbalanced current when the circuit is not balanced.

$I_{Neutral} = \sqrt{(N_{Line\ 1}{}^2 = N_{Line\ 2}{}^2 = N_{Line\ 3}{}^2) - (N_{Line\ 1} \times N_{Line\ 2}) + (N_{Line\ 2} \times N_{Line\ 3}) + (N_{Line\ 1} \times N_{Line\ 3})}$

$N_{Line\ 1}$ = Neutral Current Line 1

$N_{Line\ 2}$ = Neutral Current Line 1

$N_{Line\ 3}$ = Neutral Current Line 1

❑ Balanced Circuits

What is the neutral current for a 4-wire, 208Y/120-volt feeder where L_1 = 100 ampere, L_2 = 100 ampere, and L_3 = 100 ampere (Figure 2–27)?

(a) 50 ampere (b) 100 ampere (c) 125 ampere (d) 0 ampere

• Answer: (d) 0 ampere

$I_{Neutral} = \sqrt{(N_{Line\ 1}{}^2 \times N_{Line\ 2}{}^2 + N_{Line\ 3}{}^2) - (N_{Line\ 1} \times N_{Line\ 2}) + (N_{Line\ 2} \times N_{Line\ 3}) + (N_{Line\ 1} \times N_{Line\ 3})}$

$I_{Neutral} = \sqrt{(100\ Amps^2 + 100\ Amps^2 + 100\ Amps^2) - [(100A \times 100A) + (100A \times 100A) + (100A \times 100A)]} = \sqrt{0}$

$I_{Neutral} = 0$ amperes

❑ **Unbalanced Circuits**

What is the neutral current for a 4-wire, 208Y/120-volt feeder where L_1 = 100 ampere, L_2 = 100 ampere, and L_3 = 50 ampere (Figure 2–28)?

(a) 50 ampere (b) 100 ampere (c) 125 ampere (d) 0 ampere

• Answer: (a) 50 ampere

$I_{Neutral} = \sqrt{(N_{Line\ 1}{}^2 \times N_{Line\ 2}{}^2 + N_{Line\ 3}{}^2) - (N_{Line\ 1} \times N_{Line\ 2}) + (N_{Line\ 2} \times N_{Line\ 3}) + (N_{Line\ 1} \times N_{Line\ 3})}$

$I_{Neutral} = \sqrt{(100A^2 + 100A^2 + 50A^2) - [(100A \times 100A) + (100A \times 50A) + (100A \times 50A)]}$

$= \sqrt{2{,}500}$

$I_{Neutral} = 50$ amperes

Nonlinear Load Neutral Current

The neutral conductor of a balanced wye, 4-wire circuit will carry current when supplying power to nonlinear loads, such as computers, copy machines, laser printers, fluorescent lighting, etc. The current can be as much as two times the phase current, depending on the harmonic content of the loads.

❑ **Nonlinear Loads**

What is the neutral current for a 4-wire, 208Y/120-volt feeder supplying power to a nonlinear load, where L_1 = 100 ampere, L_2 = 100 ampere, and L_3 = 100 ampere? Assume that the harmonic content results in neutral current equal to 200 percent of the phase current (Figure 2–29).

(a) 80 ampere (b) 100 ampere (c) 125 ampere (d) 200 ampere

• Answer: (d) 200 ampere

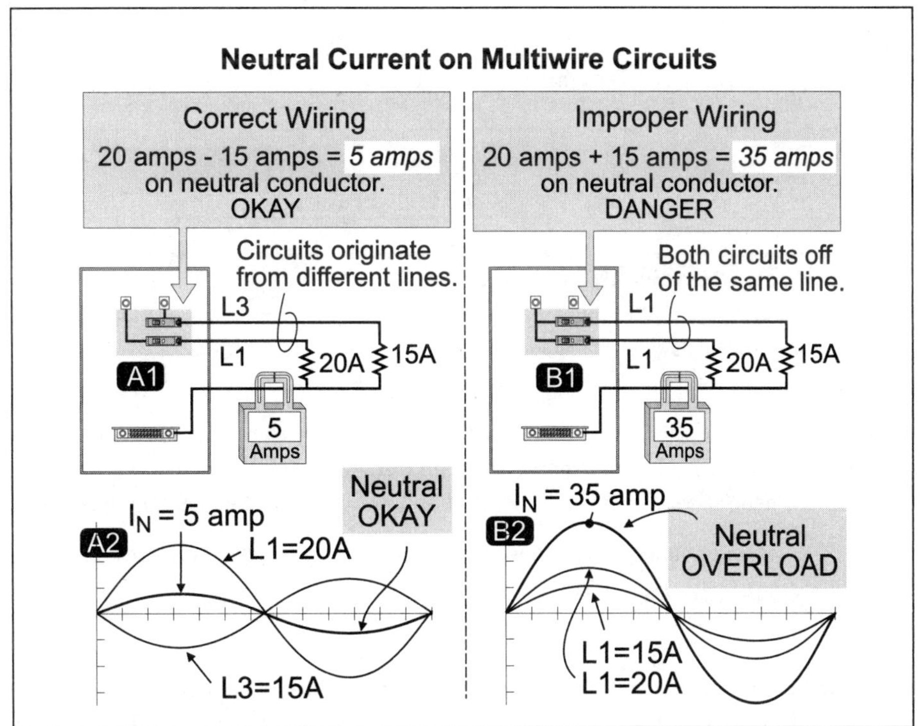

Figure 2–30
Neutral Current on Multiwire Circuits

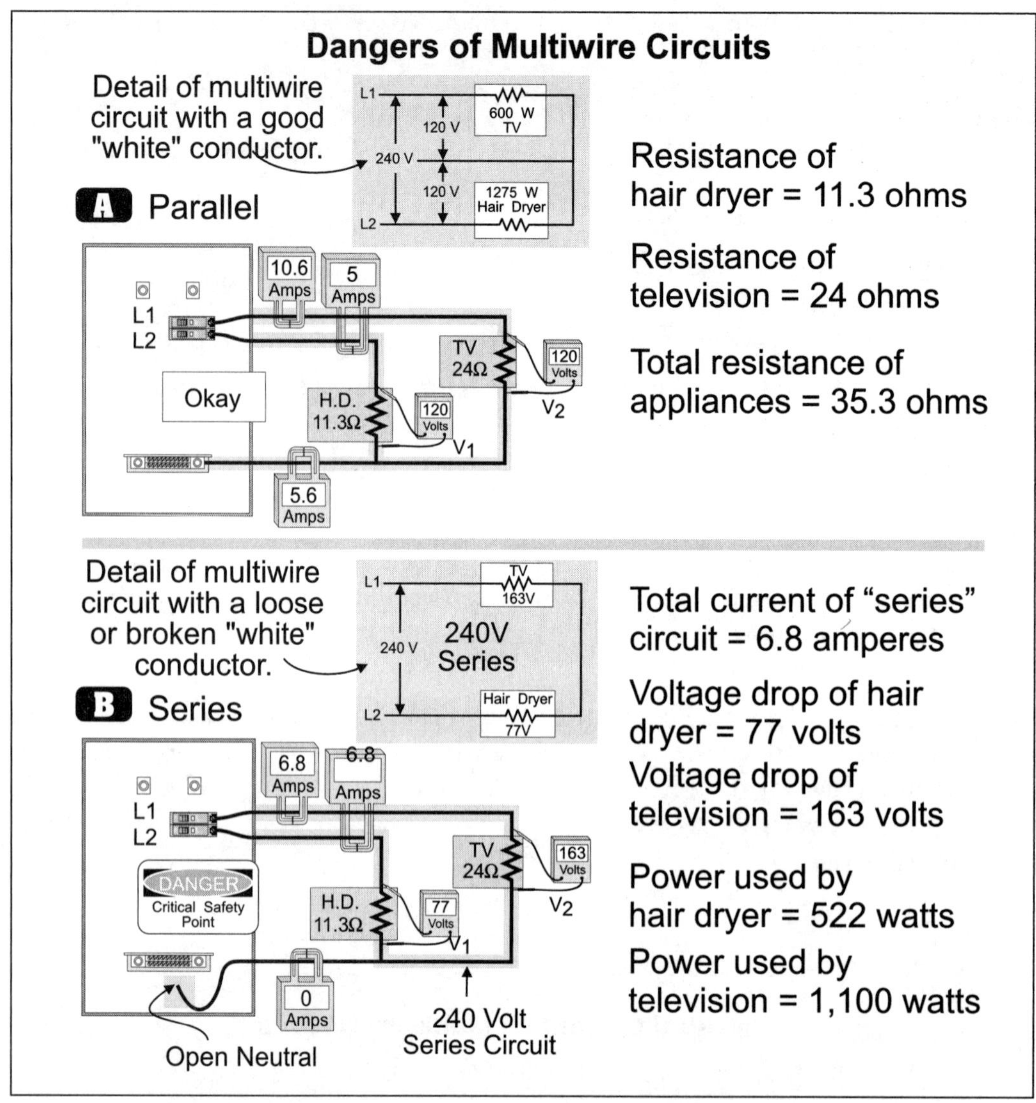

Figure 2–31
Neutral Current on Multiwire Circuits

2–10 DANGERS OF MULTIWIRE BRANCH CIRCUIT

Improper wiring, or mishandling of multiwire branch circuits, can cause excessive neutral current (overload) or destruction of electrical equipment because of overvoltage if the neutral conductor is opened.

Overloading the Neutral

If the ungrounded conductors (hot wires) of a multiwire branch circuit are connected to different phases, the current on the grounded (neutral) conductor will cancel. If the ungrounded conductors are not connected to different phases, the current from each phase will add on the neutral conductor. This can result in an overload of the neutral conductor (Figure 2–30).

Note: Overloading of the neutral conductor will cause the insulation to look discolored due to excessive heat. Now you know why the neutral wires sometimes look burned.

Overvoltage

If the neutral conductor of a multiwire branch circuit is opened, the multiwire branch circuit changes from a parallel circuit into a series circuit. Instead of two 120-volt circuits, there is now one 240-volt circuit, which can result in fires and the destruction of the electric equipment because of overvoltage (Figure 2–31).

To determine the operating voltage of each load in an open multiwire branch circuit, use the following steps:

Step 1: ➛ Determine the resistance of each appliance, $R = E^2/P$.
E = Appliance voltage nameplate rating
P = Appliance power nameplate rating

Step 2: ➛ Determine the circuit resistance, $R_T = R_1 + R_2$.

Step 3: ➛ Determine the current of the circuit, $I = E_S/R_T$.
E_S = Voltage Source
R_T = Resistance total, from Step 2

Step 4: ➛ Determine the voltage for each appliance, $E = I_T \times R$.
I_T = Current of the circuit
R = Resistance of each resistor

Step 5: ➛ Determine the power consumed of each appliance, $P = E^2/R$.
E = Voltage the appliance operates at (squared)
R = Resistance of the appliance

❑ **At what voltage does each of the loads operate if the neutral is opened (Figure 2–31)?**

Step 1: ➛ Determine the resistance of each appliance, $R = E^2/P$.
Hair dryer rated 1,275 watts at 120 volts.
R = 120 volts²/1,275 watts = 11.3 ohms
Television rated 600 watts at 120 volts.
R = 120 volts²/600 watts = 24 ohm

Step 2: ➛ Determine the circuit resistance, $R_T = R_1 + R_2$.
R_T = 11.3 ohm + 24 ohm = 35.3 ohm

Step 3: ➛ Determine the current of the circuit, $I = E_S/R_T$.
E_S = Voltage Source, R_T = Resistance Total

$$I = \frac{240 \text{ volts}}{35.3 \text{ ohms}} = 6.8 \text{ ampere}$$

Step 4: ➛ Determine the voltage for each appliance, $E = I_T \times R$.
I_T = Current of the circuit, R = Resistance of each resistor
Hair dryer: 6.8 ampere × 11.3 ohm = 76.84 volts
Television: 6.8 ampere × 24 ohm = 163.2 volts

Step 5: ➛ Determine the power consumed of each appliance, $P = E^2/R$.
E^2 = Voltage the appliance operates at (squared)
R = Resistance of the appliance
Hair Dryer: P = 76.8 volts²/11.3 ohm = 522 watts
Television: P = 163.2 volt²/24 ohm = 1,110 watts

Note: The 600 watt, 120-volt rated TV operates at 163 volts and consumes 1,110 watts. You can kiss this TV good-bye. Because of the dangers associated with multiwire branch circuits, do not use them for sensitive or expensive equipment such as computers, stereos, etc.

Unit 2 – Electrical Circuits Summary Questions

Part A – Series Circuits

Introduction to Series Circuits

1. A closed-loop circuit is a circuit in which a specific amount of current leaves the voltage source and flows through every electrical device in a single path before it returns to the voltage source.
(a) True (b) False

2. Closed-loop circuits are typically used for signal and control circuits.
(a) True (b) False

2–1 Understanding Series Calculations

3. Resistance opposes the flow of electrons. In a series circuit, the total circuit resistance is equal to the sum of all the series resistors.
(a) True (b) False

4. The opposition to current flow results in _____ .
(a) current (b) voltage (c) voltage drop (d) none of these

5. In a series circuit, the current is _____ through the transformer, the conductors, and the appliance.
(a) proportional (b) distributed (c) additive (d) constant

6. • When power supplies are connected in series, the voltage remains the same, provided that all the polarities are connected properly.
(a) True (b) False

7. The power consumed in a series circuit is equal to the power of the largest resistor in the series circuit.
(a) True (b) False

Part B – Parallel Circuits

Introduction to Parallel Circuits

8. A _____ circuit is a circuit in which current leaves the voltage source, branches through different parts of the circuit in different magnitudes, and then branches back to the voltage source.
(a) series (b) parallel (c) series-parallel (d) multiwire

2–4 Understanding Parallel Calculations

9. • The power supply provides the pressure needed to move the electrons; however, the _____ oppose(s) the current flow.
(a) power supply (b) conductors (c) appliances (d) all of these

10. When power supplies are connected in parallel, the amp-hour capacity remains the same.
(a) True (b) False

11. The total current of a parallel circuit is equal to the sum of the branch currents. The current in each branch can be calculated by the formula: $I = E/R$.
(a) True (b) False

12. When current flows through a resistor, power is consumed. The power consumed of each branch can be determined by the formula: $P = I^2 \times R$. The total power consumed in a parallel circuit is equal to the largest branch power.
(a) True (b) False

2–5 Parallel Circuit Resistance Calculations

13. The basic method(s) of calculating total resistance of a parallel circuit is/are _____ .
(a) equal resistor method (b) product of the sum method
(c) reciprocal method (d) all of these

14. The total resistance of three 6 ohm resistors in parallel is _____ .
(a) 6 ohm (b) 12 ohm (c) 18 ohm (d) none of these

15. The circuit resistance of a 600-watt coffee pot and a 1,000-watt skillet is _____ when connected to a 120-volt parallel circuit.
(a) 24 ohm (b) 14.4 ohm (c) 38.4 ohm (d) 9 ohm

16. The resistance total of a 20 ohm, 20 ohm, and 10 ohm resistor in parallel is _____ .
(a) 5 ohm (b) 20 ohm (c) 30 ohm (d) 50 ohm

2–6 Parallel Circuit Summary

17. • Which of the following statements is/are true about parallel circuits?
I. The total resistance of a parallel circuit is less than the smallest resistor of the circuit.
II. Current total is equal to the sum of the branch currents.
III. The power of all resistors is equal to the sum of the branch powers.
(a) I and II (b) I, II, and III (c) II and III (d) I and III

Part C – Series-Parallel and Multiwire Branch Circuits

Introduction to Series-Parallel Circuits

18. A _____ is a circuit that contains some resistors in series and some resistors in parallel to each other.
(a) parallel circuit (b) series circuit (c) series-parallel circuit (d) none of these

Part D – Multiwire Branch Circuits

Introduction to Multiwire Branch Circuits

19. A multiwire branch circuit has two or more ungrounded conductors having a potential difference between them and an equal difference of potential between each ungrounded conductor and the grounded (neutral) conductor.
(a) True (b) False

2–9 Neutral Current Calculations

20. • The current on the grounded (neutral) conductor of a 2-wire circuit will be _____ of the current on the ungrounded conductor.
(a) 50% (b) 70% (c) 80% (d) 100%

Three-Wire Circuits

21. • A balanced 120/240 volt, single-phase, 3-wire circuit is connected so the ungrounded conductors are from different transformer phases (Line 1 and Line 2). The current on the grounded (neutral) conductor will be _____ of the ungrounded conductor current.
(a) 0% (b) 70% (c) 80% (d) 100%

22. A 3-wire, 120/240 volt circuit will carry 10 ampere unbalanced neutral current if: Line 1 = 20 ampere and Line 2 = 10 ampere.
(a) True (b) False

23. • What is the neutral current for a 20 ampere, 3-wire, 208Y/120-volt circuit?
(a) 0 ampere (b) 10 ampere (c) 20 ampere (d) 40 ampere

Four-Wire Circuits

24. The neutral of a 4-wire, 208Y/120- or 480Y/277-volt circuit will carry the unbalanced current when the circuit is balanced.
(a) True (b) False

25. What is the neutral current for a 4-wire, 208Y/120-volt circuit, L1 = 20 ampere, L2 = 20 ampere, L3 = 20 ampere?
(a) 0 ampere (b) 10 ampere (c) 20 ampere (d) 40 ampere

26. • The grounded (neutral) conductor of a balanced wye 4-wire circuit will carry no current when supplying power to balanced nonlinear loads.
(a) True (b) False

27. • A 3-phase, 4-wire multiwire branch circuit has 20 ampere on each 120-volt phase. The neutral conductor could carry as much as _____ .
(a) 0 amperes (b) 10 amperes (c) 15 amperes (d) 40 amperes

2–10 Dangers of Multiwire Branch Circuit

28. Improper wiring or mishandling of multiwire branch circuits can cause _____ connected to the circuit.
(a) overloading of the ungrounded conductors
(b) overloading of the grounded (neutral) conductors
(c) destruction of equipment because of overvoltage
(d) b and c

29. • Because of the dangers associated with the open neutral (grounded conductor), the continuity of the _____ conductor cannot be dependent on the receptacle [*Section 300-13(b)*].
(a) ungrounded (b) grounded (c) a and b (d) none of these

☆ Challenge Questions

Part A – Series Circuits

30. • Two resistors, one rated 4 ohm and one rated 8 ohm, are connected in series. If the voltage drop across both resistors is 12 volts, then the current that would pass through the 4 ohm resistor is _____ .
(a) 1 ampere (b) 2 ampere
(c) 4 ampere (d) 8 ampere

31. • A series circuit has four 40 ohm resistors and the power supply is 120 volts. The voltage drop of each resistor would be _____ .
(a) one-quarter of the source voltage
(b) 30 volts
(c) the same across each resistor
(d) all of these

32. • The power consumed in a series circuit is _____ .
(a) the sum of the power consumed of each load
(b) determined by the formula $P_T = I^2 \times R_T$
(c) determined by the formula $P_T = E \times I$
(d) all of these

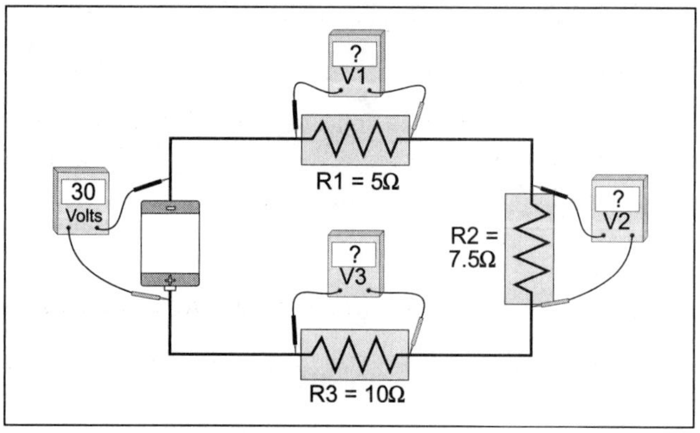

Figure 2–32

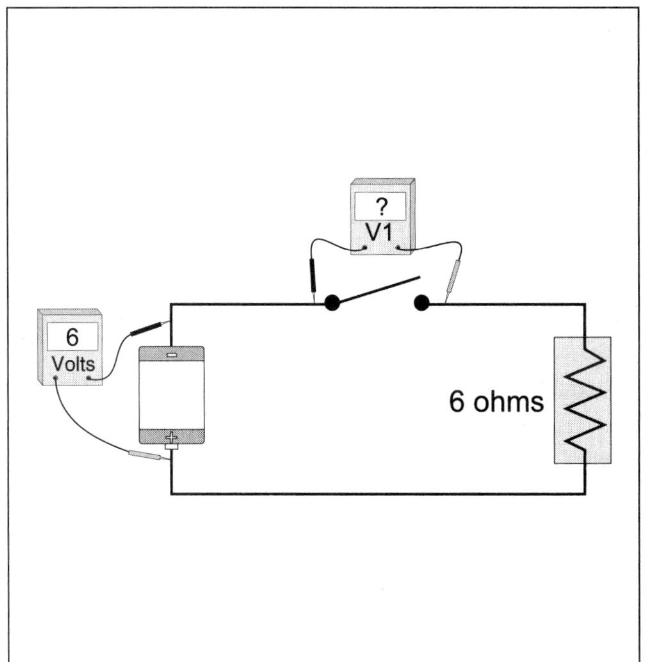

Figure 2–33

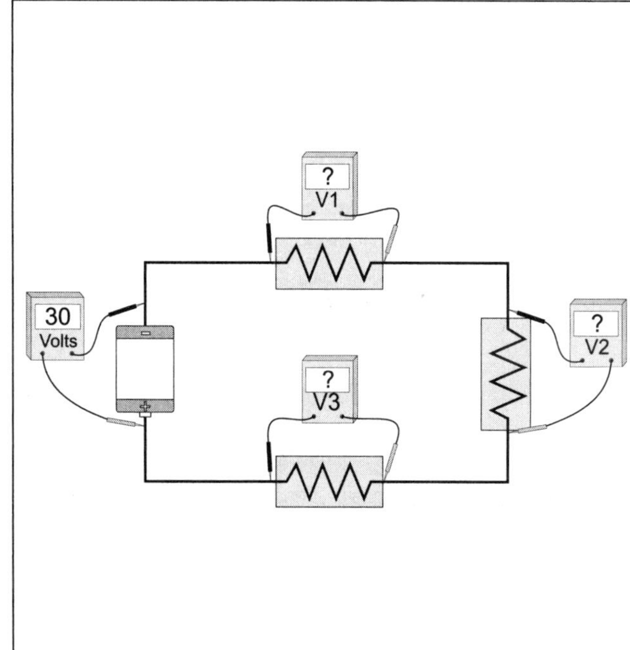

Figure 2–34

33. • The reading on voltmeter 2 (V_2) is _____ (Figure 2–32).
(a) 5 volts (b) 7 volts (c) 10 volts (d) 6 volts

34. The voltmeter connected across the switch would read _____ (Figure 2–33).
(a) 3 volts (b) 12 volts (c) 6 volts (d) 18 volts

Part B – Parallel Circuits

35. • In general, when multiple light bulbs are wired in a single fixture, they are connected in _____ to each other.
(a) series (b) series-parallel (c) parallel (d) order of wattage

36. • A single-phase, dual-rated, 120/240-volt motor will have its winding connected in _____ when supplied by 120 volts.
(a) series (b) parallel (c) series-parallel (d) parallel-series

37. • The voltmeters shown in Figure 2–34 are connected _____ each of the loads.
(a) in series to (b) across (c) in parallel to (d) b and c

38. • If the supply voltage remains constant, resistors will consume the most power when they are connected _____ .
(a) all in series
(b) all in parallel
(c) with two parallel pairs in series
(d) with one pair in parallel and the other two in series

Circuit Resistance

39. A parallel circuit has three resistors. One resistor is rated 2 ohm, one is 3 ohm, and the other is 7 ohm. The total resistance of the parallel circuit is _____ . Remember the total resistance of any parallel circuit is always less than the smallest resistor.
(a) 12 ohm (b) 1 ohm (c) 42 ohm (d) 1.35 ohm

Figure 2–35 applies to the next three questions.

40. • The total current of the circuit can be measured by ammeter _____ (Figure 2–35).
(a) 1 (b) 2 (c) 3 (d) none of these

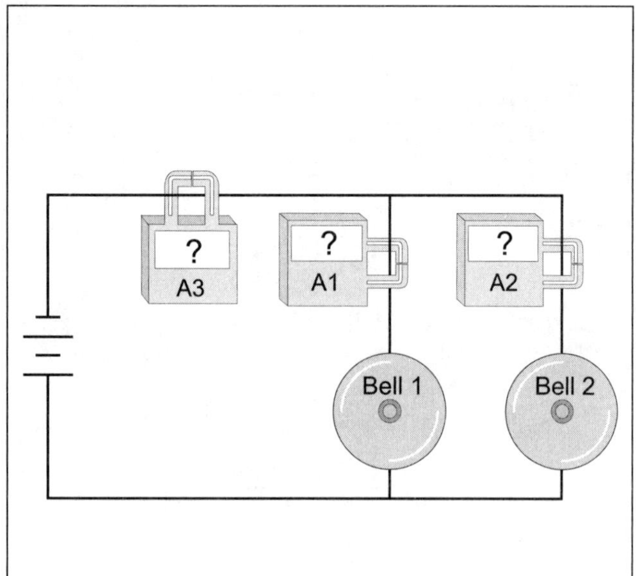

Figure 2–35

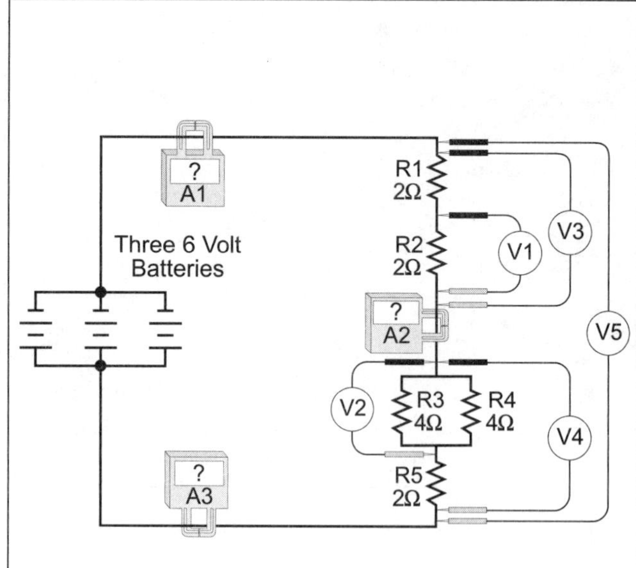

Figure 2–36

41. • If Bell 2 consumed 12 watts of power when supplied by two 12-volt batteries (connected in series), the resistance of this bell would be _____ (Figure 2–35).
(a) 9.6 ohm (b) 44 ohm (c) 576 ohm (d) 48 ohm

42. Determine the total circuit resistance of the parallel circuit based on the following facts (Figure 2–35):
1. The current on ammeter 1 reads 0.75 ampere.
2. The voltage of the circuit is 30 volts.
3. Bell 2 has a resistance of 48 ohm.
Tip: Total resistance of a parallel circuit is always less than the smallest resistor.
(a) 22 ohm (b) 48 ohm (c) 1920 ohm (d) 60 ohm

Part C – Series-Parallel Circuits

Figure 2–36 applies to the next three questions:

43. • The total current of this circuit can be read on _____ (Figure 2–36).
(a) ammeter 1 (b) ammeter 2 (c) ammeter 3 (d) all of these

44. • The reading of V_2 is _____ (Figure 2–36).
(a) 1.5 volts (b) 4 volts (c) 4 volts (d) 8 volts

45. • The reading of V_4 is _____ (Figure 2–36).
(a) 1.5 volts (b) 3 volts (c) 5 volts (d) 8 volts

Figure 2–37 applies to the next two questions:

46. Resistor R_1 has a resistance of 5 ohm, resistors R_2, R_3, and R_4 have a resistance of 15 ohm each. The total resistance of this series-parallel circuit is _____ (Figure 2–37).
(a) 50 ohm (b) 35 ohm (c) 25 ohm (d) 10 ohm

47. • What is the voltage drop across R_1? R_1 is 5 ohm and total resistance of R_2, R_3, and R_4 is 5 ohm (Figure 2–37).
(a) 60 volts (b) 33 volts (c) 40 volts (d) 120 volts

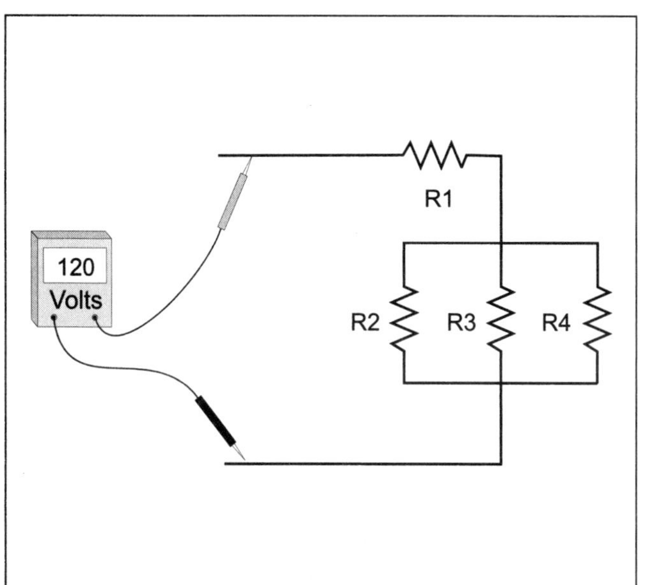

Figure 2–37

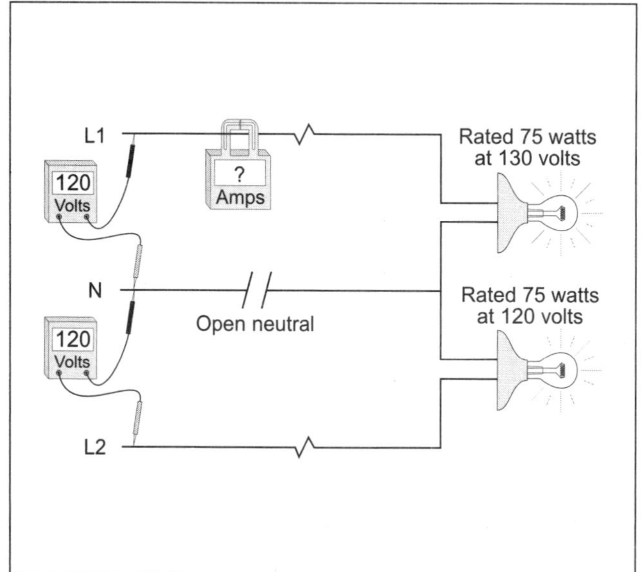

Figure 2–38

Part D – Multiwire Branch Circuits

48. • If the neutral of the circuit in the diagram is opened, the circuit becomes one series circuit of 240 volts. Under this condition, the current of the circuit is _____ . Tip: Determine the total resistance (Figure 2–38).
(a) 0.67 ampere (b) 0.58 ampere (c) 2.25 ampere (d) 0.25 ampere

Unit 3

Understanding Alternating Current

OBJECTIVES

After reading this unit, the student should be able to briefly explain the following concepts:

Part A – Alternating Current Fundamentals

Alternating current
Armature turning frequency
Current flow
Generator
Magnetic cores
Phase differences in degrees
Phase – in and out
Values of alternating current
Waveform

Part B and C – Induction and Capacitance

Charge, discharging and testing of capacitors
Conductor impedance
Conductor shape
Induced voltage and applied current
Magnetic cores
Use of capacitors

Part D – Power Factor and Efficiency

Apparent power (volt-amperes)
Efficiency
Power factor
True power (watts)

After reading this unit, the student should be able to briefly explain the following terms:

Part A – Alternating Current Fundamentals

Ampere-turns
Armature speed
Coil
Conductor cross-sectional area
Effective (RMS)
Effective to peak
Electromagnetic field
Frequency
Impedance
Induced voltage
Magnetic field
Magnetic flux lines
Peak
Peak to effective
Phase relationship
RMS
Root-mean-square
Self-inductance
Skin effect
Waveform

Part B and C – Induction and Capacitance

Back-EMF
Capacitance
Capacitive reactance
Capacitor
Coil
Counterelectromotive force
Eddy currents
Farads
Frequency
Henrys
Impedance
Induced voltage
Induction
Inductive reactance
Phase relationship
Self-inductance
Skin effect

Part D – Power Factor and Efficiency

Apparent power
Efficiency
Input watts
Output watts
Power factor
True power
Volts-amperes
Wattmeter

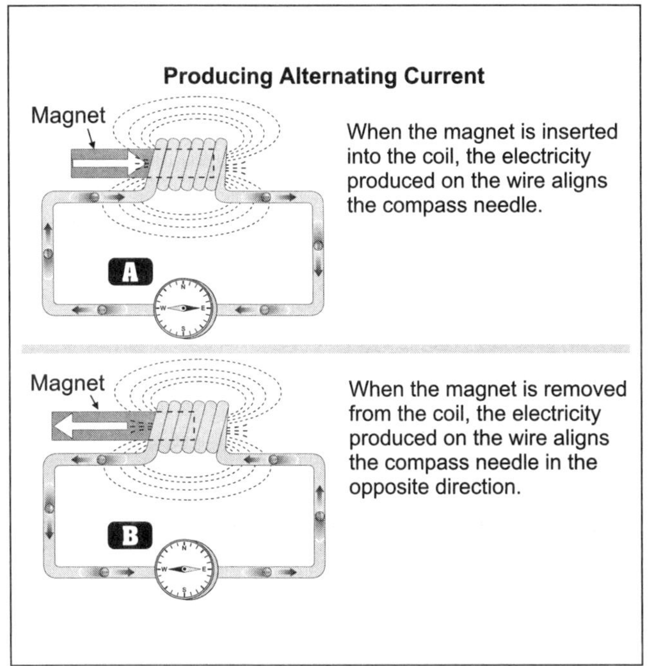

Figure 3–1
Producing Alternating Current

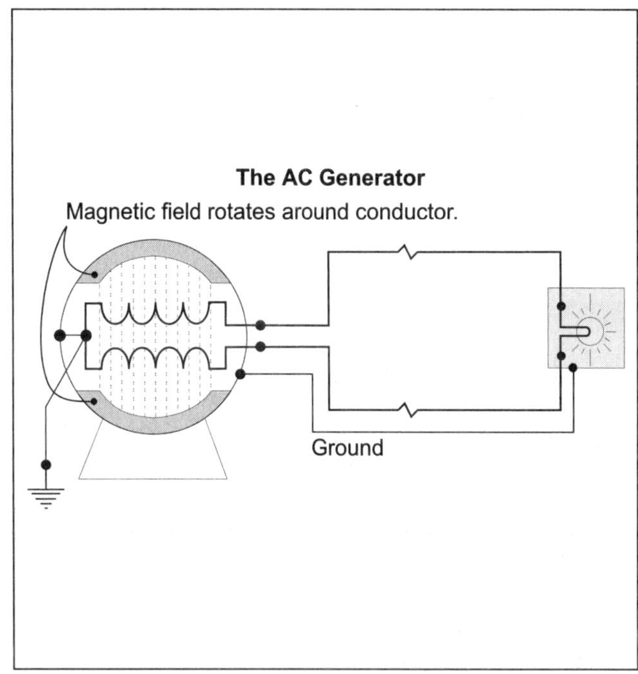

Figure 3–2
The AC Generator

PART A – ALTERNATING CURRENT FUNDAMENTALS

3–1 CURRENT FLOW

For current to flow in a circuit, the circuit must be a closed loop and the power supply must push the electrons through the completed circuit. The transfer of electrical energy can be accomplished by electrons flowing in one constant direction (direct current) or by electrons flowing in one direction and then reversing in the other direction (alternating current).

3–2 ALTERNATING CURRENT

Alternating current is generally used instead of direct current for electrical systems and building wiring. This is because alternating current can easily have voltage variations (transformers). In addition, alternating current can be transmitted inexpensively at high voltage over long distances resulting in reduced voltage drop of the power distribution system as well as smaller power distribution wire and equipment. Alternating current can be used for certain applications for which direct current is not suitable.

Alternating current is produced when electrons in a conductor are forced to move because there is a moving magnetic field. The lines of force from the magnetic field cause the electrons in the wire to flow in a specific direction. When the lines of force of the magnetic field move in the opposite direction, the electrons in the wire are forced to flow in the opposite direction (Figure 3–1). Electrons will flow only when there is relative motion between the conductors and the magnetic field.

3–3 ALTERNATING CURRENT GENERATOR

An alternating current generator consists of many loops of wire that rotate between the flux lines of a magnetic field. Each conductor loop travels through the magnetic lines of force in opposite directions, causing the electrons within the conductor to move in a specific direction (Figure 3–2). The force on the electrons caused by the magnetic flux lines is called voltage, or electromotive force (EMF), and is abbreviated as V or E.

The magnitude of the electromotive force is dependent on the number of turns (wraps) of wire, the strength of the magnetic field, and the speed at which the coil rotates. The rotating conductor loop mounted on a shaft is called a rotor or armature. Slip, or collector rings, and carbon brushes are used to connect the output voltage from the generator to an external circuit.

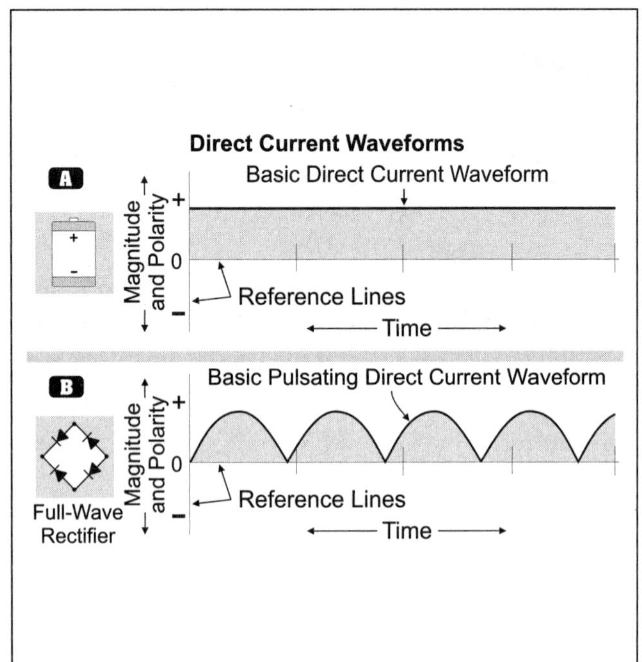

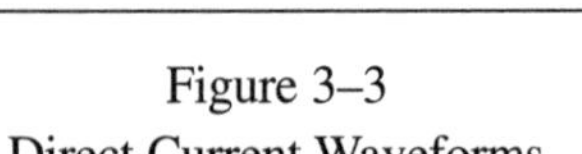

Figure 3–3
Direct Current Waveforms

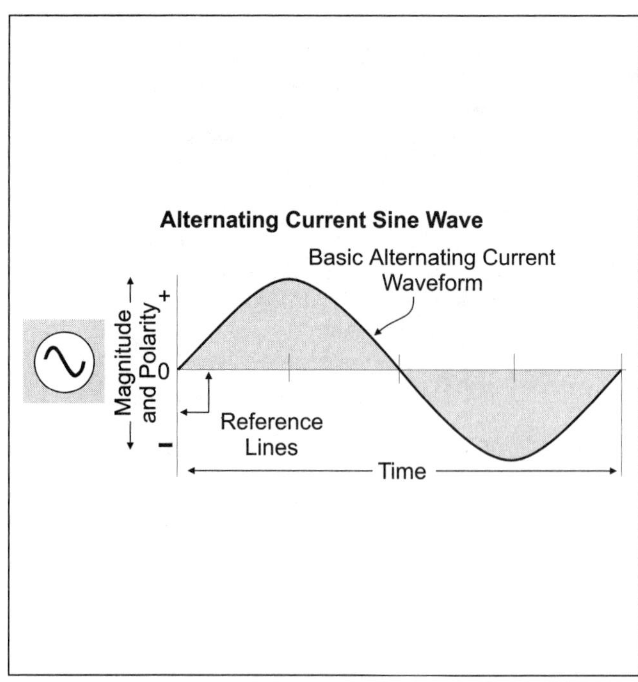

Figure 3–4
Alternating Current Sine Wave

3–4 WAVEFORM

A waveform is a pictorial view of the shape and magnitude of the level and direction of current or voltage over a period of time. The waveform represents the magnitude and direction of the current or voltage over time in relationship with the generators armature.

Direct Current Waveform

The polarity of the voltage of direct current is positive, and the current flows through the circuit in the same direction at all times. In general, direct current voltage and current remain the same magnitude, particularly when supplied from batteries or rectifiers (Figure 3–3).

Alternating Current Waveform

The waveform for alternating current displays the level and direction of the current and voltage for every instant of time for one full revolution of the armature. When the waveform of an alternating current circuit is symmetrical, with positive polarity above and negative below the zero reference level, the waveform is called a sine wave (Figure 3–4).

3–5 ARMATURE TURNING FREQUENCY

Frequency is a term used to indicate the number of times a generators armature turns one full revolution (360°) in one second. This is expressed as hertz, or cycles per second, and is abbreviated as Hz or cycles per second (cps). In the United States, frequency is not a problem for most electrical work because it remains constant at 60 hertz (Figure 3–5).

Armature Speed

A current or voltage that is produced when the generators armature makes one cycle (360°) in 1/60th of a second is called 60 Hz. This is because the armature travels at

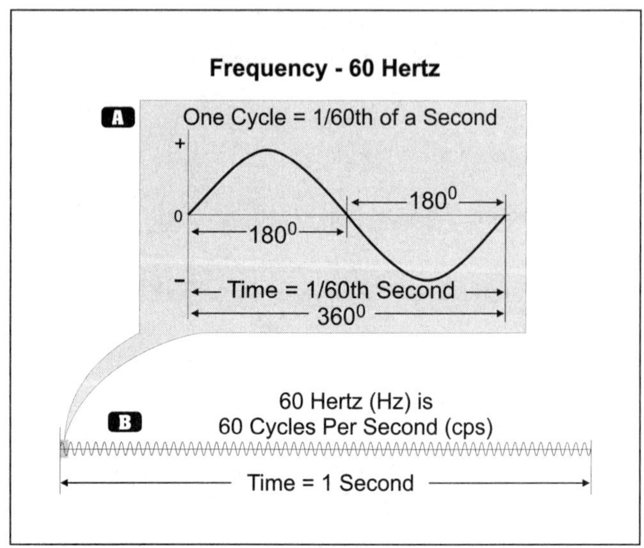

Figure 3–5
Frequency – 60 Hertz

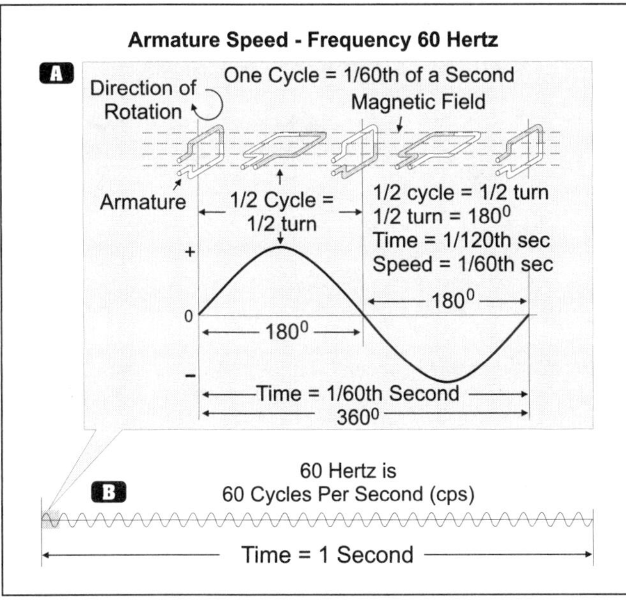

Figure 3–6
Armature Speed

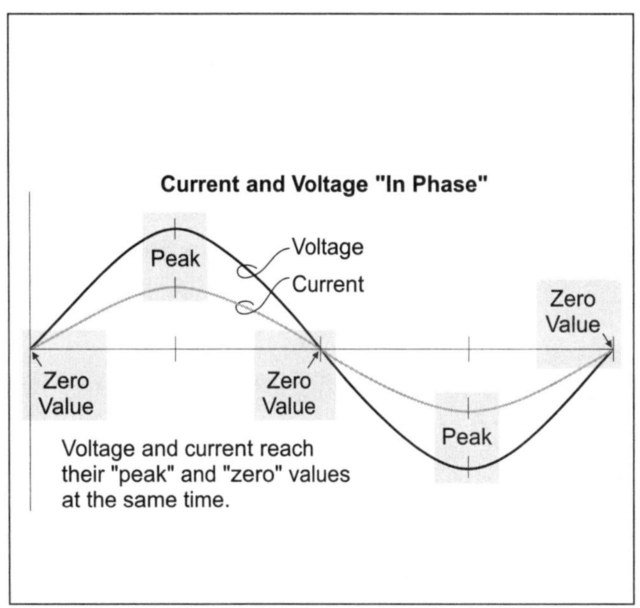

Figure 3–7
Current and Voltage "In Phase"

a speed of one cycle every 1/60th of a second. A 60 Hz circuit takes 1/60th of a second for the armature to complete one full cycle (360°), and it will take 1/120th of a second to complete $^1/_2$ cycle (180°) (Figure 3–6).

3–6 PHASE – IN AND OUT

Phase is a term used to indicate the time relationship between two waveforms, such as voltage to current, or voltage to voltage. *In phase* means that two waveforms are in step with each other, that is, they cross the horizontal zero axis at the same time (Figure 3–7). The terms lead and lag are used to describe the relative position, in degrees, between two waveforms. The waveform that is ahead *leads* and the waveform behind *lags* (Figure 3–8).

3–7 PHASE DIFFERENCES IN DEGREES

Phase differences are expressed in *degrees,* with one full revolution of the armature equal to 360°. Single-phase, 120/240 volt power is out of phase by 180°, and three-phase power is out of phase by 120° (Figure 3–9).

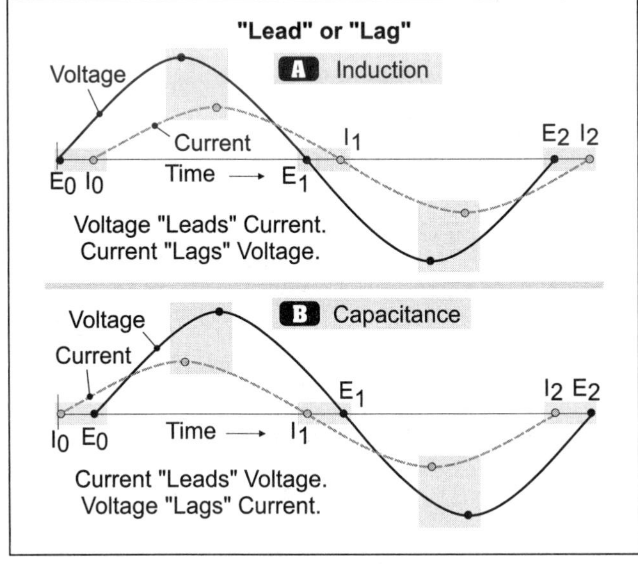

Figure 3–8
Lead or Lag

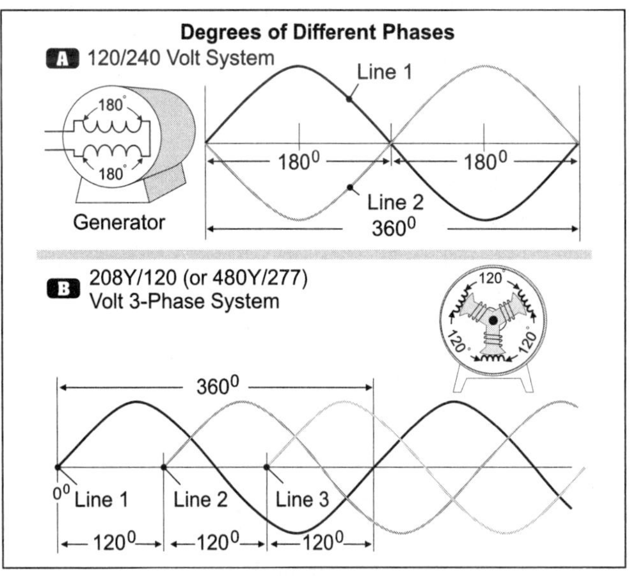

Figure 3–9
Degrees of Different Phases

3–8 VALUES OF ALTERNATING CURRENT

There are many different types of values in alternating current, the most important are peak (maximum) and effective. In alternating current systems, effective is the value that will cause the same amount of heat to be produced, as in a direct current circuit containing only resistance. Another term for effective is RMS, which is root-mean-square. In the field, whenever you measure voltage, you are always measuring effective voltage. Newer types of clamp-on ammeter have the ability to measure peak as well as effective (RMS) current (Figure 3–10).

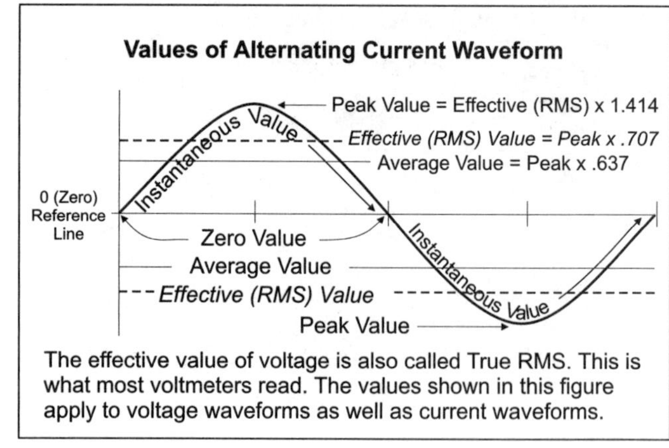

Figure 3–10
Values of Alternating Current Waveform

Note: The values of direct current generally remains constant and are equal to the effective values of alternating current.

❑ **Current**

Twelve ampere effective alternating current will have the same heating effect as _____ ampere direct current.

(a) 7 (b) 9 (c) 11 (d) 12

• Answer: (d) 12 ampere

❑ **Voltage**

One hundred and twenty volt effective alternating current will produce the same heating effect as _____ volts direct current.

(a) 70 (b) 90 (c) 120 (d) 170

• Answer: (c) 120 volts

Effective to Peak

To convert effective to peak, or vice versa, the following formulas should be helpful:

Peak = Effective (RMS) × 1.414 or

Peak = Effective (RMS)/0.707

❑ **Effective to Peak**

What is the peak voltage of a 120-volt effective circuit?

(a) 120 volts (b) 170 volts (c) 190 volts (d) 208 volts

• Answer: (b) 170 volts

Peak Volts = Effective (RMS) volts × 1.414

Peak Volts = 120 volts × 1.414 = 170 volts

Peak-to-Effective

Peak to effective can be determined by the formula:

Effective (RMS) = Peak × 0.707

❑ **Peak-to-Effective**

What is the effective current of 60 ampere peak?

(a) 42 ampere (b) 60 ampere (c) 72 ampere (d) none of these

• Answer: (a) 42 ampere

Effective Current = Peak Current × 0.707

Effective Current = 60 ampere × 0.707 = 42 ampere

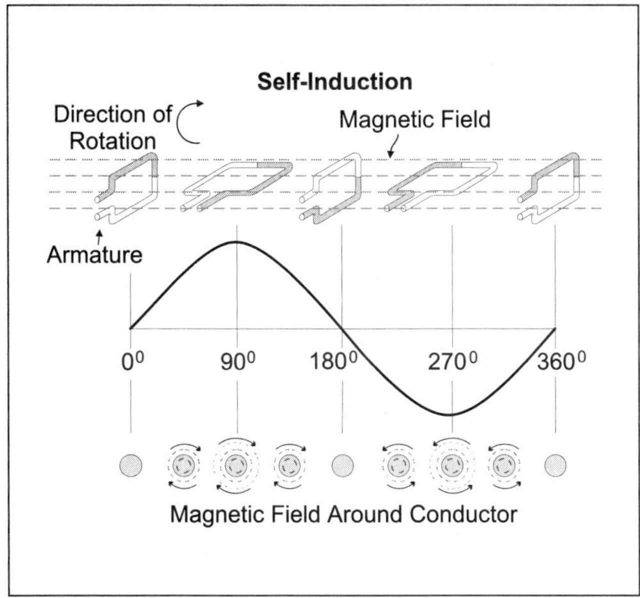

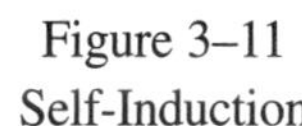
Figure 3–11
Self-Induction

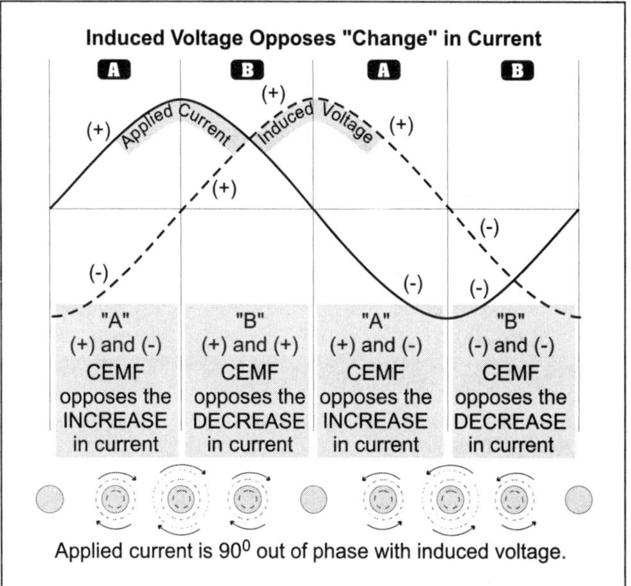

Figure 3–12
Induced Voltage

PART B – INDUCTION

INDUCTION INTRODUCTION

When a magnetic field moves through a conductor, it causes the electrons in the conductor to move. The movement of electrons caused by electromagnetism is called *induction.* In an alternating current circuit, the current movement of the electrons increases and decreases with the rotation of the *armature* through the magnetic field. As the current flowing through the conductor increases or decreases, it causes an expanding and collapsing electromagnetic field within the conductor (Figure 3–11). This varying electromagnetic field within the conductor causes the electrons in the conductor to move at 90° to the flowing electrons within the conductor. This movement of electrons because of the conductor's electromagnetic field is called self-induction. Self-induction (induced voltage) is commonly called *counterelectromotive-force (CEMF),* or back-EMF.

3–9 INDUCED VOLTAGE AND APPLIED CURRENT

The induced voltage in the conductor because of the conductor's counterelectromotive-force (CEMF) is always 90° out of phase with the applied current (Figure 3–12). When the current in the conductor increases, the induced voltage tries to prevent the current from increasing by storing some of the electrical energy in the magnetic field around the conductor. When the current in the conductor decreases, the induced voltage attempts to keep the current from decreasing by releasing the energy from the magnetic field back into the conductor. The opposition to any change in current because of induced voltage is called inductive reactance, which is measured in ohms, abbreviated as X_L, and known as *Lenz's Law.* Inductive reactance changes proportionally with frequency and can be found by the formula:

$$\mathbf{X_L = 2\pi fL}$$

$\pi = 3.14$

f = frequency

L = Inductance in henrys

❑ Inductive Reactance

What is the inductive reactance of a conductor in a transformer that has an inductance of 0.001 henrys? Calculate at 60 Hz, 180 Hz, and 300 Hz (Figure 3–13).

• Answers:

$X_L = 2\pi fL$

X_L at 60 Hz, $= 2 \times 3.14 \times 60 \text{ Hz} \times 0.001 \text{ henrys} = 0.377 \text{ ohm}$

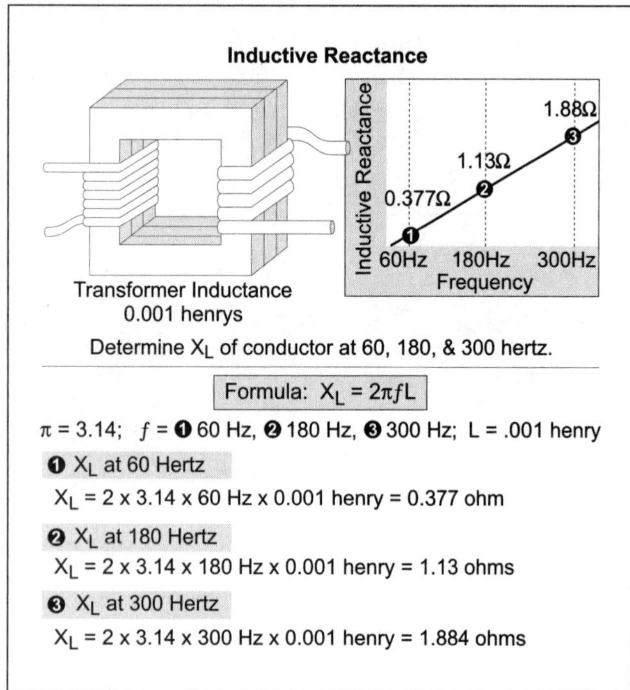

Figure 3–13
Inductive Reactance

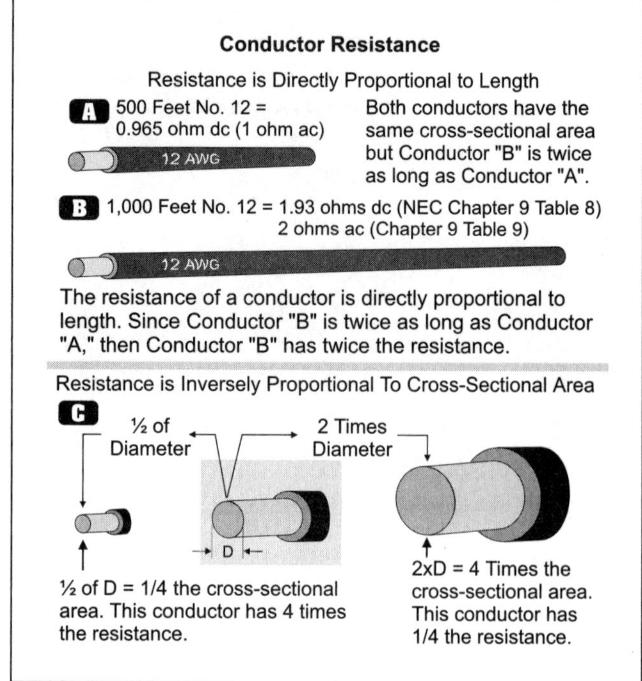

Figure 3–14
Conductor Resistance

X_L at 180 Hz, = 2 × 3.14 × 180 Hz × 0.001 henrys = 1.13 ohm

X_L at 300 Hz, = 2 × 3.14 × 300 Hz × 0.001 henrys = 1.884 ohm

3–10 CONDUCTOR IMPEDANCE

The resistance of a conductor is a physical property of the conductors that oppose the flow of electrons. This resistance is directly proportional to the conductor length and is inversely proportional to the conductor cross-sectional area. This means that the longer the wire, the greater the conductor resistance; the smaller the cross-sectional area of the wire, the greater the conductor resistance (Figure 3–14).

The opposition to alternating current flow due to inductive reactance and conductor resistance is called impedance (Z). Impedance is only present in alternating current circuits and is measured in ohms. This opposition to alternating current (Z) is greater than the resistance (R) of a direct current circuit because of *counterelectromotive-force, eddy currents*, and *skin effect.*

Eddy Currents and Skin Effect

Eddy currents are small independent currents that are produced as a result of the expanding and collapsing magnetic field of alternating current flow (Figure 3–15). The expanding and collapsing magnetic field of an alternating current circuit also induces a counterelectromotive-force in the conductors that repel the flowing electrons towards the surface of the conductor (Figure 3–16). The counterelectromotive-force causes the circuit current to flow near the surface of the conductor rather than at the conductor's center. The flow of electrons near the surface is known as skin effect. Eddy currents and skin effect decrease the effective conductor cross-sectional area for current flow, which results in an increase in the conductor's impedance which is measured in ohms.

3–11 INDUCTION AND CONDUCTOR SHAPE

The shape into which the conductor is configured affects the amount of *self-induced voltage* in the conductor. If a conductor is coiled into adjacent loops *(winding),* the electromagnetic field *(flux lines)* of each conductor loop adds together to create a stronger overall magnetic field (Figure 3–17). The amount of self-induced voltage created within a conductor is directly proportional to the current flow, the length of the conductor, and the frequency at which the magnetic fields cut through the conductors (Figure 3–18).

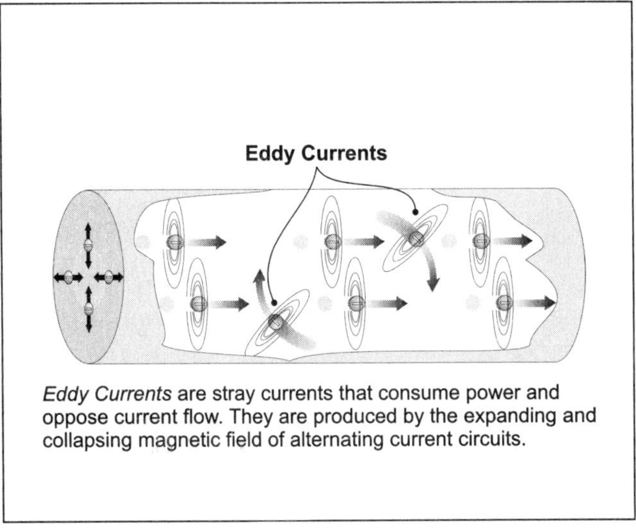

Figure 3–15
Eddy Currents

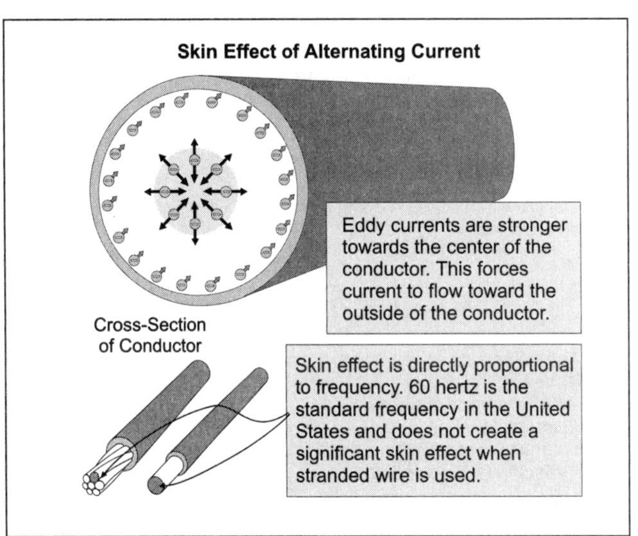

Figure 3–16
Skin Effect of Alternating Current

3–12 INDUCTION AND MAGNETIC CORES

Self-induction in a *coil* of conductors is affected by the winding current magnitude, the core material, the number of conductor loops *(turns),* and the spacing of the winding (Figure 3–19). An *iron core* within the conductor winding permits an easy path for the electromagnetic flux lines which produce a greater counterelectromotive-force than air core windings. In addition, conductor coils are measured in *ampere-turns*, that is, the current in ampere times the number of conductor loops or turns.

Ampere-Turns = Coil Ampere × Number of Coil Turns

❑ **Ampere-Turns**

A magnetic coil in a transformer has 1,000 turns and carries 5 ampere. What is the ampere-turns rating of this transformer (Figure 3–20)?

(a) 5,000 ampere-turns (b) 7,500 ampere-turns (c) 10,000 ampere-turns (d) none of these

• Answer: (a) 5,000 ampere-turns

Ampere-Turns = Ampere × Number of Coil Turns

Ampere-Turns = 5 ampere × 1,000 turns = 5,000

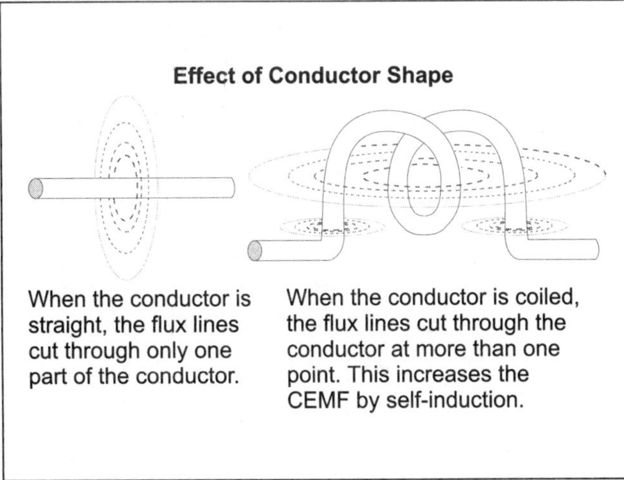

Figure 3–17
Effect of Conductor Shape

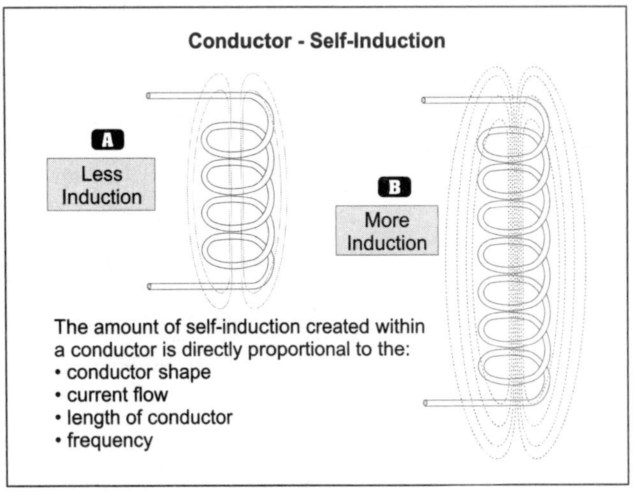

Figure 3–18
Conductor – Self-Induction

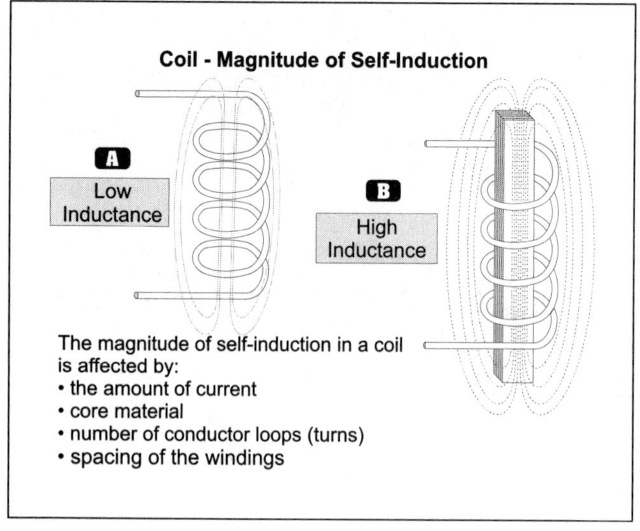

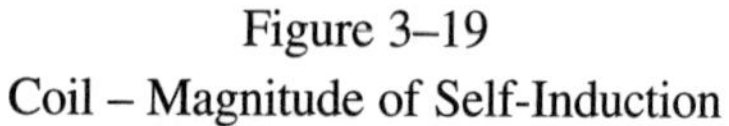
Figure 3–19
Coil – Magnitude of Self-Induction

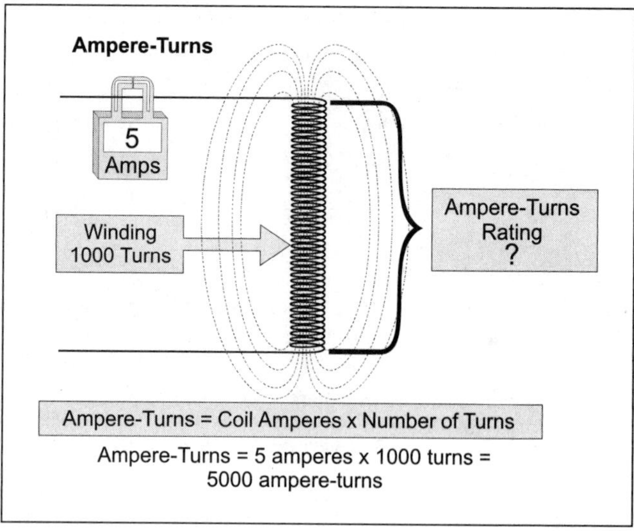

Figure 3–20
Ampere Turns

PART C – CAPACITANCE

CAPACITANCE INTRODUCTION

Capacitance is the property of an electric circuit that enables the storage of electric energy by means of an electrostatic field, much as a spring stores mechanical energy (Figure 3–21). Capacitance exists whenever an insulating material separates two objects that have a difference of potential. A capacitor is two metal foils separated by insulating material (wax paper) rolled together. Devices that intentionally introduce capacitance into circuits are called capacitors or condensers.

3–13 CHARGE, TESTING, AND DISCHARGING

When a capacitor has a potential difference between the plates, it is said to be *charged.* One plate has an excess of free electrons (–), and the other plate has a lack of electrons (+). Because of the insulation *(dielectric)* between the capacitor foils, electrons cannot flow from one plate to the other. However, there are electric lines of force between the plates, and this force is called the *electric field* (Figure 3–22).

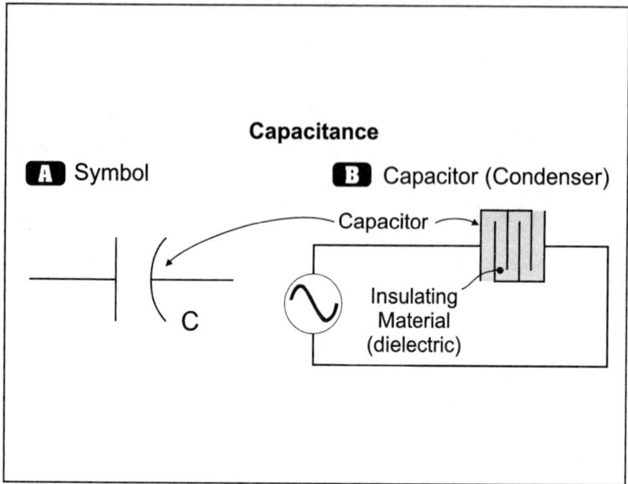

Figure 3–21
Capacitance

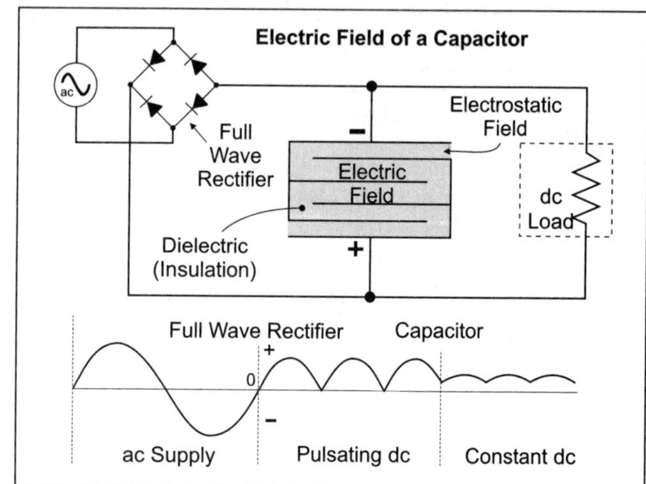

Figure 3–22
Electric Field of a Capacitor

Factors that determine capacitance are: the surface area of the plates, the distance between the plates, and the dielectric insulating material between the plates. Capacitance is measured in *farads,* but the opposition offered to the flow of current by a capacitor is called *capacitive reactance*, which is measured in *ohms*. Capacitive reactance is abbreviated X_C, is inversely proportional to frequency and can be calculated by the equation:

$$X_C = \frac{1}{2\pi fC}$$

$\pi = 3.14$, f = frequency, C = Capacitance measure in farads

❑ **Capacitive Reactance**

What is the capacitive reactance of a 0.000001 farad capacitor supplied by 60 Hz, 180 Hz, and 300 Hz (Figure 3–23)?

Answer: $X_C = \frac{1}{2\pi fC}$.

$$X_C \text{ at 60 Hz} = \frac{1}{2 \times 3.14 \times 60 \text{ Hz} \times 0.000001 \text{ farads}} = 2{,}654 \text{ ohm}$$

$$X_C \text{ at 180 Hz} = \frac{1}{2 \times 3.14 \times 180 \text{ Hz} \times 0.000001 \text{ farads}} = 885 \text{ ohm}$$

$$X_C \text{ at 300 Hz} = \frac{1}{2 \times 3.14 \times 300 \text{ Hz} \times 0.000001 \text{ farads}} = 530.46 \text{ ohm}$$

Capacitor Short and Discharge

The more the capacitor is charged, the stronger the electric field. If the capacitor is overcharged, the electrons from the negative plate could be pulled through the dielectric insulation to the positive plate, resulting in a short of the capacitor. To discharge a capacitor, all that is required is a conducting path between the capacitor plates.

Testing a Capacitor

If a capacitor is connected in series with a direct current power supply and a test lamp, the lamp will be continuously illuminated if the capacitor is shorted, but will remain dark if the capacitor is good (Figure 3–24).

3–14 USE OF CAPACITORS

Capacitors (condensers) are used to prevent arcing across the *contacts* of low ampere switches and direct current power supplies for electronic loads such as a computer (Figure 3–25). In addition, capacitors cause the current to lead the voltage by

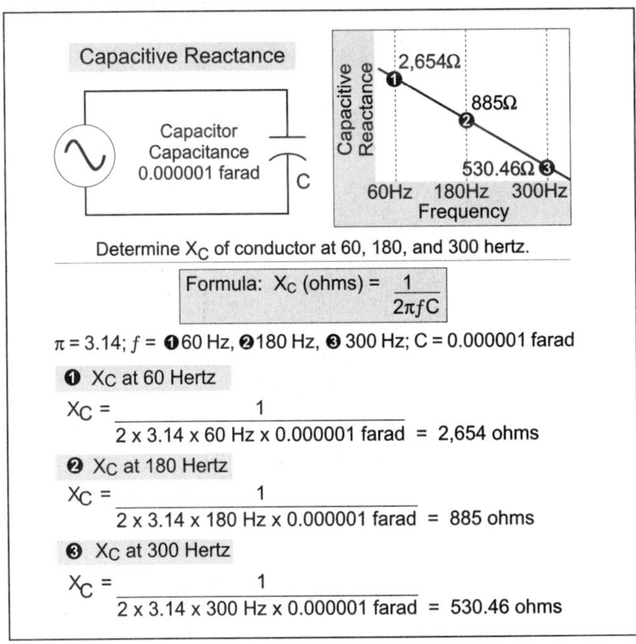

Figure 3–23
Capacitive Reactance

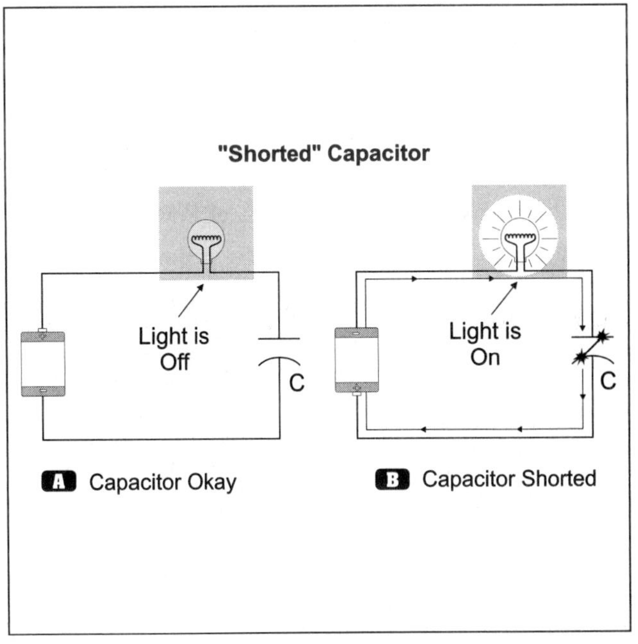

Figure 3–24
Shorted Capacitor

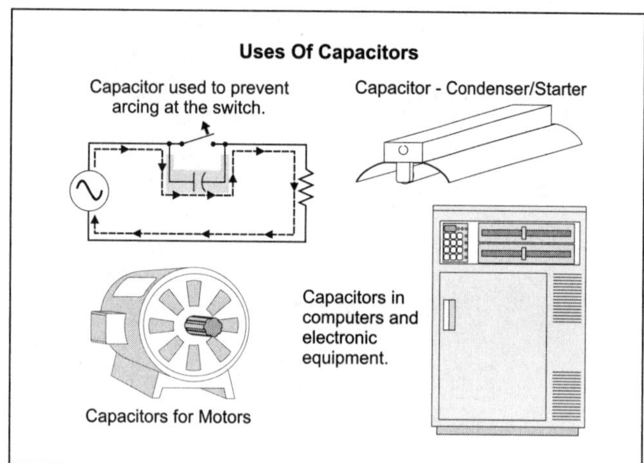

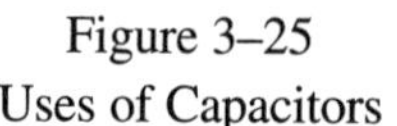

Figure 3–25
Uses of Capacitors

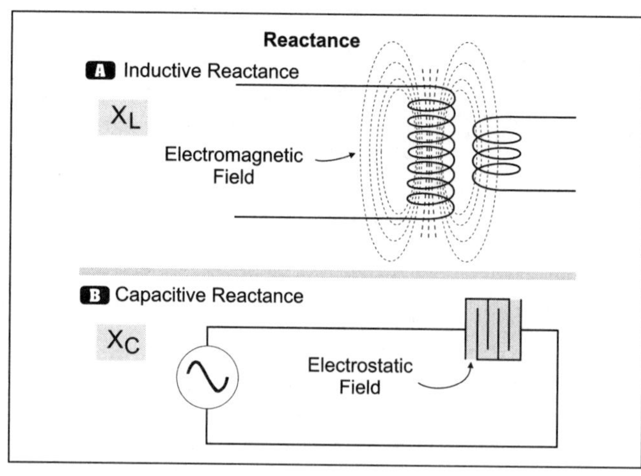

Figure 3–26
Reactance

as much as 90° and can be used to correct poor power factor due to inductive reactive loads such as motors. Poor power factor, due to induction, causes current to lag voltage by as much as 90°. Correction of the power factor to *unity* (100%) is known as *resonance* and occurs when inductive reactance is equal to capacitive reactance in a series circuit $X_L = X_C$.

PART D – POWER FACTOR AND EFFICIENCY

POWER FACTOR INTRODUCTION

Alternating current circuits develop inductive and capacitive reactance which causes some power to be stored temporarily in the electromagnetic field of the inductor or in the electrostatic field of the capacitor (Figure 3–26). Because of the temporary storage of energy, the voltage and current of inductive and capacitive circuits are not in phase. This out of phase relationship between voltage and current is called *power factor* (Figure 3–27). Power factor affects the calculation of power in alternating current circuits, but not in direct current circuits.

3–15 APPARENT POWER (VOLT-AMPERE)

In alternating current circuits, the value from multiplying volts times ampere equals *apparent power* and is commonly referred to as *volt-ampere*, VA, or *kilovolt-amperes* kVA. Apparent power (VA) is equal to or greater than *true power* (watts) depending on the *power factor*. When sizing circuits or equipment, always size the circuit equipment according to the apparent power (volt-ampere), not the true power (watt).

Apparent Power = Volts × Ampere (single-phase)

Apparent Power = Volts × Ampere × $\sqrt{3}$ (three-phase)

Apparent Power = True Power/Power Factor

❑ **Apparent Power**

What is the apparent power of a 16 ampere load operating at 120 volts (Figure 3–28)?

(a) 2,400 VA (b) 1,920 VA (c) 1,632 VA (d) none of these

• Answer: (b) 1,920 VA

❑ **Apparent Power**

What is the apparent power of a 250-watt fixture that has a power factor of 80 percent (Figure 3–29)?

(a) 200 VA (b) 250 VA (c) 313 VA (d) 375 VA

• Answer: (c) 313 VA

Apparent Power = True Power/Power Factor

Apparent Power = 250 watts/0.8, = 313 volt-ampere

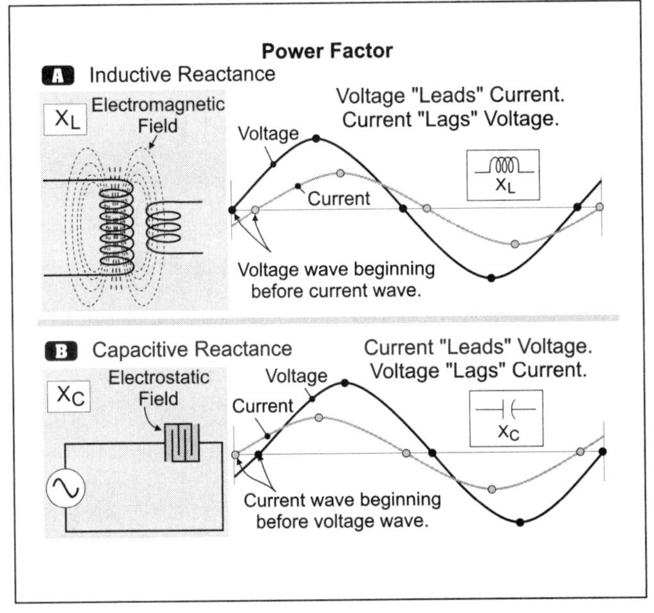

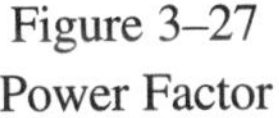

Figure 3–27
Power Factor

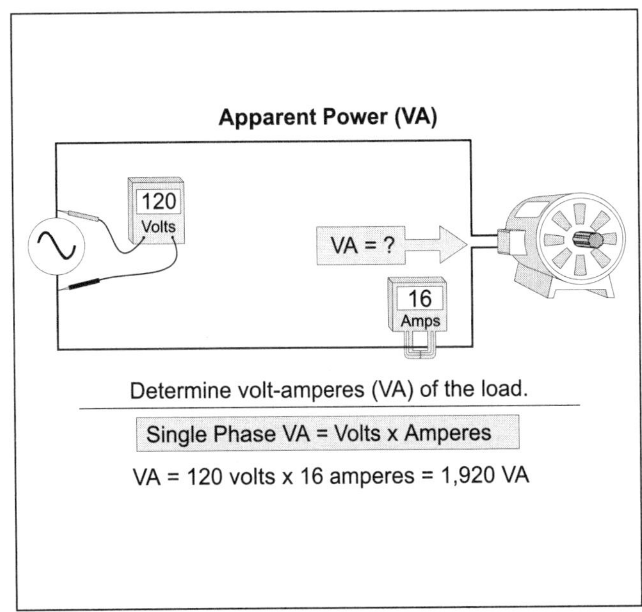

Figure 3–28
Apparent Power (VA)

3–16 POWER FACTOR

Power factor is the ratio of active power (watts) to apparent power (volt-ampere) expressed as a percent that does not exceed 100 percent. Power factor is a form of measurement of how far the voltage and current are out of phase with each other. In an alternating current circuit that supplies power to resistive loads such as incandescent lighting, heating elements, etc., the circuit voltage and current will be *in phase*. The term in phase means that the voltage and current reach their zero and peak values at the same time, resulting in a power factor of 100 percent or *unity*. Unity can occur if the circuit only supplies resistive loads or capacitive reactance (X_C) is equal to inductive reactance (X_L).

The formulas for determining power factor are:

Power Factor = True Power/Apparent Power

Power Factor = Watts/Volt-Ampere

Power Factor = Resistance (R)/Impedance (Z)

The formulas for current:

I = Watts/(Volts Line to Line × Power Factor) – Single-Phase

I = Watts/(Volts Line to Line × 1.732 × Power Factor) – Three-Phase

❑ **Power Factor**

What is the power factor of a fluorescent lighting fixture that produces 160 watts of light and has a ballast rated for 200 volt-ampere (Figure 3–30)?

(a) 80 percent (b) 90 percent (c) 100 percent (d) none of these

• Answer: (a) 80 percent

$$\text{Power Factor} = \frac{\text{True Power}}{\text{VA}} = \frac{160 \text{ watts}}{200 \text{ volt-ampere}} = 0.80 \text{ or } 80\%$$

❑ **Current**

What is the current flow for three 5 kW, 230-volt, single-phase loads that have a power factor of 90 percent?

(a) 72 ampere (b) 90 ampere (c) 100 ampere (d) none of these

• Answer: (a) 72 ampere

I = Watts/(Volts × Power Factor) = 15,000 Watts/(230 volts × 0.9) = 72 ampere

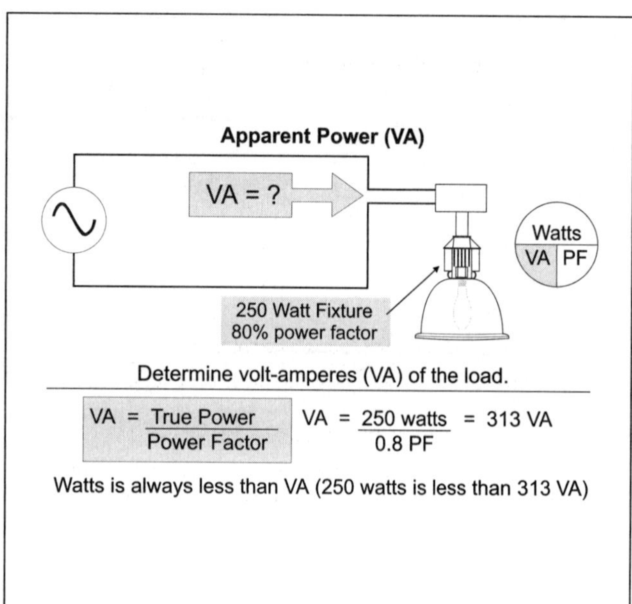

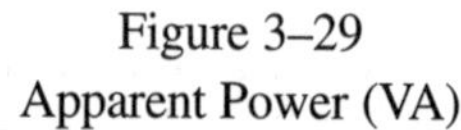

Figure 3–29
Apparent Power (VA)

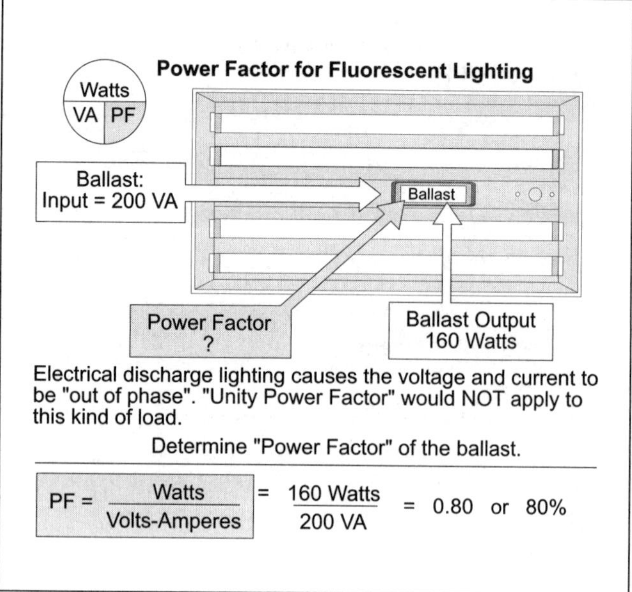

Figure 3–30
Power Factor for Fluorescent Lighting

3–17 TRUE POWER (WATTS)

ALTERNATING CURRENT

True power is the energy consumed for work, expressed by the unit called the watt. Power is measured by a wattmeter which is connected series-parallel [series (measures ampere) and parallel (measures volts)] to the circuit conductors. The true power of a circuit that contains inductive or capacitance reactance can be calculated by the use of one of the following formulas:

True Power (alternating current) = Volts × Ampere × Power Factor (single-phase)

True Power (alternating current) = Volts × Ampere × Power Factor × $\sqrt{3}$ (three-phase)

Note: True power for a direct current circuit is calculated as Volts × Ampere.

❑ **True Power**

What is the true power of a 16-ampere load operating at 120 volts with a power factor of 85 percent (Figure 3–31)?

(a) 2,400 watts (b) 1,920 watts (c) 1,632 watts (d) none of these

• Answer: (c) 1,632 watts

True Power = Volt-ampere × Power Factor

True Power = (120 volts × 16 ampere) × 0.85, = 1,632 watts per hour

Note: True Power is always equal to or less than apparent power.

DIRECT CURRENT

In direct current or alternating current circuits at unity power factor:

True Power = Volts × Ampere

❑ **True Power**

What is true power of a 30 ampere, 240-volt (single-phase) resistive load that has unity power factor (Figure 3–32)?

(a) 4,300 watts (b) 7,200 watts (c) 7,200 VA (d) none of these

• Answer: (b) 7,200 watts

True Power = Volts × Ampere

True Power = 240 volts × 30 ampere = 7,200 watts

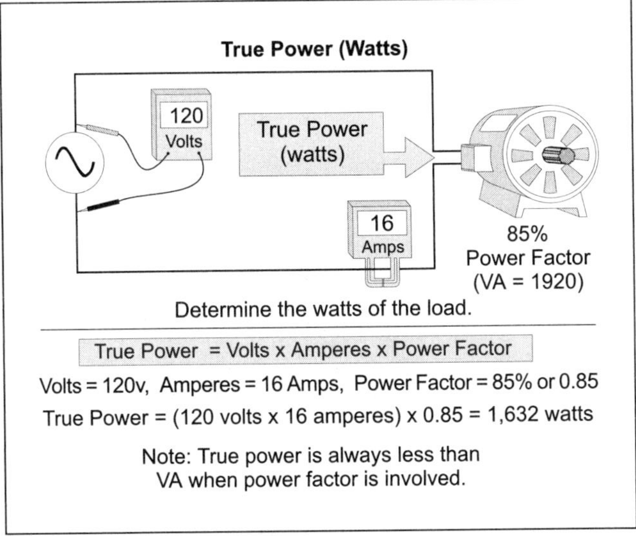

Figure 3–31
True Power (Watts)

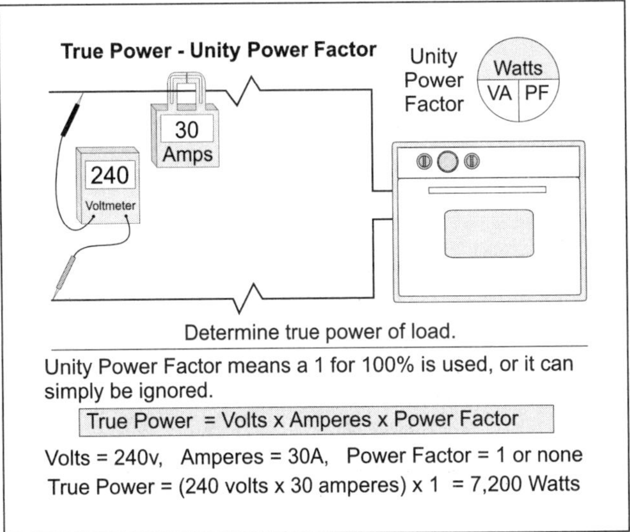

Figure 3–32
True Power – Unity Power Factor

3–18 EFFICIENCY

Efficiency has nothing to do with power factor. Efficiency is the ratio of output power to input power, where power factor is the ratio of true power (watts) to apparent power (volt-ampere). Energy that is not used for its intended purpose is called power loss. Power losses can be caused by conductor resistance, friction, mechanical loading, etc. Power losses of equipment are expressed by the term efficiency, which is the ratio of the input power to the output power (Figure 3–33). The formulas for efficiency are:

Efficiency = Output watts/Input watts

Input watts = Output watts/Efficiency

Output watts = Input watts × Efficiency

❑ **Efficiency**

If the input of a load is 800 watts and the output is 640 watts, what is the efficiency of the equipment (Figure 3–34)?

(a) 60 percent (b) 70 percent (c) 80 percent (d) 100 percent

• Answer: (c) 80 percent

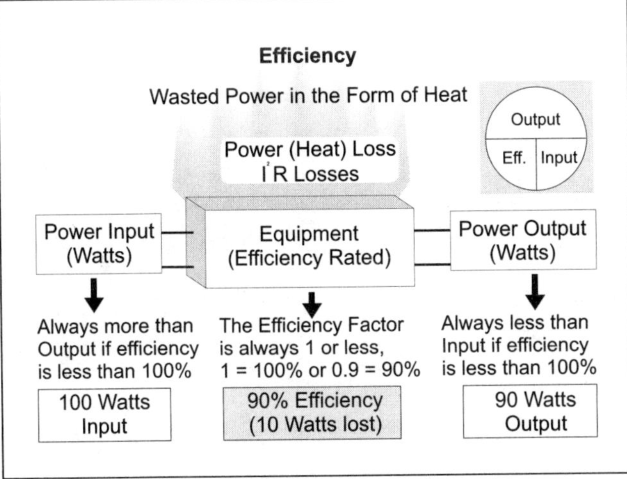

Figure 3–33
Efficiency

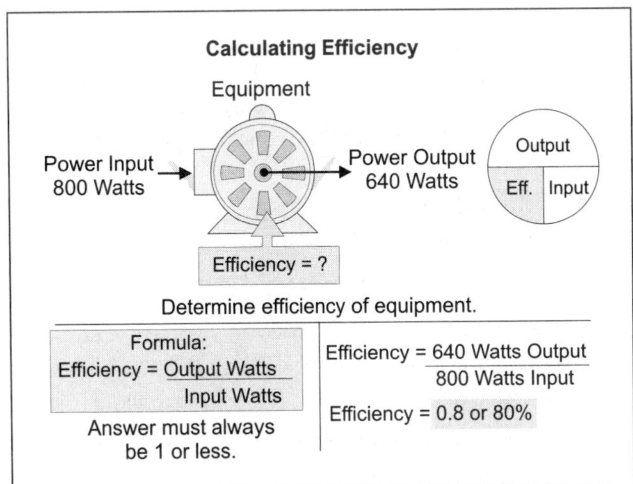

Figure 3–34
Calculating Efficiency

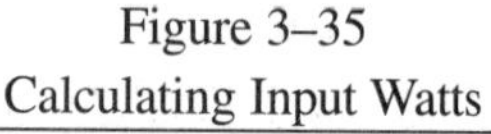

Figure 3–35
Calculating Input Watts

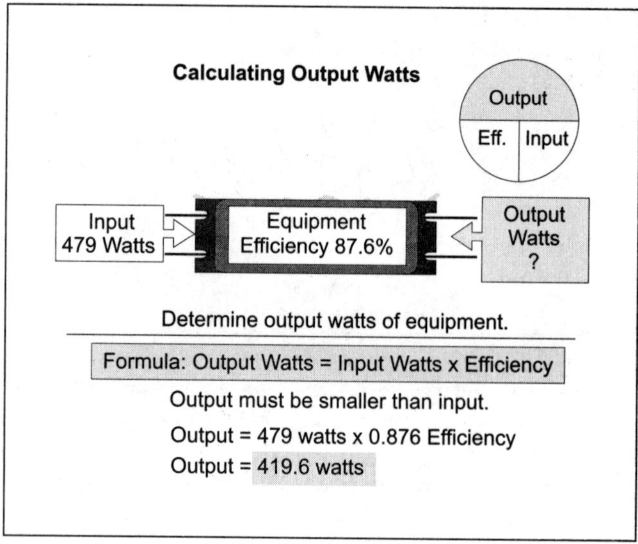

Figure 3–36
Calculating Output Watts

$$\text{Efficiency} = \frac{\text{Output}}{\text{Input}}$$

$$\text{Efficiency} = \frac{640 \text{ watts}}{800 \text{ watts}} = 0.8 \text{ or } 80\%$$

Note: Efficiency is always less than 100 percent.

❑ Input

If the output of a load is 250 watts and the equipment is 88 percent efficient, what are the input watts (Figure 3–35)?

(a) 200 watts (b) 250 watts (c) 285 watts (d) 325 watts

- Answer: (c) 285 watts

$$\text{Input} = \frac{\text{Output}}{\text{Efficiency}}$$

$$\text{Input} = \frac{250 \text{ watts}}{0.88 \text{ efficiency}}$$

Input = 284 watts

Note: Input is always greater than the output.

❑ Output

If a load is 87.6 percent efficient, for every 479 watts of input there will be _____ watts of output (Figure 3–36).

(a) 440 (b) 480 (c) 390 (d) 420

- Answer: (d) 420 watts

Output = Input $\times$ Efficiency

Output = 479 $\times$ .876

Output = 419.6 watts

Note: Output is always less than input.

Unit 3 – Understanding Alternating Current Summary Questions

Part A – Alternating Current Fundamentals

3–1 Current Flow

1. The effects of electron movement are the same regardless of the direction of the current flow.
(a) True (b) False

3–2 Alternating Current

2. Alternating current is primarily used because it can be transmitted inexpensively and because it can be used for certain applications for which direct current is not suitable.
(a) True (b) False

3. Faraday's experiments revealed that when a magnetic field moves through a coil of wire, the lines of force of the magnetic field makes the electrons in the wire flow in a specific direction. When the magnetic field moves in the opposite direction, electrons in the wire flow continue to flow in the same direction.
(a) True (b) False

3–3 Alternating Current Generator

4. A simple alternating current generator consists of a rotating loop of wire between the lines of force of opposite poles of a magnet. The magnitude of the voltage is dependent on the _____ .
(a) number of turns of wire (b) strength of the magnetic field
(c) speed at which the coil rotates (d) all of these

3–4 Waveform

5. • A waveform is used to display the level and direction of current and voltage. The waveform for _____ circuits displays the level and direction of the current and voltage for every instant of time for one full revolution of the armature.
(a) direct current (b) alternating current (c) a and b (d) none of these

6. For alternating current circuits, the waveform is called the _____ .
(a) frequency (b) cycle (c) degree (d) none of these

3–5 Armature Turning Frequency

7. The number of times the armature turns in one second is called the frequency. Frequency is expressed as _____ or cycles per second.
(a) degrees (b) sign wave (c) phase (d) hertz

3–6 Phase – In and Out

8. When two waveforms are in step with each other, they are said to be in phase. In a purely resistive alternating current circuit, the current and voltage are in phase. This means that they both reach their zero and _____ values at the same time.
(a) peak (b) effective (c) average (d) none of these

3–7 Phase Differences in Degrees

9. Phase differences are expressed in _____ .
(a) sign (b) phase (c) hertz (d) degrees

10. The terms _____ and _____ are used to describe the relative positions in time of two waveforms (voltage or current).
(a) hertz, phase (b) frequency, phase (c) sine, degrees (d) lead, lag

3–8 Values of Alternating Current

11. • _____ is the value of the voltage or current at any one particular moment of time.
(a) Peak (b) Root-mean-square (c) Effective (d) Instantaneous

12. _____ is the maximum value that alternating current or voltage reaches.
(a) Peak (b) Root-mean-square (c) Instantaneous (d) None of these

13. • "Effective" is the alternating current voltage or current value that produces the same amount of heat in a resistor that would be produced by the same amount of direct current voltage or current. _____ is the same as effective.
(a) Peak (b) Root-mean-square (c) Instantaneous (d) None of these

Part B – Induction

Induction Introduction

14. The movement of electrons because of electromagnetism is called _____ .
(a) flux lines (b) voltage (c) induction (d) magnetic field

15. The induction of voltage in a conductor because of expanding and collapsing magnetic fields is known as _____ .
(a) flux lines (b) power (c) self-induced voltage (d) magnetic field

16. Change in current produces a magnetic field through the conductor which produces an induced voltage that always opposes the change in current. The induced voltage that opposes the change in current is called _____ .
(a) CEMF
(b) counterelectromotive-force
(c) back-EMF
(d) all of these

3–9 Induced Voltage and Applied Current

17. When the conductor current increases, the direction of the induced voltage (CEMF) in the conductor is opposite to the direction of the conductor current and tries to prevent the current from _____ .
(a) decreasing (b) increasing

18. When the current in the conductor decreases, the direction of the induced voltage in the conductor attempts to keep the current from decreasing by releasing the energy from the magnetic field back into the conductor.
(a) True (b) False

3–10 Conductor Impedance

19. In direct current circuits, the only property that affects current and voltage flow is _____, which is a physical property of the conductors that oppose current flow.
(a) voltage (b) CEMF (c) back-EMF (d) none of these

20. Conductor resistance is directly proportional to the conductor length and inversely proportional to conductor cross-sectional area.
(a) True (b) False

21. Alternating currents produce a CEMF that is set up inside the conductor which increases the effective resistance of the conductor because of eddy currents and skin effect.
(a) True (b) False

22. The opposition to current flow in a conductor because of resistance and induction is called _____ .
(a) resistance (b) capacitance (c) induction (d) impedance

23. • Eddy currents are small independent currents that are produced as a result of direct current. Eddy currents flow erratically through a conductor, consume power, and increase the effective conductor resistance by opposing the current flow.
(a) True (b) False

24. The expanding and collapsing magnetic field of the conductors caused by the current flow in a conductor induces a voltage in the conductors which repels the flowing electrons towards the surface of the conductor. This has the effect of decreasing the effective conductor cross-sectional area, which causes an increase in the conductor impedance. The flow of electrons near the surface is known as _____ .
(a) eddy currents (b) induced voltage (c) impedance (d) skin effect

3–11 Induction and Conductor Shape

25. The amount of the self-induced voltage created within the conductor is directly proportional to the current flow, the length of the conductor, and frequency at which the magnetic fields cut through the conductors.
(a) True (b) False

3–12 Induction and Magnetic Cores

26. • Self-inductance (CEMF) in a coil is effected by the _____ .
(a) winding length and shape (b) core material
(c) frequency (d) all of these

Part C – Capacitance

Capacitance Introduction

27. _____ is a property of an electric circuit that enables it to store electric energy by means of an electrostatic field and to release this energy at a latter time.
(a) Capacitance (b) Induction (c) Self-induction (d) None of these

3–13 Charge, Testing, and Discharging

28. When a capacitor has a potential difference between the plates, it is said to be _____ . One plate has an excess of free electrons, and the other plate has fewer electrons.
(a) induced (b) charged (c) discharged (d) shorted

29. If the capacitor is overcharged, the electrons from the negative plate may be pulled through the insulation to the positive plate. The capacitor is said to have _____ .
(a) charged (b) discharged (c) induced (d) shorted

30. To discharge a capacitor, all that is required is a _____ path between the capacitor plates.
(a) conducting (b) insulating (c) isolating (d) none of these

3–14 Uses of Capacitors

31. What helps prevent arcing across the contacts of electric switches?
(a) springs (b) condenser (c) inductor (d) resistor

32. The current in a purely capacitive circuit _____ .
(a) leads the applied voltage by 90° (b) lags the applied voltage by 90°
(c) leads the applied voltage by 180° (d) lags the applied voltage by 180°

33. Circuits containing inductive or capacitive reactance temporarily store power in the electromagnetic field of induction and the electrostatic field of capacitors.
(a) True (b) False

Part D – Power Factor and Efficiency

3–15 Apparent Power (Volt-Ampere)

34. If you measure voltage and current in an inductive or capacitive circuit and then multiply them together, you would obtain the circuit's _____ .
(a) true power (b) power factor (c) apparent power (d) power loss

35. Apparent power is equal to or greater than true power depending on the power factor.
(a) True (b) False

36. When sizing circuits or equipment, always size the circuit components and transformers according to the apparent power, not the true power.
(a) True (b) False

3–16 Power Factor

37. Power factor is a measurement of how far the voltage and current are out of phase with each other. Power factor is the ratio between true power (resistive load) to apparent power (reactive load). Power factor can be expressed by the formula: _____ .
(a) $PF = P/E$ (b) $PF = R/Z$ (c) $PF = I^2R$ (d) $PF = Z/R$

38. • When current and voltage are in phase, the power factor is _____ .
(a) 100 percent (b) unity (c) 90° (d) a and b

39. In an alternating current circuit that supplies power to resistive loads such as incandescent lighting, heating elements, etc., the circuit voltage and current are said to be _____, resulting in a power factor of unity.
(a) out of phase (b) leading by 90°
(c) 90° out of phase (d) none of these

3–17 True Power (Watts)

40. True power is the energy consumed for work expressed by the term watts. To determine the true power of a circuit that contains inductive or capacitance reactance, we must multiply the volts times the current times the _____ .
(a) efficiency (b) sine wave (c) power factor (d) none of these

41. In direct current circuits, the voltage and current are constant and the true power is simply volts times ampere.
(a) True (b) False

42. What does it cost per year (9 cents per kWH) for ten 150-watt recessed fixtures to operate if they are on 6 hours per day?
(a) $153 (b) $296 (c) $235 (d) $180

Power Factor

43. A 2 × 4 recessed fixture contains four 34-watt lamps, and the ballast is rated 1.5 ampere at 120 volts. What is the power factor of the ballast assuming 100 percent efficiency?
(a) 55% (b) 65% (c) 70% (d) 75%

44. • What is the apparent power of a 20-ampere load operating at 120 volts with a power factor of 85 percent?
(a) 2,400 watts (b) 1,920 watts (c) 1,632 watts (d) 2,400 VA

45. • What is the true power of a 20-ampere load operating at 120 volts with a power factor of 85 percent?
(a) 2,400 watts (b) 1,920 watts (c) 1,632 watts (d) none of these

46. • Since power factor cannot be greater than 100 percent, true power is equal to or less than the apparent power. Because of power factor, the VA of the load is greater than the watts, which results in less loads per circuit, more circuits, and larger kVA transformers.
(a) True (b) False

47. • What is the true power of a 10-ampere circuit operating at 120 volts with unity (100%) power factor?
(a) 1,200 VA (b) 2,400 VA (c) 1,200 watts (d) 2,400 watts

48. What size transformer is required for a 125 ampere, 240-volt, single-phase load?
(a) 3 kVA (b) 30 kVA (c) 12.5 kVA (d) 15 kVA

49. What size transformer is required for a 30 kW-load that has a power factor of 85 percent?
(a) 12.5 kVA (b) 35 kVA (c) 7.5 kVA (d) 15 kVA

50. • How many 20 ampere, 120-volt circuits are required for forty-two, 300-watt recessed fixtures (noncontinuous load)? Note: Only so many fixtures are permitted on a circuit.
(a) 3 circuits (b) 4 circuits (c) 5 circuits (d) 6 circuits

51. • How many 20 ampere, 120-volt circuits are required for forty-two, 300-watt recessed fixtures (noncontinuous load) with a power factor of 85 percent?
(a) 5 circuits (b) 6 circuits (c) 7 circuits (d) 8 circuits

3–18 Efficiency

52. Efficiency is the ratio of the output to input power.
(a) True (b) False

53. If the output is 1,320 watts and the input is 1,800 watts, what is the efficiency of the equipment?
(a) 62 percent (b) 73 percent (c) 0 percent (d) 100 percent

54. If the output is 160 watts and the equipment is 88 percent efficient, what is the input ampere at 120 volts?
(a) 0.75 ampere (b) 1.500 ampere (c) 2.275 ampere (d) 3.250 ampere

55. A transformer that is 97 percent efficient produces _____ watts output for every 1 kW input.
(a) 970 watts (b) 1,000 watts (c) 1,030 watts (d) 1,300 watts

☆ Challenge Questions

Part A – Alternating Current Fundamentals

3–2 Alternating Current

56. The primary reason(s) for high voltage transmission lines is (are) _____ .
(a) reduced voltage drop (b) smaller wire
(c) smaller equipment (d) all of these

57. One of the advantages of a higher-voltage system as compared to a lower-voltage system (for the same wattage loads) is _____ .
(a) reduced voltage drop (b) reduced power use
(c) large currents (d) lower electrical pressure

58. The advantage of alternating current over direct current is that alternating current provides for _____ .
(a) better speed control
(b) ease of voltage variation
(c) lower resistance at high currents
(d) none of these

3–4 Waveform

59. A waveform represents _____ .
(a) the magnitude and direction of current or voltage
(b) how current or voltage can vary with time
(c) how output voltage can vary with the generator armature
(d) all of these

3–5 Armature Turning Frequency

60. Frequency of an alternating current waveform is the number of times the current or voltage goes through 360° in _____ .
(a) $^1/_{10th}$ second (b) 5 seconds (c) 1 second (d) 60 seconds

61. How much time does it take for 60 Hz alternating current to travel through 180°?
(a) $^1/_{120}$ second (b) $^1/_{40}$ second (c) $^1/_{180}$ second (d) none of these

3–8 Values of Alternating Current

62. • The heating effects of 10 ampere alternating current as compared with 10 ampere direct current is _____ .
(a) the same (b) less (c) greater (d) none of these

63. If the maximum value of an alternating current system is 50 ampere, the RMS value would be approximately _____ .
(a) 25 ampere (b) 30 ampere (c) 35 ampere (d) 40 ampere

64. • The maximum value of 120-volt direct current is equal to the maximum value of an equivalent alternating current.
(a) True (b) False

65. • You are getting a 120 volt reading on your voltmeter. This is an indication of the _____ value of the voltage source.
(a) average (b) peak (c) effective (d) instantaneous

Part B – Induction

3–9 Induced Voltage and Applied Current

66. _____ Law states that a change in current produces a counterelectromotive-force whose direction is such that it opposes the change in current.
(a) Kirchoff's Second (b) Kirchoff's First (c) Lenz's (d) Hertz

67. Inductive reactance is abbreviated as _____ .
(a) I^2R (b) L_X (c) X_L (d) Z

68. Inductive reactance changes proportionally with frequency.
(a) True (b) False

69. Inductive reactance is measured in _____ .
(a) farads (b) watts (c) ohms (d) coulombs

70. • If the frequency is constant, the inductive reactance of a circuit will _____ .
(a) remain constant regardless of the current and voltage changes
(b) vary directly with the voltage
(c) vary directly with the current
(d) not affect the impedance

3–10 Conductor Impedance

71. The total opposition to current flow in an alternating current circuit is expressed in ohms and is called _____ .
(a) impedance (b) conductance (c) reluctance (d) none of these

72. Impedance is present in _____ type circuit(s).
(a) resistance (b) direct current (c) alternating current (d) none of these

73. Conductor resistance to alternating current flow is _____ the resistance to direct current.
(a) higher than (b) lower than (c) the same as (d) none of these

Part C – Capacitance

3–13 Charge, Testing, and Discharging

74. If a test lamp is placed in series with a capacitor with a direct current voltage source and the lamp is continuously illuminated, it is an indication that the capacitor is _____ .
(a) fully charged (b) shorted (c) fully discharged (d) open-circuit

75. The insulating material between the surface plates of a capacitor is called the _____ .
(a) inhibitor (b) electrolyte (c) dielectric (d) regulator

76. Capacitors are measured in _____ .
(a) watts (b) volts (c) farads (d) henrys

77. Capacitive reactance is measured in _____ .
(a) ohms (b) volts (c) watts (d) henrys

3–14 Uses of Capacitors

78. • In a circuit that has only capacitive reactance (X_C), the voltage and current are said to be out of phase to each other because the voltage _____ .
(a) leads the current by 90°
(b) lags the current by 90°
(c) leads the current
(d) none of these

79. Resonance occurs when _____ .
(a) only resistance occurs in the system
(b) the power factor is equal to 0
(c) $X_L = X_C$ are in a series circuit
(d) when R = Z

Part D – Power Factor and Efficiency

3–15 Apparent Power (Volt-Ampere)

80. If you multiply the voltage times the current in an inductive or capacitive circuit, the answer you obtain will be the _____ of the circuit.
(a) watts (b) true power (c) apparent power (d) all of these

81. The apparent power of a 19.2 ampere, 120-volt load is _____ .
(a) 2,304 kVA (b) 2.3 kVA (c) 2.3 VA (d) 230 kVA

3–16 Power Factor

82. Power factor in an alternating current circuit will be unity (100%), if the circuit contains only _____ .
(a) induction motors (b) transformers (c) reactance coils (d) resistive loads

83. • Three 8 kW electric discharge lighting bank circuits, which have a 92 percent power factor, are connected to a 230-volt, 3-phase source. The current flow of these lights is _____ .
(a) 37 ampere (b) 50 ampere (c) 65 ampere (d) 75 ampere

3–17 True Power (Watts)

84. • A wattmeter is connected in _____ in the circuit.
(a) series (b) parallel (c) series-parallel (d) none of these

85. • True power is always voltage times current for _____ .
(a) all alternating current circuits
(b) direct current circuits
(c) alternating current circuits at unity power factor
(d) b and c

86. Power consumed in either a single-phase alternating current or direct current system is always equal to _____ .
(a) $E \times I$ (b) $E \times R$ (c) $I^2 \times R$ (d) $E/(I \times R)$

87. • The power consumed on a 76 ampere, 208-volt, 3-phase circuit that has a power factor of 89 percent is _____ .
(a) 27,379 watts (b) 35,808 watts (c) 24,367 watts (d) 12,456 watts

88. • The true power of a single-phase, 2.1 kVA load with a power factor of 91 percent is _____ .
(a) 2.1 kW (b) 1.91 kW (c) 1.75 kW (d) 1,911 kW

3–18 Efficiency

89. • Motor efficiency can be determined by which of the following formulas?
(a) hp × 746
(b) hp × 746/VA Input
(c) hp × 746/Watts Input
(d) hp × 746/kVA Input

90. The efficiency ratio of a 4,000 VA transformer with a secondary VA of 3,600 VA is _____ .
(a) 80% (b) 70% (c) 90% (d) 110%

Unit 4

Motors and Transformers

OBJECTIVES

After reading this unit, the student should be able to briefly explain the following concepts:

Part A – Motors
Alternating current motors
Dual voltage motors
Horsepower/watts
Motor speed control
Nameplate ampere
Reversing DC motors
Reversing AC motors
Volt-ampere calculations

Part B – Transformers
Transformer primary vs. secondary
Transformer current
Transformer kVA rating
Transformer power losses
Transformer turns ratio

After reading this unit, the student should be able to briefly explain the following terms:

Part A – Motors
Armature winding
Commutator
Dual voltage
Field winding
Horsepower ratings
Magnetic field
Motor full load current
Nameplate ampere
Torque
Watts rating

Part B – Transformers
Auto transformers
Circuit impedance
Conductor losses
Core losses
Counter-electromotive force
Current transformers (CT)
Eddy current losses
Excitation current
Flux leakage loss
Hysteresis losses
Kilo volt-ampere
Line current
Ratio
Self-excited
Step-down transformers
Step-up transformers

PART A – MOTORS

MOTOR INTRODUCTION

The *electric motor* operates on the principle of the attracting and repelling forces of magnetic fields. One magnetic field (permanent or electromagnetic) is *stationary* and the other magnetic field (called the *armature* or *rotor*) rotates between the poles of the stationary magnet (Figure 4–1). The turning or repelling forces between the magnetic fields are called *torque.* Torque is dependent on the strength of the stationary magnetic field, the strength of magnetic field of the armature, and the physical construction of the motor.

The repelling force of like magnetic polarities, and the attraction force of unlike polarities, causes the armature to rotate in the electric motor. For a direct current motor, a device called a *commutator* is placed on the end of the conductor loop or armature. The purpose of the commutator is to maintain the proper polarity of the loop, so as to keep the armature or rotor turning.

The *armature winding* of a motor carries a starting of current when voltage is first applied to the motor windings. As the armature turns, it cuts the lines of force of the *field winding* resulting in an increase in counterelectromotive-force, (CEMF) (Figure 4–2). The increased CEMF results in an increase in inductive reactance, which increases the circuit impedance. The increased circuit impedance causes a decrease in the motor armature running current (Figure 4–3).

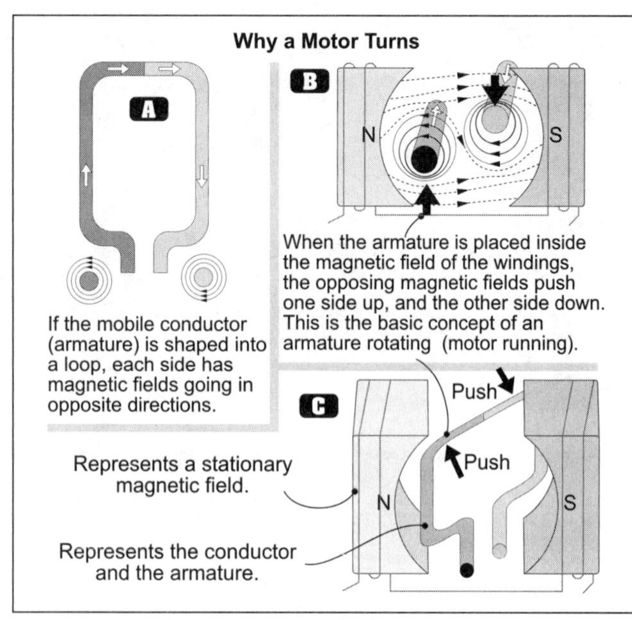

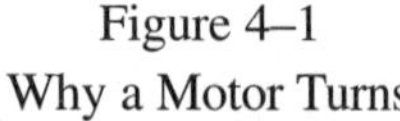

Figure 4–1
Why a Motor Turns

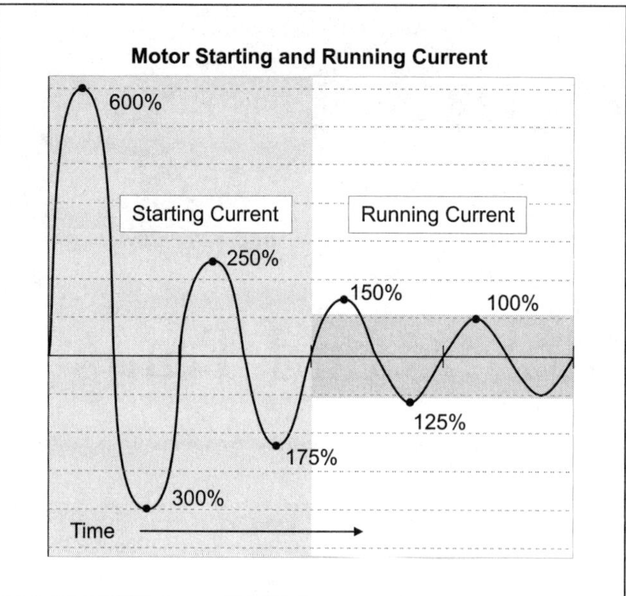

Figure 4–2
Motor Starting and Running Current

4–1 MOTOR SPEED CONTROL

One of the advantages of *direct current motor* over the alternating current motor is the motor's ability to maintain a constant speed. But series direct current motors are susceptible to run away (increase rotational speed) when not connected to a load.

If the speed of a direct current motor is increased, the armature winding is cut by the magnetic field at an increasing rate, resulting in an increase of the armatures CEMF. The increased CEMF in the armature acts to cut down on the armature current, resulting in the motor slowing down. Placing a load on a direct current motor causes the motor to slow down, which reduces the rate at which the armature winding is cut by the magnetic field flux lines. A reduction of the armature flux lines results in a decrease in the armature's CEMF and an increase in motor speed.

4–2 REVERSING A DIRECT CURRENT MOTOR

To reverse a direct current motor, you must reverse either the *magnetic field* of the field winding or the magnetic field of the armature. This is accomplished by reversing either the field or armature current flow (Figure 4–4). Because most direct current motors have the field and armature winding connected to the same direct current power supply, reversing the polarity of the power supply changes both the field and armature simultaneously. To reverse the rotation of a direct current motor, you must reverse either the field or armature leads, but not both.

4–3 ALTERNATING CURRENT MOTORS

Fractional horsepower motors that can operate on either alternating or direct current are called universal motors. A motor that will not operate on direct current is called an induction motor.

The induction motor is the purest form of an alternating current motor, with no physical connection between its rotating member *(rotor)*, and stationary member *(stator)*. Two common types of alternating current induction motors are synchronous and wound rotor motors.

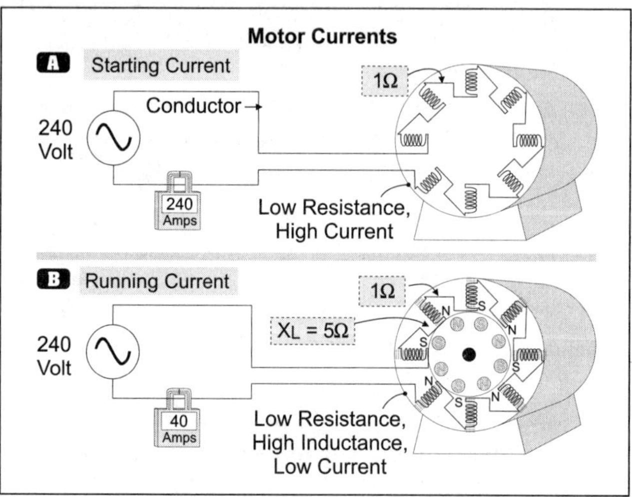

Figure 4–3
Motor Currents

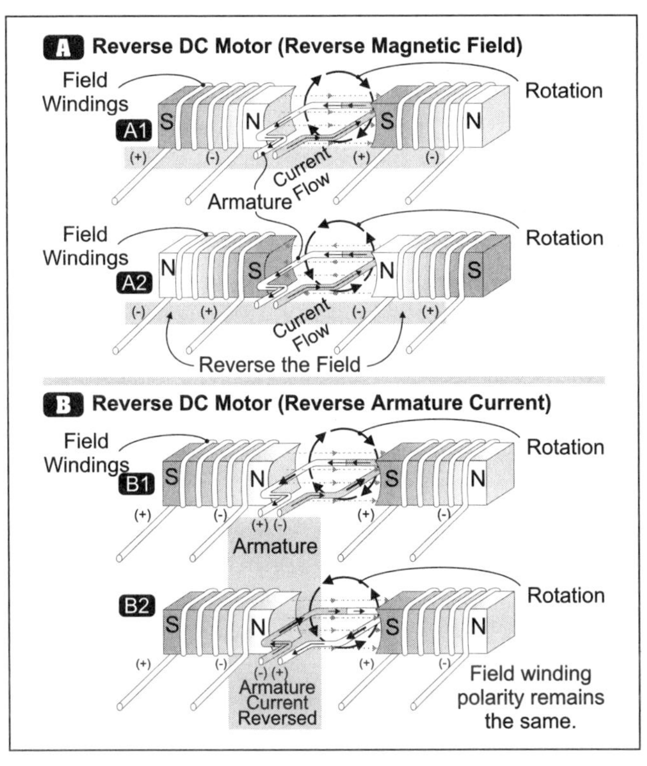

Figure 4–4
Reverse DC Motor

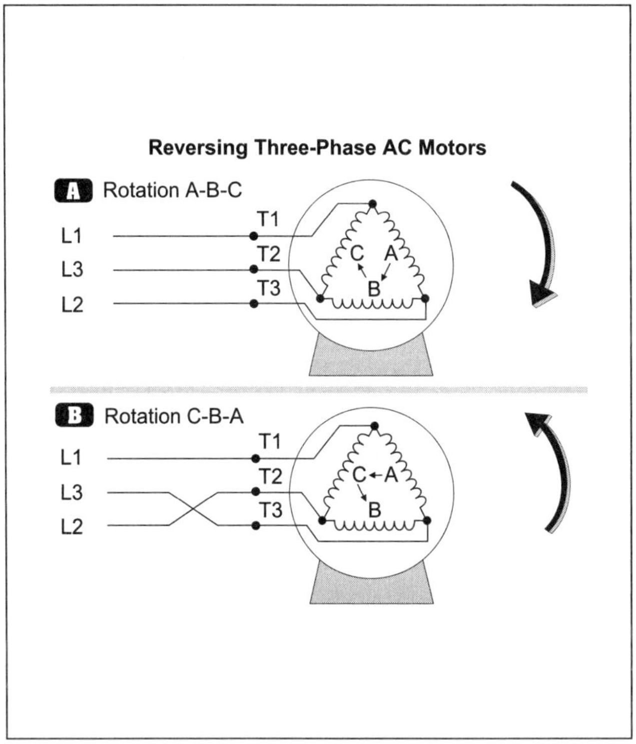

Reversing Three-Phase AC Motors

A Rotation A-B-C

L1 L3 L2 T1 T2 T3 C A B

B Rotation C-B-A

L1 L3 L2 T1 T2 T3 C A B

Figure 4–5
Reversing Three-Phase AC Motors

Synchronous Motors

The synchronous motor's rotor is locked in step with the rotating stator field. Synchronous motors maintain their speed with a high degree of accuracy and are used for electric clocks and other timing devices.

Wound Rotor Motors

Because of their high starting torque requirements, wound rotor motors are used only in special applications. They only operate on three-phase alternating current.

4–4 REVERSING ALTERNATING CURRENT MOTORS

Three-phase alternating current motors can be reversed by reversing any two of the three line conductors that supply the motor (Figure 4–5).

4–5 MOTOR VOLT-AMPERE CALCULATIONS

Dual voltage motors are made with two field windings, each rated for the lower voltage marked on the nameplate of the motor. When the motor is wired for the lower voltage, the field windings are connected in parallel; when wired for the higher voltage, the motor windings are connected in series (Figure 4–6).

Motor Input VA

Regardless of the voltage connection, the *power* consumed by a motor is the same at either voltage. To determine the *motor input* apparent power (VA), use the following formulas:

Motor VA (single-phase) = Volts × Ampere

Motor VA (three-phase) = Volts × Ampere $\times \sqrt{3}$

❏ **Motor VA Single-Phase**

What is the motor VA of a single-phase 115/230-volt, 1 horsepower motor that has a current rating of 16 ampere at 115 volts and 8 ampere at 230 volts (Figure 4–7)?

(a) 1,450 VA (b) 1,600 VA (c) 1,840 VA (d) 1,920 VA

• Answer: (c) 1,840 VA

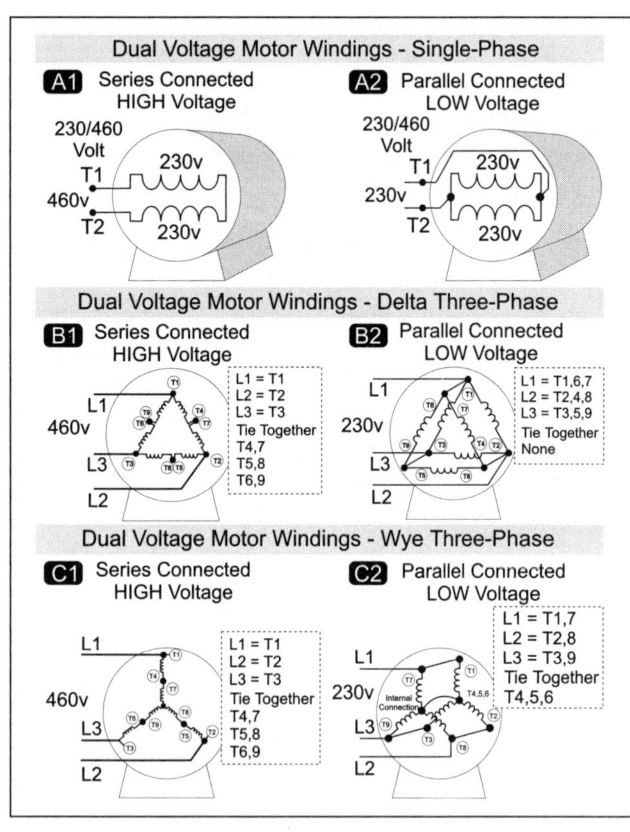

Figure 4–6
Dual Voltage Motor Windings – Single-Phase

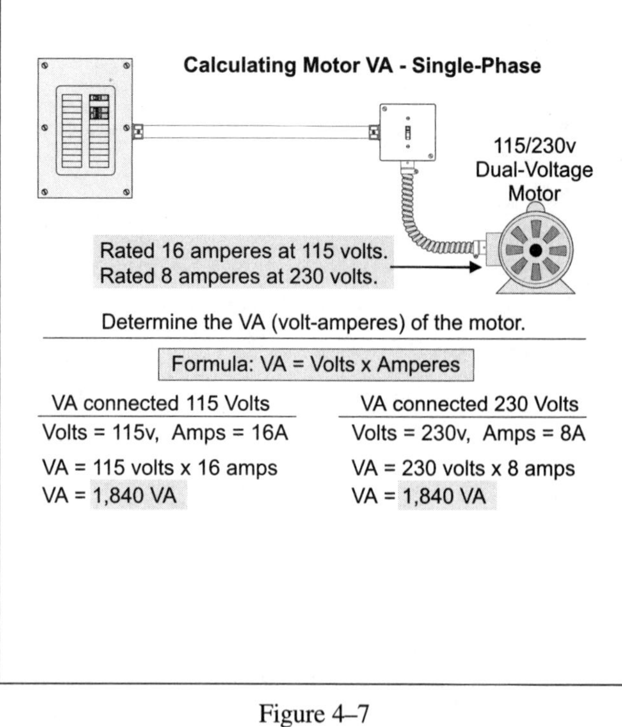

Figure 4–7
Calculating Motor VA – Single-Phase

Motor VA = Volts × Ampere

Volts = 115 or 230, Ampere = 16 or 8 ampere

Motor VA = 115 volts × 16 ampere, or = 230 volts × 8 ampere, = 1,840 VA

❑ Motor VA Three-Phase

What is the motor VA of a 3-phase, 230/460 volt, 30 horsepower motor that has a current rating of 40 ampere at 460 volts (Figure 4–8)?

(a) 41,450 VA (b) 31,600 VA (c) 21,840 VA (d) 31,869 VA

• Answer: (d) 31,869 VA

Motor VA = Volts × Ampere × $\sqrt{3}$

Volts = 460 volts, Ampere = 40 ampere, $\sqrt{3} = 1.732$

Motor VA = 460 volts × 40 ampere × 1.732, = 31,869 VA

4–6 MOTOR HORSEPOWER/WATTS

The mechanical work (output) of a motor is rated in horsepower and can be converted to electrical energy as 746 watts per horsepower.

Horsepower = Output Watts/746 Watts

Motor Output Watts = Horsepower × 746

❑ Motor Horsepower

What size horsepower motor is required to produce 15 kW output (Figure 4–9)?

(a) 5 horsepower (b) 10 horsepower (c) 20 horsepower (d) 30 horsepower

• Answer: (c) 20 horsepower

$$\text{Horsepower} = \frac{\text{Output Watts}}{746} = \frac{15{,}000\text{ watts}}{746\text{ watts}} = 20\text{ horsepower}$$

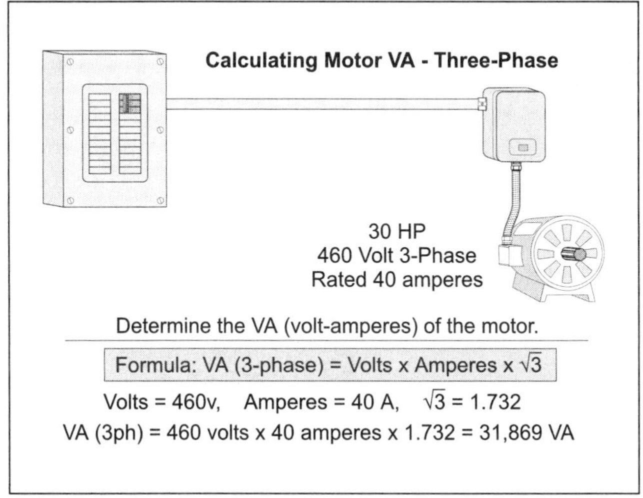

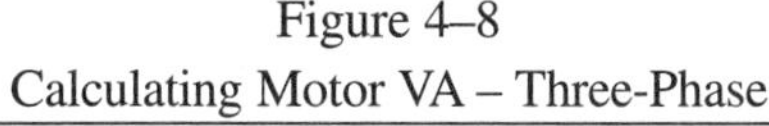
Figure 4–8
Calculating Motor VA – Three-Phase

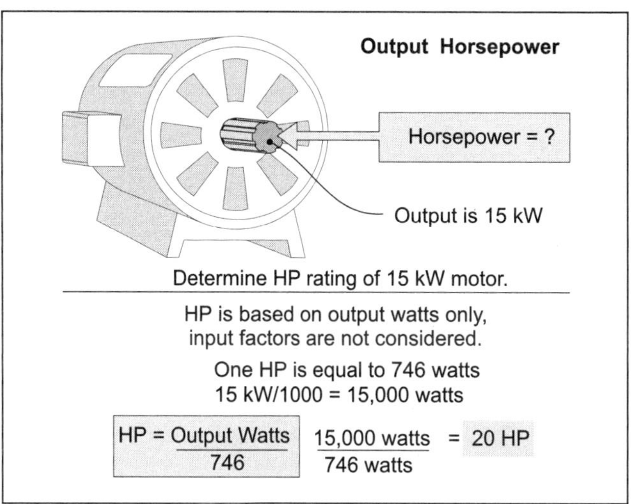

Figure 4–9
Output Horsepower

❑ **Motor Output Watt**

What is the output watt rating of a 10 horsepower motor (Figure 4–10)?

(a) 3 kW (b) 2.2 kVA (c) 3.2 kW (d) 7.5 kW

• Answer: (d) 7.5 kW

Output Watts = 10 horsepower × 746 Watts, = 7,460 watts

$$\text{Output watts} = \frac{7{,}460 \text{ watts}}{1{,}000} = 7.46 \text{ kW}$$

4–7 MOTOR NAMEPLATE AMPERE

The motor nameplate indicates the motor operating voltage and current. The actual current drawn by the motor depends on how much the motor is loaded. It is important not to overload a motor above its rated horsepower because the current of the motor will increase to a point that the motor winding will be destroyed from excess heat. The nameplate current can be calculated by the following formulas:

$$\textbf{Motor Nameplate (single-phase) Ampere} = \frac{\textbf{Motor Horsepower} \times \textbf{746 Watts}}{(\textbf{Volts} \times \textbf{Efficiency} \times \textbf{Power Factor})}$$

$$\textbf{Motor Nameplate (three-phase) Ampere} = \frac{\textbf{Motor Horsepower} \times \textbf{746 Watts}}{(\textbf{Volts} \times \sqrt{3} \times \textbf{Efficiency} \times \textbf{Power Factor})}$$

❑ **Nameplate Ampere – Single-Phase**

What is the nameplate ampere for a 7.5 horsepower motor rated 240 volts, single-phase? The efficiency is 87 percent and the power factor is 93 percent (Figure 4–11).

(a) 19 ampere (b) 24 ampere (c) 19 ampere (d) 29 ampere

• Answer: (d) 29 ampere

$$\text{Nameplate ampere} = \frac{\text{Horsepower} \times 746 \text{ Watts}}{(\text{Volts} \times \text{Efficiency} \times \text{Power Factor})}$$

$$\text{Nameplate ampere} = \frac{7.5 \text{ Horsepower} \times 746 \text{ watts}}{(240 \text{ volts} \times 0.87 \text{ Efficiency} \times 0.93 \text{ Power Factor})} = 28.81 \text{ ampere}$$

❑ **Nameplate Ampere – Three-Phase**

What is the nameplate ampere of a 40 horsepower motor rated 208 volts, 3-phase? The efficiency is 80 percent and the power factor is 90 percent (Figure 4–12).

(a) 85 ampere (b) 95 ampere (c) 105 ampere (d) 115 ampere

• Answer: (d) 115 ampere

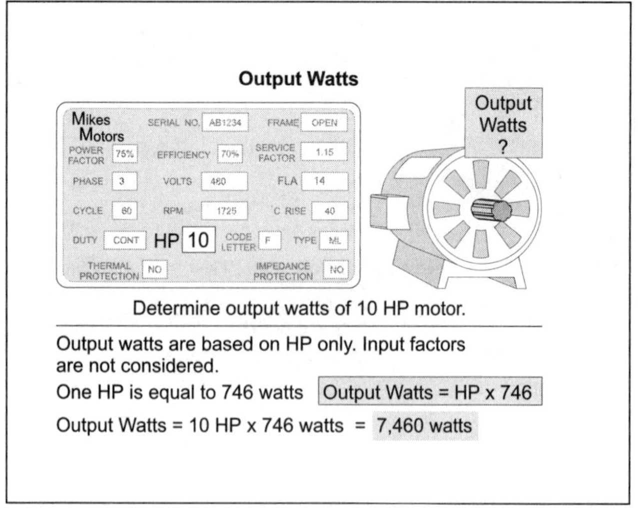

Figure 4–10
Output Watts

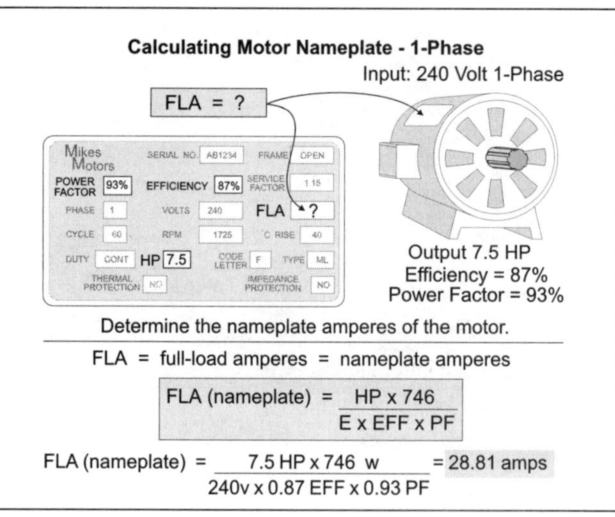

Figure 4–11
Calculating Motor Nameplate

$$\text{Nameplate ampere} = \frac{\text{Horsepower} \times 746 \text{ Watts}}{(\text{Volts} \times \sqrt{3} \times \text{Efficiency} \times \text{Power Factor})}$$

$$\text{Nameplate} = \frac{40 \text{ Horsepower} \times 746 \text{ Watts}}{(208 \text{ volts} \times 1.732 \times 0.8 \text{ Efficiency} \times 0.9 \text{ Power Factor})} = 29{,}840/259.4 = 115 \text{ ampere}$$

PART B – TRANSFORMER BASICS

TRANSFORMER INTRODUCTION

A *transformer* is a stationary device used to raise or lower voltage. Transformers have the ability to transfer electrical energy from one circuit to another by *mutual induction* between two conductor coils. Mutual induction occurs between two conductor coils *(windings)* when electromagnetic lines of force within one winding induces a voltage into a second winding (Figure 4–13).

Current transformers use the circuit conductors as the primary winding and step the current down for metering. Often, the *ratio* of the current transformer is 1,000 to 1. This means that if there are 400 ampere on the phase conductors, the current transformer steps the current down to 0.4 ampere for the meter.

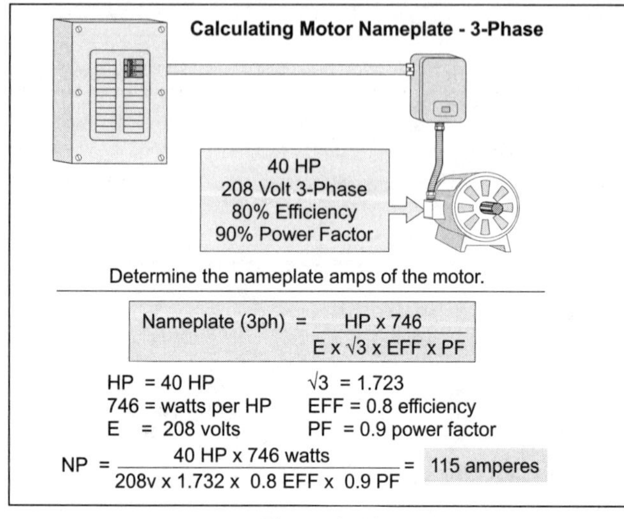

Figure 4–12
Calculating Motor Nameplate – 3-Phase

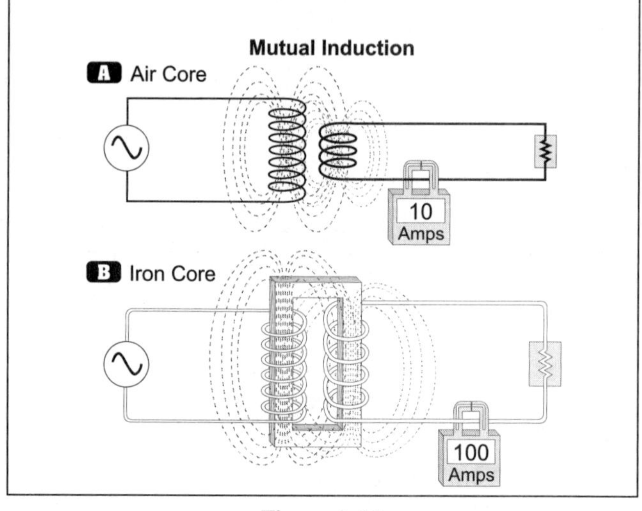

Figure 4–13
Mutual Induction

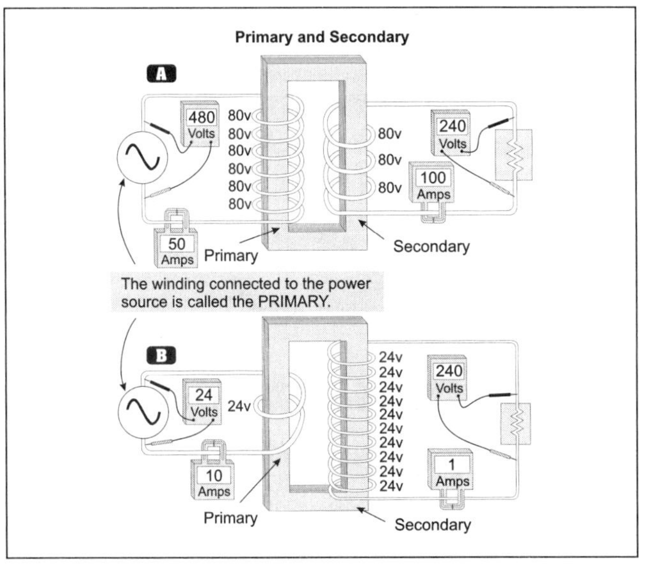

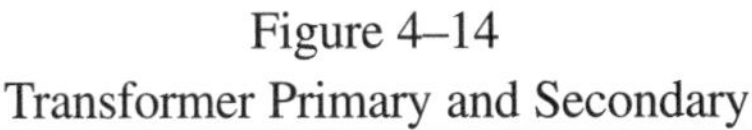

Figure 4–14
Transformer Primary and Secondary

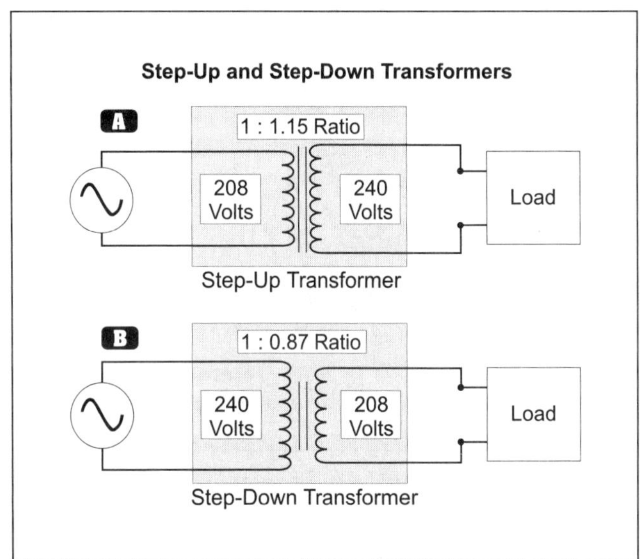

Figure 4–15
Step-Up and Step-Down Trransformers

The *magnetic coupling (magnetomotive force, MMF)* between the primary and secondary winding can be increased by increasing the winding *ampere-turns*. Ampere-turns can be increased by increasing the number of coils and/or the current through each coil. When the current in the core of a transformer is raised to a point where there is high *flux density*, additional increases in current will produce few additional flux lines. The transformer iron core is said to be *saturated.*

4–8 TRANSFORMER PRIMARY AND SECONDARY

The transformer *winding* connected to the source is called the *primary winding*. The transformer winding connected to the load is called the *secondary winding*.

4–9 TRANSFORMER SECONDARY AND PRIMARY VOLTAGE

Voltage induced in the secondary winding of a transformer is equal to the sum of the voltages induced in each *loop* of the secondary winding. The voltage induced in the secondary of a transformer depends on the number of secondary conductor turns cut by the primary magnetic flux lines. The greater the number of secondary conductor loops, the greater the secondary voltage (Figure 4–14).

Step-Up and Step-Down Transformers

The secondary winding of a *step-down transformer* has fewer turns than the primary winding, resulting in a lower secondary voltage. The secondary winding of a *step-up transformer* has more turns than the primary winding, resulting in a higher secondary voltage (Figure 4–15).

4–10 AUTOTRANSFORMERS

Autotransformers are transformers that use a common winding for both the primary and the secondary. The disadvantage of an auto transformer is the lack of isolation between the primary and secondary conductors, but they are often used because they are less expensive (Figure 4–16).

4–11 TRANSFORMER POWER LOSSES

When current flows through the winding of a transformer, power is dissipated in the form of heat. This loss is referred to as conductor I^2R loss. In addition, losses include flux leakage, core loss from eddy currents, and hysteresis heating losses (Figure 4–17).

Conductor Resistance Loss

Transformer windings are made of many turns of wire. The resistance of the conductors is directly proportional to the length of the conductors and inversely proportional to the cross-sectional area of the conductor. The more turns there are, the longer the conductor is, and the greater the conductor resistance. Conductor losses can be determined by the formula: $P = I^2R$.

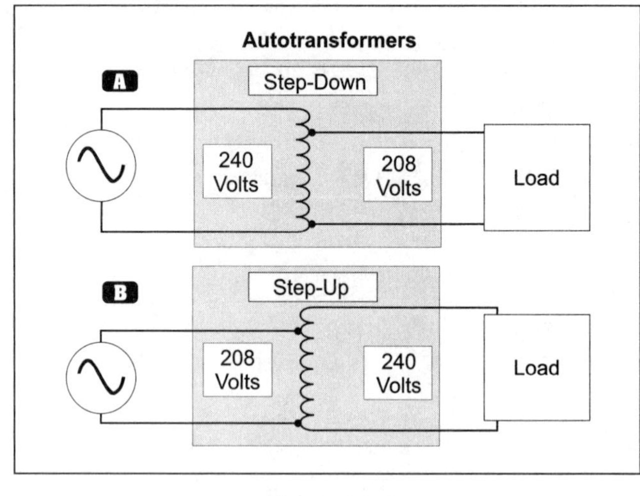

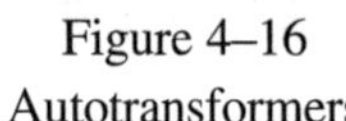
Figure 4–16
Autotransformers

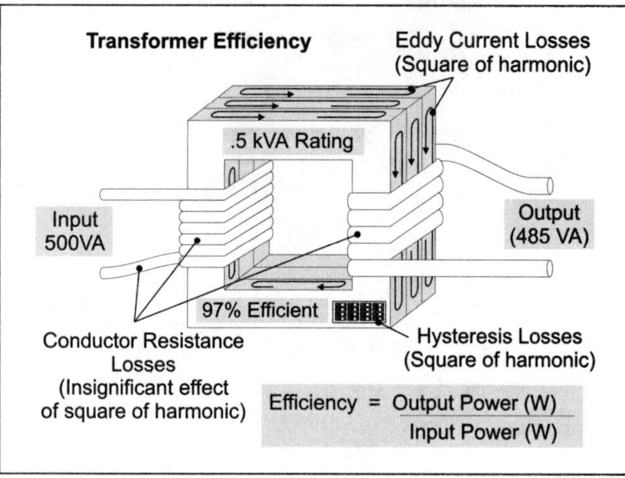

Figure 4–17
Transformer Efficiency

Flux Leakage Loss

The *flux leakage loss* represents the electromagnetic flux lines between the primary and secondary winding that are not used to convert electrical energy from the primary to the secondary. They represent wasted energy.

Core Losses

Iron is the only metal used for transformer cores because it offers low resistance to magnetic flux lines *(low reluctance).* Iron cores permit more flux lines between the primary and secondary winding, thereby increasing the magnetic coupling between the primary and secondary windings. However, alternating currents produce electromagnetic fields within the windings that induce a circulating current in the iron core. These circulating currents *(eddy currents)* flow within the iron core producing power losses that cannot be transferred to the secondary winding. Transformer iron cores are laminated to have a small cross-sectional area to reduce the eddy currents and their associated losses.

Hysteresis Losses

Each time the primary magnetic field expands and collapses, the transformer's iron core molecules realign themselves to the changing polarity of the electromagnetic field. The energy required to realign the iron core molecules to the changing electromagnetic field is called hysteresis losses.

Heating by the Square of the Frequency

Hysteresis and eddy current losses are affected by the square of the alternating current frequency. For this reason, care must be taken when iron core transformers are used in applications involving high frequencies and nonlinear loads. If the transformer operates at the third harmonic (180 Hz), the losses will be the square of the multiple of the fundamental frequency. The third harmonic frequency (180 Hz) is three times the fundamental frequency (60 Hz).

❑ Transformer Losses by Square of Frequency

What is the effect on a 60 Hz rated transformer of third harmonic loads (180 Hz)?

(a) heating increases three times
(b) heating increases six times
(c) heating increases nine times
(d) no significant heating

- Answer: (c) heating increases nine times (3^2)

4–12 TRANSFORMER TURNS RATIO

The relationship of the primary winding voltage to the secondary winding voltage is the same as the relation between the number of primary turns as compared to the number of secondary turns. This relationship is called turns ratio or voltage ratio.

❑ Delta Winding Turns Ratio

What is the turns ratio of a delta-delta transformer? The primary winding is 480 volts, and the secondary winding voltage is 240 volts (Figure 4–18)?

(a) 4:1 (b) 1:4 (c) 2:1 (d) 1:2

- Answer: (c) 2:1

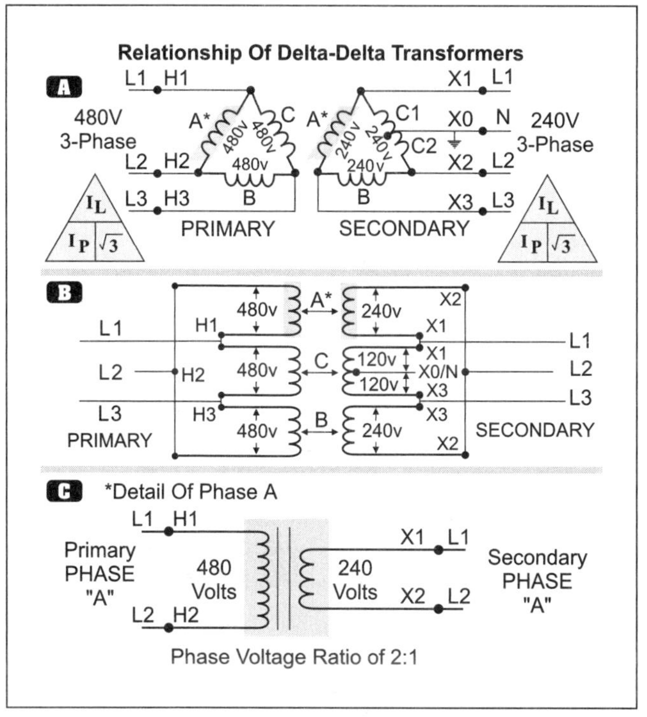

Figure 4–18
Relationships of Delta-Delta Transformers

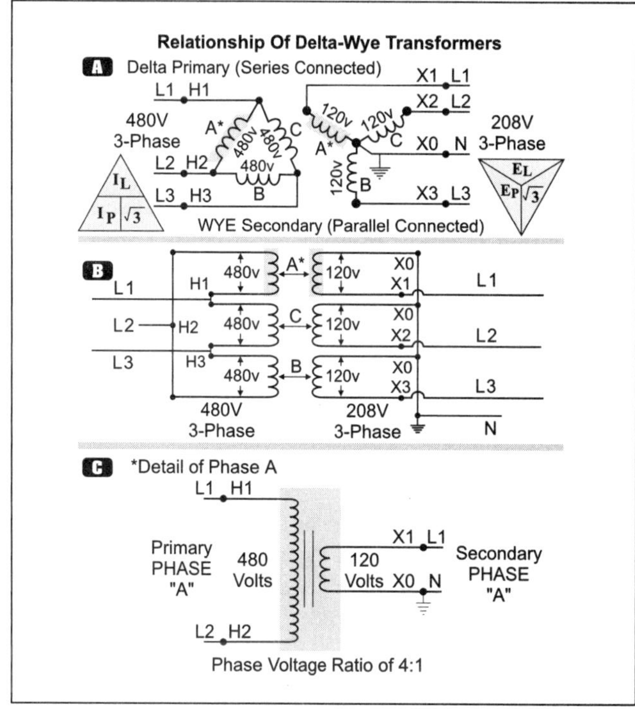

Figure 4–19
Relationships of Delta-Wye Transformers

The primary winding voltage is 480 and the secondary winding voltage is 240.

This results in a ratio of 480:240 or 2:1.

❑ **Wye Winding Ratio**

What is the turns ratio of a delta-wye transformer? The primary winding is 480 volts, and the secondary winding is 120 volts (Figure 4–19).

(a) 4:1 (b) 1:4 (c) 2.3:1 (d) 1:2.3

• Answer: (a) 4:1

The primary winding voltage is 480, and the secondary winding voltage is 120.

This results in a ratio of 480:120 or 4:1.

4–13 TRANSFORMER kVA RATING

Transformers are rated in kilovolt-ampere, abbreviated as kVA.

4–14 TRANSFORMER CURRENT

Whenever the number of primary turns is greater than the number of secondary turns, the secondary voltage is less than the primary voltage. This results in secondary current being greater than the primary current because the power remains the same, but the voltage changes. Since the secondary voltage of most transformers is less than the primary voltage, the secondary conductors carry more current than the primary (Figure 4–20). Primary and secondary line current can be calculated by:

Current (single-phase) = Volt-Ampere/Volts

Current (three-phase) = Volt-Ampere/(Volts × 1.732)

❑ **Transformer Current – Single-Phase**

What is the primary and secondary line current for a single-phase 25 kVA transformer, rated 480 volts primary and 240 volts secondary (Figure 4–21)?

(a) 52/104 ampere (b) 104/52 ampere (c) 104/208 ampere (d) 208/104 ampere

• Answer: (a) 52/104 ampere

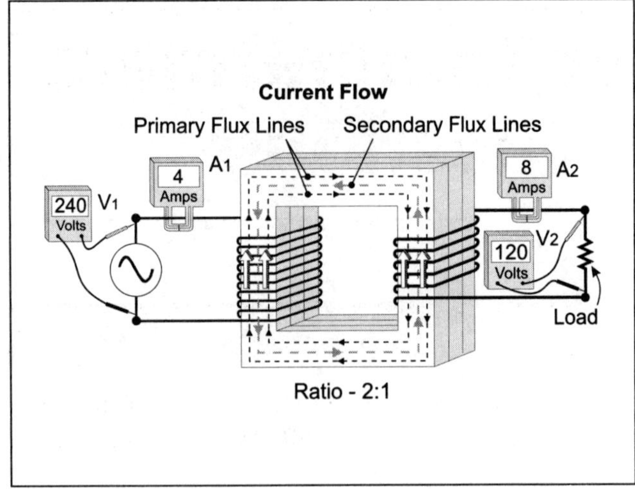

Figure 4–20
Transformer Current Flow

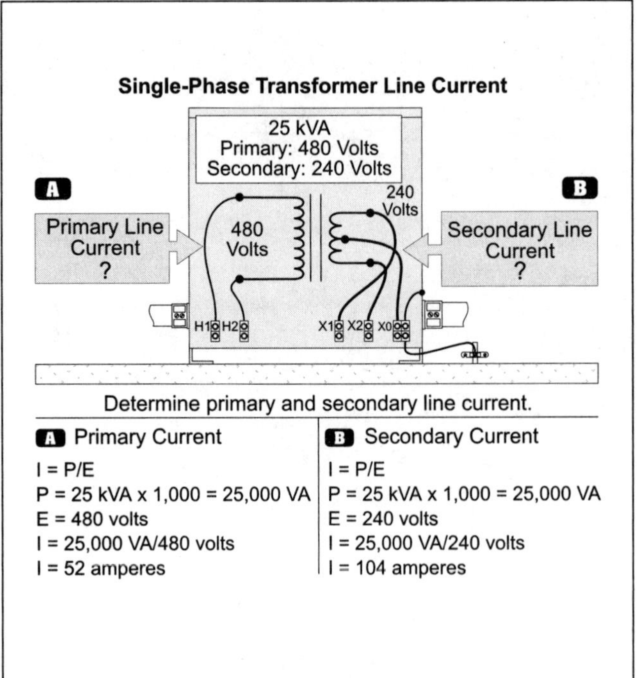

Figure 4–21
Transformer Line Current

Primary current = VA/E

$$\text{Primary current} = \frac{25{,}000 \text{ VA}}{480 \text{ volts}}$$

Primary current = 52 ampere

Secondary current = VA/E

$$\text{Secondary current} = \frac{25{,}000 \text{ VA}}{240 \text{ volts}}$$

Secondary current = 104 ampere

❑ Transformer Current – Three-Phase

What is the primary and secondary line current for a 3-phase 37.5 kVA transformer, rated 480 volts primary and 208 volts secondary (Figure 4–22)?

(a) 45/104 ampere (b) 104/40 ampere
(c) 208/140 ampere (d) 140/120 ampere

• Answer: (a) 45/104 ampere

$$\text{Primary current} = \frac{\text{VA}}{\text{E} \times \sqrt{3}}$$

$$\text{Primary current} = \frac{37{,}500 \text{ VA}}{(480 \text{ volts} \times 1.732)}$$

Primary current = 45 ampere

$$\text{Secondary current} = \frac{\text{VA}}{\text{E} \times \sqrt{3}}$$

$$\text{Secondary current} = \frac{37{,}500 \text{ VA}}{(208 \text{ volts} \times 1.732)}$$

Secondary current = 104 ampere

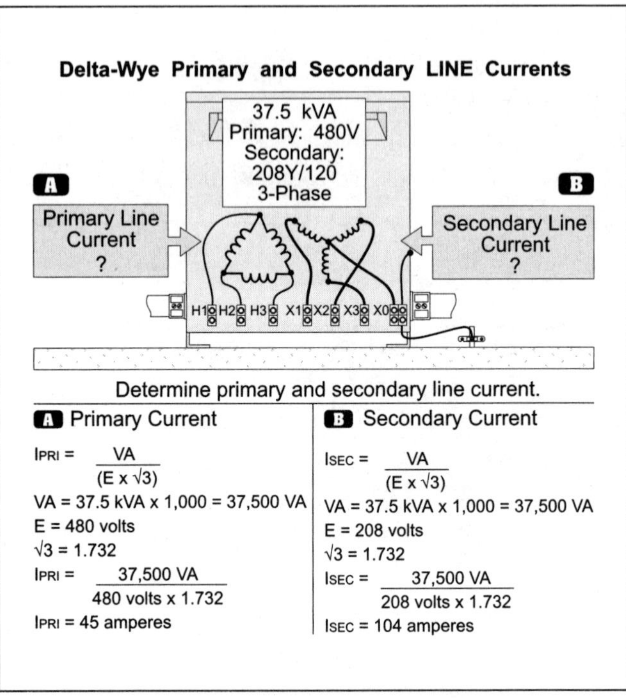

Figure 4–22
Delta-Wye Transformers Primary and Secondary Line Currents

Unit 4 – Motors and Transformers Summary Questions

Part A – Motors

Motor Introduction

1. When voltage is applied to the motor's armature, short-circuit current will flow and the armature will begin to turn. As the armature starts turning, it cuts the lines of force of the field winding, resulting in induced CEMF in the armature conductor, which reduces the short-circuit current.
(a) True (b) False

2. An electric motor works because of the effects a _____ has against a wire carrying an electric current.
(a) magnetic field (b) commutator (c) voltage source (d) none of these

3. The rotating part of a direct current motor or generator is called the _____ .
(a) shaft (b) rotor (c) capacitor (d) field

4. For a direct current motor, a device called a _____ is placed on the end of the conductor loop (armature). The polarity of the loop is maintained with the proper magnetic field to keep the opposing magnetic fields pushing each other in the same direction. This in turn keeps the loop turning.
(a) coil (b) resistor (c) commutator (d) none of these

4–1 Motor Speed Control

5. • One of the great advantages of the direct current motor is the motor's ability to maintain a constant speed. If the speed of a direct current motor is increased, the armature will cut through the field winding magnetic flux at an increasing rate resulting in a lower CEMF that acts to cut down on the increased armature current, which slows the motor back down.
(a) True (b) False

6. • Placing a load on a direct current motor causes the motor to slow down, which increases the rate at which the field flux lines are cut by the armature. As a result, the armature CEMF increases, resulting in an increase in the applied armature voltage and current. The increase in current results in an increase in motor speed.
(a) True (b) False

4–2 Reversing a Direct Current Motor

7. To reverse a direct current motor, you must reverse the direction of the _____ .
(a) field current (b) armature current (c) a or b (d) a and b

4–3 Alternating Current Motors

8. In a(n) _____ motor, the rotor is actually locked in step with the rotating stator field and is dragged along at the synchronous speed of the rotating magnetic field. _____ motors maintain their speed with a high degree of accuracy and are used for electric clocks and other timing devices.
(a) alternating current (b) universal (c) wound rotor (d) synchronous

9. _____ rotor motors are used only as special applications because of their high starting torque requirements and only operate on 3-phase alternating current power.
(a) Alternating current (b) Universal (c) Wound rotor (d) Synchronous

10. _____ motors are fractional horsepower motors that operate equally well on alternating current and direct current and are used for vacuum cleaners, electric drills, mixers, and light household appliances.
(a) Alternating current (b) Universal (c) Wound rotor (d) Synchronous

4–4 Reversing Alternating Current Motors

11. • Three-phase alternating current motors can be reversed by changing the wiring from the line wiring from ABC phase configuration to_____ .
(a) BCA (b) CAB (c) CBA (d) ABC

4–5 Motor Volt-Ampere Calculations

12. Dual-voltage 277/480 volt motors are made with two field windings, each rated at 277 volts. The field windings are connected in parallel for _____ volt operation and in series for _____ volt operation.
(a) 277, 480 (b) 480, 277 (c) 277, 277 (d) 480, 480

4–6 Motor Horsepower/Watts

13. What size motor is required to produce 30 kW output?
(a) 20 horsepower (b) 30 horsepower (c) 40 horsepower (d) 50 horsepower

14. What are the output watts of a 15 horsepower motor?
(a) 11 kW (b) 15 kVA (c) 22 kW (d) 31 kW

15. What are the output watts of a 5 horsepower motor, 3-phase 480 volt, efficiency 75 percent and power factor 70 percent?
(a) 3.75 kW (b) 7.5 kVA (c) 7.5 kW (d) 10 kW

4–7 Motor Nameplate Ampere

16. • For practical purposes you will not need to calculate motor nameplate current; but you should understand how it is calculated.
(a) True (b) False

17. What are the nameplate ampere for a 5 horsepower motor, 240 volt, single-phase, efficiency at 90 percent and power factor of 80 percent?
(a) 19.3 ampere (b) 21.6 ampere (c) 28.2 ampere (d) 31.1 ampere

18. What are the nameplate ampere of a 20 horsepower 208 volt, 3-phase motor, power factor 90 percent, and efficiency of 80 percent?
(a) 50 ampere (b) 58 ampere (c) 65 ampere (d) 80 ampere

Part B – Transformer Basics

Transformer Introduction

19. A _____ is a stationary device used to raise or lower the voltage and has the ability to transfer electrical energy from one circuit to another, with no physical connection between the two.
(a) capacitor (b) motor (c) relay (d) transformer

20. Transformers operate on the principle of _____ .
(a) magnetoelectricity (b) triboelectric effect (c) thermocouple (d) mutual induction

4–8 Primary and Secondary Transformer

21. The transformer winding that is connected to the source is called the _____ winding and the transformer winding that is connected to the load is called the _____ . Transformers are reversible; that is, either winding can be used as the primary or secondary.

(a) secondary, primary (b) primary, secondary (c) depends on the wiring (d) none of these

22. Voltage induced in the secondary winding of a transformer is dependent on the number of secondary turns as compared to the number of primary turns.
(a) True (b) False

23. The secondary winding of a step-down transformer has _____ turns than the primary, resulting in a _____ secondary voltage as compared to the primary.
(a) less, higher (b) more, lower (c) less, lower (d) more, higher

24. The secondary winding of a step-up transformer has _____ turns than the primary, resulting in a _____ secondary voltage as compared to the primary.
(a) less, higher (b) more, lower (c) less, lower (d) more, higher

4–10 Autotransformers

25. Autotransformers use the same winding for both the primary and secondary. The disadvantage of an autotransformer is the lack of _____ between the primary and secondary conductors.
(a) power (b) voltage (c) isolation (d) grounding

4–11 Transformer Power Losses

26. The most common causes of power losses for transformers are _____ .
(a) conductor resistance (b) eddy currents (c) hysteresis (d) all of these

27. The leakage of the electromagnetic flux lines between the primary and secondary winding represents wasted energy.
(a) True (b) False

28. • The expanding and collapsing electromagnetic field of the transformer also induces a voltage in the transformer core. The induced voltage causes _____ to flow within the core, which removes energy from the transformer winding and represents wasted power.
(a) eddy currents (b) flux (c) inductive (d) hysteresis

29. Eddy currents can be reduced by dividing the core into many flat sections or laminations. Because the laminations have a _____ cross-sectional area, the resistance offered to the eddy currents is greatly increased.
(a) round (b) porous (c) large (d) small

30. As current flows through the transformer, the iron core is temporarily magnetized by the electromagnetic field created by the alternating current. Each time the primary magnetic field expands and collapses, the core molecules realign themselves to the changing polarity of the electromagnetic field. The energy required to realign the core molecules to the changing electromagnetic field is called the _____ loss of the core.
(a) eddy current (b) flux (c) inductive (d) hysteresis

4–12 Transformer Turns Ratio

31. The relationship of the primary winding voltage to the secondary winding voltage is the same as the relation between the number of conductor turns on the primary as compared to the secondary. This relationship is called _____ .
(a) ratio (b) efficiency (c) power factor (d) none of these

32. • The primary phase voltage is 240 and the secondary phase is 480. This results in a ratio of _____ .
(a) 1:2 (b) 2:1 (c) 4:1 (d) 1:4

4–13 Transformer kVA Rating

33. Transformers are rated in _____ .
(a) VA (b) kW (c) watts (d) kVA

4–14 Transformer Current

34. • The current flow in the secondary transformer winding creates an electromagnetic field that opposes the primary electromagnetic field resulting in less primary CEMF. The primary current automatically increases in direct proportion to the secondary current.
(a) True (b) False

35. • The transformer winding with the _____ number of turns will have the lower current and the winding with the _____ number of turns will have the higher current.
(a) lesser, greater (b) most, most (c) least, least (d) greater, lesser

✯ Challenge Questions

Part A – Motors

4–1 Motor Speed Control

36. A(n) _____ type of electric motor tends to run away if it is not always connected to its load.
(a) direct current series (b) direct current shunt
(c) alternating current induction (d) alternating current synchronous

37. • A(n) _____ motor has a wide speed range.
(a) alternating current (b) direct current (c) synchronous (d) induction

4–2 Reversing a Direct Current Motor

38. • If the two line (supply) leads of a direct current series motor are reversed, the motor will _____ .
(a) not run (b) run backwards (c) run the same as before (d) become a generator

39. • To reverse a direct current series motor, one may simply reverse the supply (power) leads.
(a) True (b) False

4–3 Alternating Current Motors

40. The _____ induction motor is used only in special applications and is always operated on 3-phase alternating current power.
(a) compound (b) synchronous (c) split phase (d) wound rotor

41. • The rotating part of a direct current motor or generator is called the _____ .
(a) shaft (b) rotor (c) armature (d) b or c

4–5 Motor Volt-Ampere Calculations

42. The input volt-ampere of a 5 horsepower (15.2 ampere), 230 volt, 3-phase motor are closest to _____ .
(a) 7,500 VA (b) 6,100 VA (c) 5,300 VA (d) 4,600 VA

43. The input volt-ampere of a 1 horsepower (16 ampere), 115 volt single-phase motor is _____ .
(a) 2,960 VA (b) 1,840 VA (c) 3,190 VA (d) 1,650 VA

Part B – Transformer Basics

4–11 Transformer Power Losses

44. • When the current in the core of a transformer has risen to a point where high flux density has been reached and additional increases in current produce few additional flux lines, the metal core is said to be _____ .
(a) maximum (b) saturated (c) full (d) none of these

45. • Magnetomotive force (MMF) can be increased by increasing the _____ .
(a) number of ampere-turns (b) current in the coils
(c) number of coils (d) all of these

4–12 Transformer Turns Ratio

46. • The secondary current of a transformer that has a turns ratio of 2:1 is _____ the primary current.
(a) higher than (b) lower than (c) the same as (d) none of these

4–13 Transformer kVA Rating

47. The primary kVA rating for a transformer that is 100 percent efficient, with a secondary of 12 volts (E) and with secondary current (I) of 5 ampere is _____ .
(a) 600 kVA (b) 30 kVA (c) 6 kVA (d) 0.06 kVA

4–14 Transformer Current

48. • The primary of a transformer has 100 turns and the secondary has 10 turns. What is the primary current of the transformer if the secondary current (I) is 5 ampere?
(a) 25 ampere (b) 10 ampere (c) 5 ampere (d) 0.5 ampere

49. • Which winding of a current transformer will carry more current?
Note: A clamp-on ammeter is a current transformer, the meter acts as the secondary.
(a) primary (b) secondary (c) interwinding (d) tertiary

50. A transformer primary winding has 900 turns and the secondary winding has 90 turns, which winding of the transformer has a larger conductor?
(a) primary (b) secondary (c) interwinding (d) none of these

51. The primary of a transformer is 480 volts and the secondary is 240 volts, which winding of this transformer has a larger conductor?
(a) tertiary (b) secondary (c) primary (d) windings are equal

52. If the transformer voltage turns ratio is 5:1 and the secondary has 10 ampere-turns, the primary current would be _____ if the secondary current was 10 ampere.
(a) 25 ampere (b) 10 ampere (c) 2 ampere (d) cannot be determined

53. The transformer primary is 240 volts, the secondary is 12 volts, and the load is two 100-watt lamps. The transformer is 92 percent efficient. The secondary current of this transformer is _____ .
(a) 1 ampere (b) 17 ampere (c) 28 ampere (d) none of these

54. The transformer primary is 240 volts, the secondary is 120 volts, and the load is 1,500 watt. The transformer is 92 percent efficient. The primary current of this transformer is _____ ampere.
(a) 6.8 (b) 8.6 (c) 9.9 (d) 7.8

55. • The primary kVA of this transformer is _____ (Figure 4–23)?
(a) 72.5 kVA (b) 42 kVA
(c) 30 kVA (d) 21 kVA

56. • The primary current of this transformer is _____ (Figure 4–23)?
(a) 90 ampere (b) 50 ampere
(c) 25 ampere (d) 12 ampere

57. • The secondary voltage of this transformer is _____ (Figure 4–24)?
(a) 6 volts (b) 12 volts
(c) 24 volts (d) 30 volts

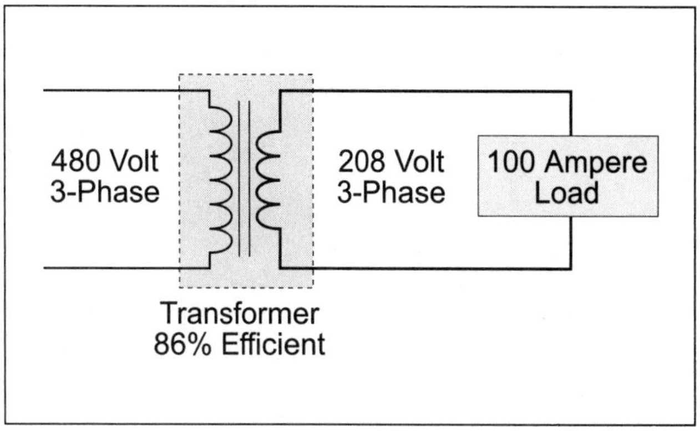

Figure 4–23

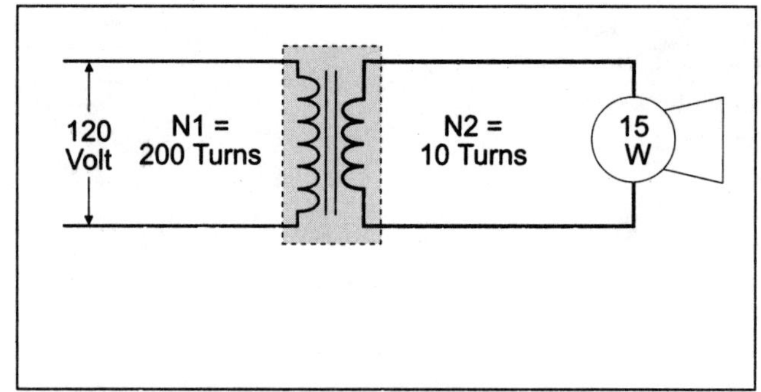

Figure 4–24

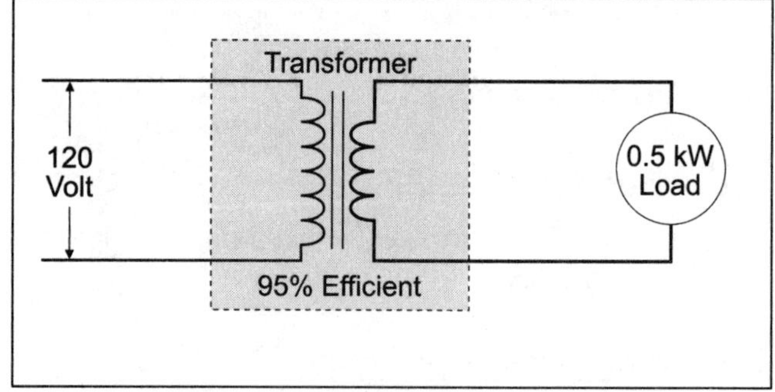

Figure 4–25

58. The primary current of the transformer is _____ (Figure 4–25)?
(a) 0.416 ampere (b) 4.38 ampere (c) 3.56 ampere (d) 41.6 ampere

59. The primary power for the transformer is _____ (Figure 4–25)?
(a) 526 VA (b) 400 VA (c) 475 VA (d) 550 VA

Transformer Miscellaneous

60. The transformer primary is 240 volts, the secondary is 12 volts, and the load is two 100-watt lamps. The transformer is 92 percent efficient. The secondary VA of this transformer _____ .
(a) is 200 VA (b) is the same as the primary VA
(c) cannot be calculated (d) none of these

61. • The secondary of a transformer is 24 volts with a load of 5 ampere. The primary is 120 volts and the transformer is 100 percent efficient. The secondary VA of this transformer _____ .
(a) is 120 VA (b) is the same as the primary VA
(c) cannot be calculated (d) a and b

62. The transformer primary is 240 volts, the secondary is 12 volts, and the load is two 100-watt lamps. The transformer is 92 percent efficient. The primary VA of this transformer _____ .
(a) is 185 VA (b) is 217 VA (c) is 0.217 VA (d) cannot be calculated

63. The output of a generator is 20 kW. The input kVA is _____ if the efficiency rating is 65 percent.
(a) 11 kVA (b) 20 kVA (c) 31 kVA (d) 33 kVA

CHAPTER 2
NEC® Calculations

Scope of Chapter 2

Unit 5

Raceway, Outlet Box, and Junction Boxes Calculations

OBJECTIVES

After reading this unit, the student should be able to briefly explain the following concepts:

Part A – Raceway Fill Calculations
- Existing raceway calculation
- Raceway sizing
- Raceway properties
- Understanding *NEC*® Chapter 9

Part B – Outlet Box Calculations
- Conductor equivalents
- Sizing box – conductors all the same size
- Volume of box

Part C – Pull and Junction Box Calculations
- Depth of box and conduit body sizing
- Pull and junction box size calculations

After reading this unit, the student should be able to briefly explain the following terms:

Part A – Raceway Fill Calculations
- Alternating current conductor resistance
- Bare conductors
- Bending radius
- Compact aluminum building wire
- Conductor properties
- Conductor fill
- Conduit bodies
- Cross-sectional area of insulated conductors
- Expansion characteristics of PVC
- Fixture wires
- Grounding conductors
- Lead-covered conductor
- *NEC*® errors
- Nipple size
- Raceway size
- Spare space area

Part B – Outlet Box Calculations
- Cable clamps
- Conductor terminating in the box
- Conductor running through the box
- Conduit bodies
- Equipment bonding jumpers
- Extension rings
- Fixture hickey
- Fixture stud
- Outlet box
- Pigtails
- Plaster rings
- Short radius conduit bodies
- Size outlet box
- Strap
- Volume
- Yoke

Part C – Pull and Junction Box Calculations
- Angle pull calculation
- Distance between raceways
- Horizontal dimension
- Junction boxes

PART A – RACEWAY FILL CALCULATIONS

5–1 UNDERSTANDING THE NATIONAL ELECTRICAL CODE®, CHAPTER 9

Chapter 9 – Tables

Table 1 – Conductor Percent Fill

The maximum percentage of conductor fill is listed in Table 1 of Chapter 9 and is based on common conditions where the length of the conductor and number of raceway bends are within reasonable limits [FPN under Table 1] (Figure 5–1).

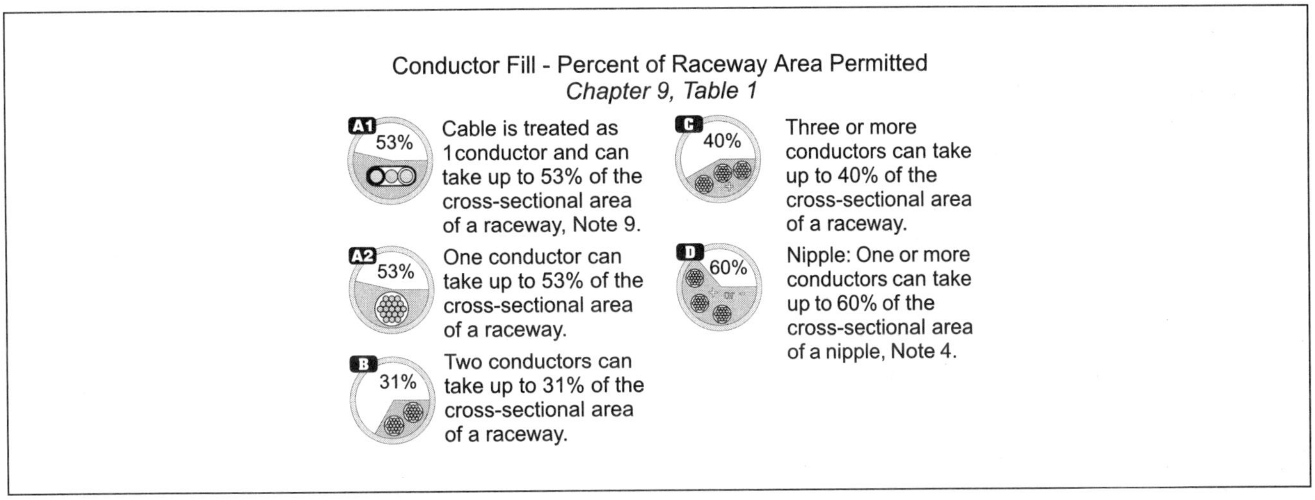

Conductor Fill - Percent of Raceway Area Permitted
Chapter 9, Table 1

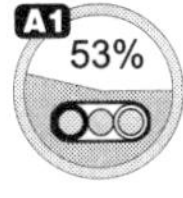

Cable is treated as 1 conductor and can take up to 53% of the cross-sectional area of a raceway, Note 9.

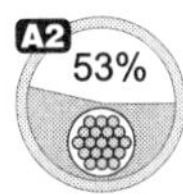

One conductor can take up to 53% of the cross-sectional area of a raceway.

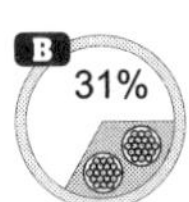

Two conductors can take up to 31% of the cross-sectional area of a raceway.

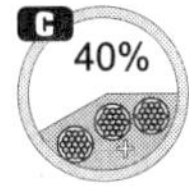

Three or more conductors can take up to 40% of the cross-sectional area of a raceway.

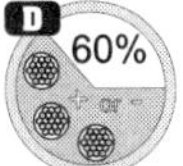

Nipple: One or more conductors can take up to 60% of the cross-sectional area of a nipple, Note 4.

Figure 5–1
Conductor Fill – Percent of Raceway Area Permitted

Table 1 of Chapter 9, Maximum Percent Conductor Fill	
Number of Conductors	**Percent Fill Permitted**
1 conductor	53% fill
2 conductors	31% fill
3 or more conductors	40% fill
Raceway 24 inches or less	60% fill Chapter 9, Note 4

Table 1, Note 1 – Conductors All the Same Size and Insulation

When all of the conductors are the same size and insulation, the number of conductors permitted in a raceway can be determined simply by looking at the tables located in Appendix C – Conduit and Tubing Fill Tables for Conductors and Fixture Wires of the Same Size.

Tables C1 through C12A are based on maximum percent fill as listed in Table 1 of Chapter 9.

Table C1 – Conductors and fixture wires in electrical metallic tubing
Table C1A – Compact conductors in electrical metallic tubing
Table C2 – Conductors and fixture wires in electrical nonmetallic tubing
Table C2A – Compact conductors in nonelectrical metallic tubing
Table C3 – Conductors and fixture wires in flexible metal conduit
Table C3A – Compact conductors in flexible metal conduit
Table C4 – Conductors and fixture wires in intermediate metal conduit
Table C4A – Compact conductors in intermediate metal conduit
Table C5 – Conductors and fixture wires in liquidtight flexible nonmetallic conduit (gray type)
Table C5A – Compact conductors in liquidtight flexible nonmetallic conduit (gray type)
Table C6 – Conductors and fixture wires in liquidtight flexible nonmetallic conduit (orange type)
Table C6A – Compact conductors in liquidtight flexible nonmetallic conduit (orange type)

Note: The appendix does not have a table for liquidtight flexible nonmetallic conduit of the black type.

Table C7 – Conductors and fixture wires in liquidtight flexible metallic conduit
Table C7A – Compact conductors in liquidtight flexible metal conduit
Table C8 – Conductors and fixture wires in rigid metal conduit

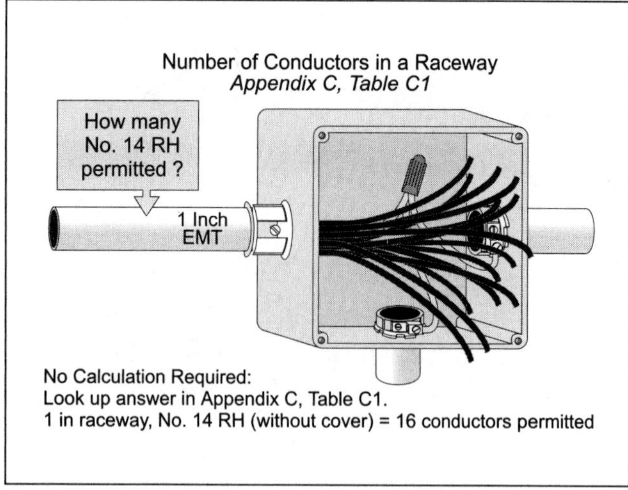

Figure 5–2
Number of Conductors in a Raceway

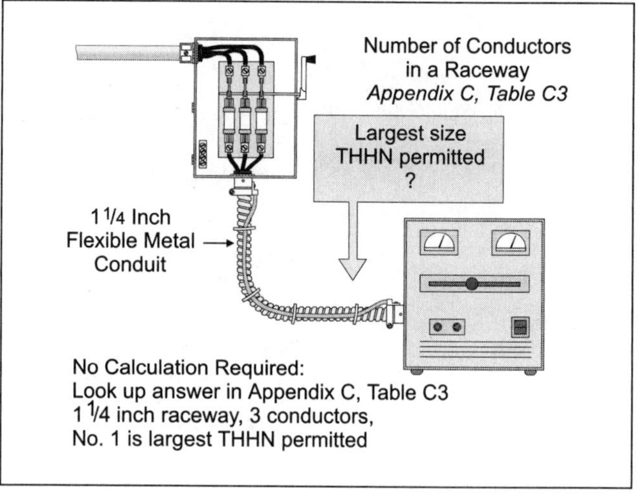

Figure 5–3
Number of Conductors in a Raceway

Table C8A – Compact conductors in rigid metal conduit
Table C9 – Conductors and fixture wires in rigid PVC conduit schedule 80
Table C9A – Compact conductors in rigid PVC conduit schedule 80
Table C10 – Conductors and fixture wires in rigid PVC conduit schedule 40
Table C10A – Compact conductors in rigid PVC conduit schedule 40
Table C11 – Conductors and fixture wires in Type A, rigid PVC conduit
Table C11A – Compact conductors in Type A, PVC conduit
Table C12 – Conductors and fixture wires in Type EB, PVC conduit
Table C12A – Compact conductors in Type EB, PVC conduit

❑ **Appendix C – Table C1**

How many No. 14 RHH conductors (without cover) can be installed in a 1 inch electrical metallic tubing (Figure 5–2)?

(a) 25 conductors (b) 16 conductors (c) 13 conductors (d) 19 conductors

• Answer: (b) 16 conductors, Appendix C, Table C1

❑ **Appendix C – Table C2A – Compact Conductor**

How many compact No. 6 XHHW conductors can be installed in a $1^{1}/_{4}$ inch nonmetallic tubing?

(a) 10 conductors (b) 6 conductors (c) 16 conductors (d) 13 conductors

• Answer: (a) 10 conductors, Appendix C, Table C2A

❑ **Appendix C – Table C3**

If $1^{1}/_{4}$ inch flexible metal conduit has three THHN conductors (not compact), what is the largest conductor permitted to be installed (Figure 5–3)?

(a) No. 1 (b) No. 1/0 (c) No. 2/0 (d) No. 3/0

• Answer: (a) No. 1, Appendix C, Table C3

❑ **Appendix C – Table C4**

How many No. 4/0 RHH conductors (with outer cover) can be installed in 2 inch intermediate metal conduit?

(a) 2 conductors (b) 1 conductor (c) 3 conductors (d) 4 conductors

• Answer: (c) 3 conductors, Appendix C, Table C4

❑ **Appendix C – Table C7 – Fixture Wire**

How many No. 18 TFFN conductors can be installed in a $^{3}/_{4}$ inch liquidtight flexible metallic conduit (Figure 5– 4)?

(a) 40 (b) 26 (c) 30 (d) 39

• Answer: (d) 39, Appendix C, Table C7

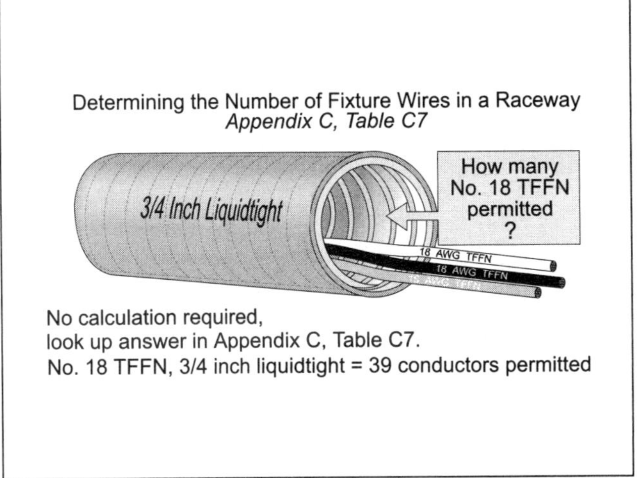

Figure 5–4
Determining the Number of Fixture Wires in a Raceway

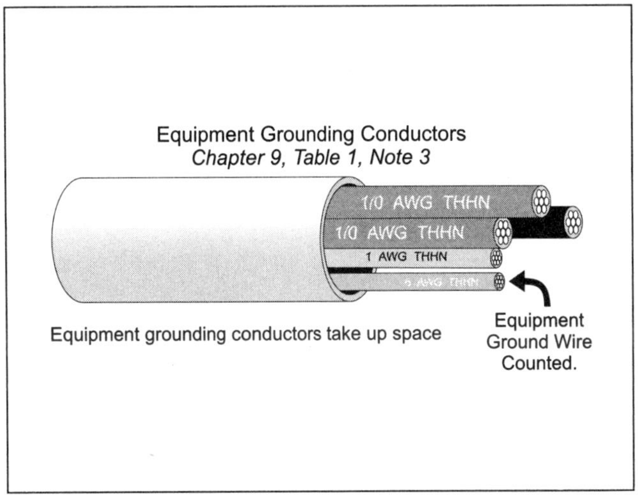

Figure 5–5
Equipment Grounding Conductors

Table 1, Note 3 – Equipment Grounding Conductors

When equipment grounding conductors are installed in a raceway, the actual area of the conductor must be used when calculating raceway fill. Chapter 9, Table 5 can be used to determine the cross-sectional area of insulated conductors, and Chapter 9, Table 8 can be used to determine the cross-sectional area of bare conductors [Note 8 of Table 1, Chapter 9] (Figure 5–5).

Table 1, Note 4 – Nipples, Raceways Not Exceeding 24 Inches

The cross-sectional areas of conduit and tubing can be found in Table 4 of Chapter 9. When a conduit or tubing raceway does not exceed 24 inches in length, it is called a nipple. Nipples are permitted to be filled to 60% of their total cross-sectional area (Figure 5–6).

Table 1, Note 7

When the calculated number of conductors (all of the same size and insulation) results in 0.8 or larger, the next whole number can be used. But be careful: this only applies when the conductors are all the same size and insulation.

Table 1, Note 8

The dimensions for bare conductor are listed in Table 8 of Chapter 9.

Chapter 9, Table 4 – Conduit and Tubing Cross-Sectional Area

Table 4 of Chapter 9 lists the dimensions and cross-sectional area for conduit and tubing. The cross-sectional area of a conduit or tubing is dependent on the raceway type (cross-sectional area of the area) and the maximum percentage fill as listed in Table 1 of Chapter 9.

❑ **Conduit Cross-Sectional Area**

What is the total cross-sectional area of $1^1/_4$ inch rigid metal conduit (Figure 5–7)?

(a) 1.063 square inches (b) 1.526 square inches
(c) 1.098 square inches (d) any of these

• Answer: (b) 1.526 square inches

Chapter 9, Table 4

Chapter 9, Table 5 – Dimensions of Insulated Conductors and Fixture Wires

Table 5 of Chapter 9 lists the cross-sectional area of insulated conductors and fixture wires.

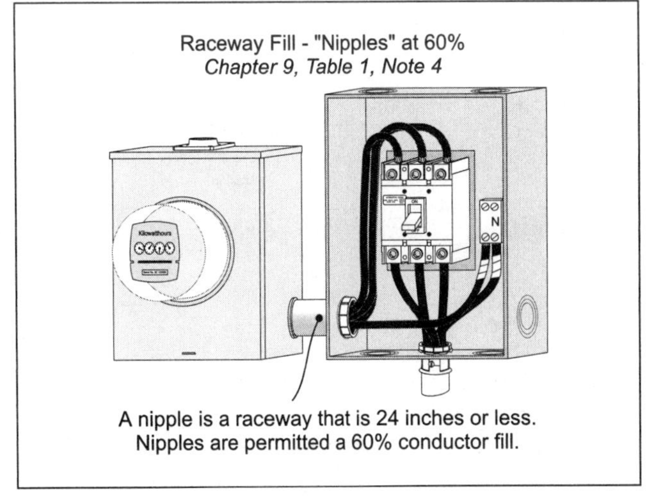

Figure 5–6
Raceway Fill – "Nipples" at 60%

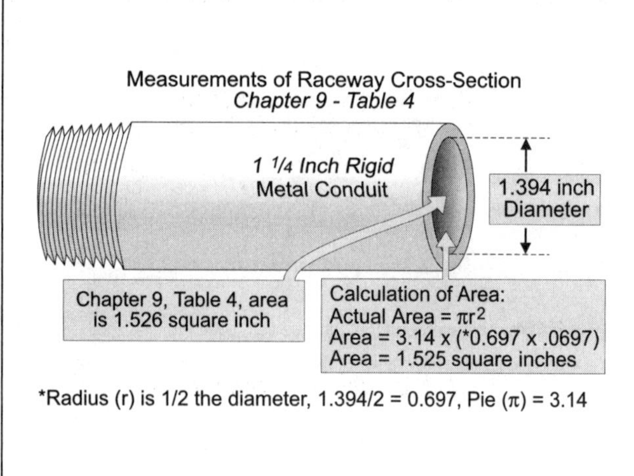

Figure 5–7
Measurements of Raceway Cross-Section

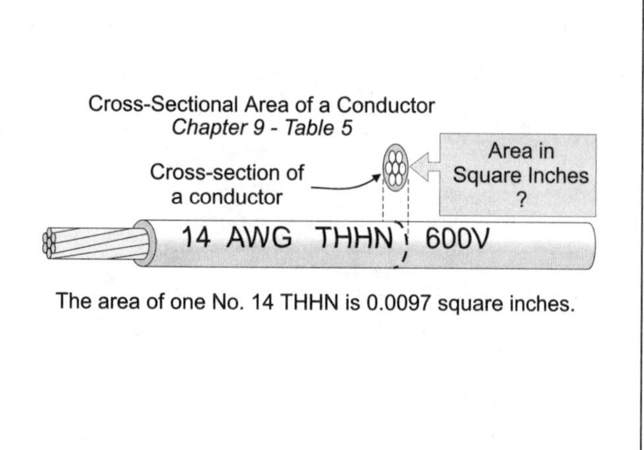

Figure 5–8
Cross-Sectional Area of a Conductor

Table 5 of Chapter 9 – Conductor Cross-Sectional Area							
	RH	RHH/RHW With *Cover*	RHH/RHW With*out* *Cover* or THW	TW	THHN THWN TFN	XHHW	BARE *Stranded* Conductors
Size AWG/kcmil	Approximate Cross-Sectional Area – Square Inches						
Column 1	Column 2	Column 3	Column 4	Column 5	Column 6	Column 7	Chapter 9, Table 8
14	.0293	.0293	.0209	.0139	.0097	.0139	**.004**
12	.0353	.0353	.0260	.0181	.0133	.0181	**.006**
10	.0437	.0437	.0333	.0243	.0211	.0243	**.011**
8	.0835	.0835	.0556	.0437	.0366	.0437	**.017**
6	.1041	.1041	.0726	.0726	.0507	.0590	**.027**
4	.1333	.1333	.0973	.0973	.0824	.0814	**.042**
3	.1521	.1521	.1134	.1134	.0973	.0962	**.053**
2	.1750	.1750	.1333	.1333	.1158	.1146	**.067**
1	.2660	.2660	.1901	.1901	.1562	.1534	**.087**
0	.3039	.3039	.2223	.2223	.1855	.1825	**.109**
00	.3505	.3505	.2624	.2624	.2233	.2190	**.137**
000	.4072	.4072	.3117	.3117	.2679	.2642	**.173**
0000	.4754	.4754	.3718	.3718	.3237	.3197	**.219**

❑ **Table 5 – THHN**

What is the cross-sectional area for one No. 14 THHN conductor (Figure 5–8)?

(a) .0206 square inch (b) .0172 square inch (c) .0097 square inch (d) .0278 square inch

• Answer: (c) .0097 square inch

❑ **Table 5 – RHW *With Outer Cover***

What is the cross-sectional area for one No. 12 RHW conductor *with outer cover*?

(a) .0206 square inch
(b) .0172 square inch
(c) .0353 square inch
(d) .0278 square inch

• Answer: (c) .0353 square inch

❑ **Table 5 – RHH *Without Outer Cover***

What is the cross-sectional area for one No. 10 RHH *without an outer cover*?

(a) .0117 square inch (b) .0333 square inch
(c) .0252 square inch (d) .0278 square inch

• Answer: (b) .0333 square inch

Chapter 9, Table 5A – Compact Aluminum Building Wire Nominal Dimensions and Areas

Tables 5A, Chapter 9 list the cross-sectional area for compact aluminum building wires. We will not use these tables for this unit.

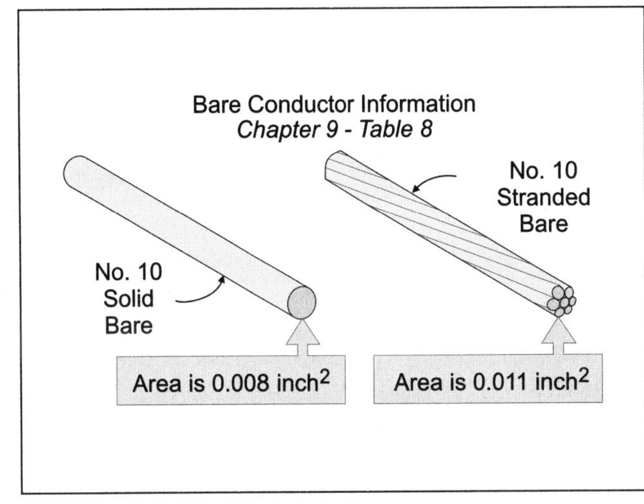

Figure 5–9
Bare Conductor Information

Chapter 9, Table 8 – Conductor Properties

Table 8 contains conductor properties such as: cross-sectional area in circular mils, number of strands per conductor, cross-sectional area in square inches for bare conductors, and conductor's resistance at 75°C for direct current for both copper and aluminum wire.

❑ **Bare Conductor – Cross-Sectional Area**

What is the cross-sectional area in square inches for one No. 10 bare conductor (Figure 5–9)?

(a) .008 solid (b) .011 stranded (c) .038 (d) a or b

• Answer: (d) .008 square inch for solid and .011 square inch for stranded

Chapter 9, Table 9 – AC Resistance for Conductors in Conduit or Tubing

Table 9 contains the alternating current resistance for copper and aluminum conductors.

❑ **Alternating Current Resistance**

What is the alternating current resistance of No. 1/0 copper installed in a metal conduit? Conductor length 1,000 feet.

(a) 0.12 ohm (b) 0.13 ohm (c) 0.14 ohm (d) 0.15 ohm

• Answer: (a) 0.12 ohm

5–2 RACEWAY AND NIPPLE CALCULATIONS

Appendix C – Tables 1 through 12 cannot be used to determine raceway sizing when conductors of different sizes (or types of insulation) are installed in the same raceway. The following Steps can be used to determine the raceway size and nipple size:

Step 1: ➼ Determine the cross-sectional area (square inches) for each conductor from Table 5 of Chapter 9 for insulated conductors, and Table 8 of Chapter 9 for bare conductors.

Step 2: ➼ Determine the total cross-sectional area for all conductors.

Step 3: ➼ Size the raceway according the percent fill as listed in Table 1 of Chapter 9:
40% for three or more conductors
60% for raceways 24 inches or less in length (nipples).

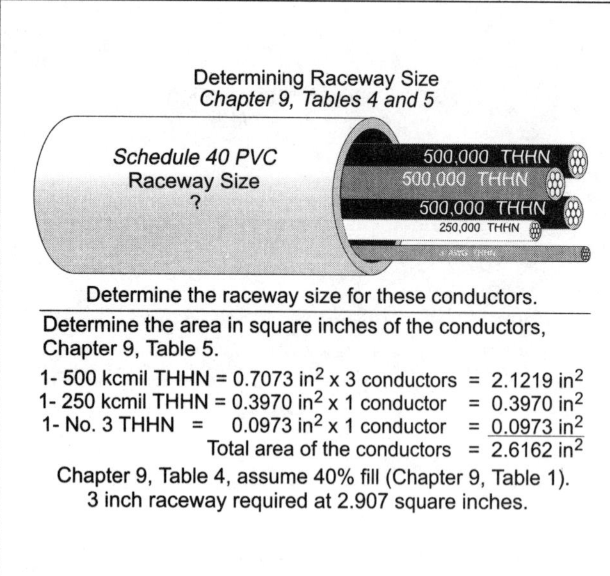

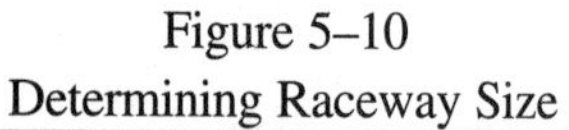
Figure 5–10
Determining Raceway Size

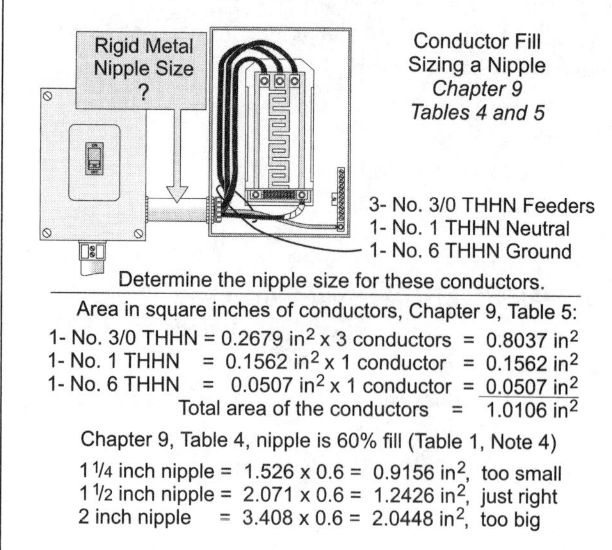

Figure 5–11
Conductor Fill – Sizing a Nipple

❑ Raceway Size

A 400 ampere feeder is installed in schedule 40 rigid nonmetallic conduit. This raceway contains three 500 kcmil THHN conductors, one 250 kcmil THHN conductor, and one No. 3 THHN conductor. What size raceway is required for these conductors (Figure 5–10)?

(a) 2 inch (b) $2^1/_2$ inch (c) 3 inch (d) $3^1/_2$ inch

- Answer: (c) 3-inch

Step 1: ➻ Determine cross-sectional area of the conductors, Table 5 of Chapter 9.
500 kcmil THHN = .7073 square inch × 3 wires = 2.1219 square inch
250 kcmil THHN = .3970 square inch × 1 wire = 0.3970 square inch
No. 3 THHN = .0973 square inch × 1 wire = 0.0973 square inch

Step 2: ➻ Total cross-sectional area of all conductors = 2.6162 square inch

Step 3: ➻ Size the conduit at 40% fill [Chapter 9, Table 1] using Table 4.
3 inch schedule 40 PVC has an cross-sectional area of 2.907 square inch for conductors

❑ Nipple Size

What size rigid metal nipple is required for three No. 3/0 THHN conductors, one No. 1 THHN conductor and one No. 6 THHN conductor (Figure 5–11)?

(a) 2 inch (b) 1 inch (c) $1^1/_2$ inch (d) none of these

- Answer: (c) $1^1/_2$ inch

Step 1: ➻ Cross-sectional area of the conductors, Table 5 of Chapter 9.
No. 3/0 THHN = 0.2679 square inch × 3 wires = 0.8037 square inch
No. 1 THHN = 0.1562 square inch × 1 wire = 0.1562 square inch
No. 6 THHN = 0.0507 square inch × 1 wire = 0.0507 square inch

Step 2: ➻ Total cross-sectional area of the conductors = 1.0106 square inches.

Step 3: ➻ Size the conduit at 60% fill [Table 1, Note 4 of Chapter 9] using Table 4.
$1^1/_4$ inch nipple = 1.5260 × 0.6 = 0.9156 square inch, too small
$1^1/_2$ inch nipple = 2.071 × 0.6 = 1.2426 square inches, just right
2 inch nipple = 3.408 × 0.6 = 2.0448 square inches, too big

5–3 EXISTING RACEWAY CALCULATIONS

There are times we need to add conductors to an existing raceway. This can be accomplished by using the following steps:

Part 1 – Determine raceway cross-sectional spare space area.

Step 1: ➵ Determine the raceway's cross-sectional area for conductor fill [Table 1 and Table 4 of Chapter 9].

Step 2: ➵ Determine the area of the existing conductors [Table 5 of Chapter 9].

Step 3: ➵ Subtract the cross-sectional area of the existing conductors (Step 2) from the area of permitted conductor fill (Step 1).

Part 2 – To determine the number of conductors permitted in spare space area:

Step 4: ➵ Determine the cross-sectional area of the conductors to be added [Table 5 of Chapter 9 for insulated conductors and Table 8 of Chapter 9 for bare conductors].

Step 5: ➵ Divide the spare space area (Step 3) by the conductors cross-sectional area (Step 4).

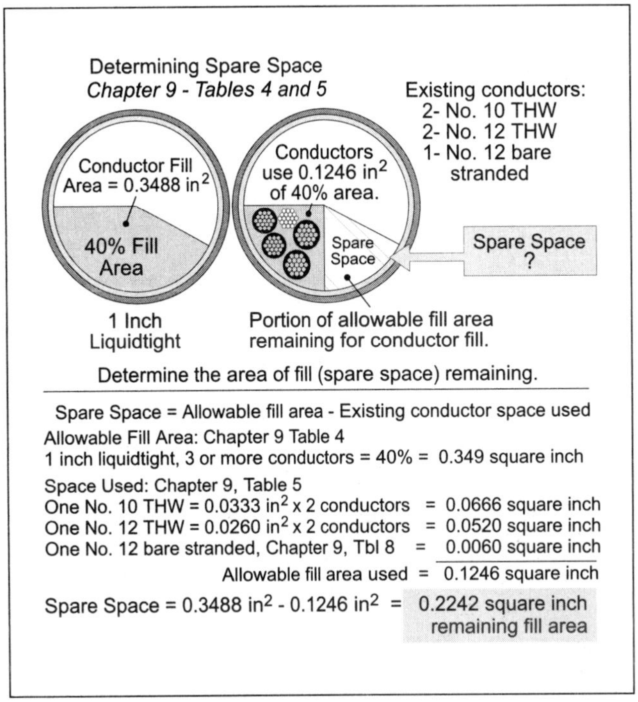

Determining Spare Space
Chapter 9 - Tables 4 and 5

Existing conductors:
2- No. 10 THW
2- No. 12 THW
1- No. 12 bare stranded

Conductor Fill Area = 0.3488 in^2

40% Fill Area

Conductors use 0.1246 in^2 of 40% area.

Spare Space

Spare Space ?

1 Inch Liquidtight

Portion of allowable fill area remaining for conductor fill.

Determine the area of fill (spare space) remaining.

Spare Space = Allowable fill area - Existing conductor space used

Allowable Fill Area: Chapter 9 Table 4
1 inch liquidtight, 3 or more conductors = 40% = 0.349 square inch

Space Used: Chapter 9, Table 5
One No. 10 THW = 0.0333 in^2 x 2 conductors = 0.0666 square inch
One No. 12 THW = 0.0260 in^2 x 2 conductors = 0.0520 square inch
One No. 12 bare stranded, Chapter 9, Tbl 8 = 0.0060 square inch
Allowable fill area used = 0.1246 square inch

Spare Space = $0.3488 \text{ in}^2 - 0.1246 \text{ in}^2$ = 0.2242 square inch remaining fill area

Figure 5–12
Determining Square Space

❑ **Spare Space Area**

An existing one inch liquidtight flexible metallic conduit contains two No. 12 THW conductors, two No. 10 THW conductors, and one No. 12 bare (stranded). What is the area remaining for additional conductors (Figure 5–12)?

(a) 0.2242 square inch (if the raceway is more than 24 inches long)
(b) 0.3986 square inch (if the raceway is less than 24 inches long)
(c) there is no spare space
(d) a and b

• Answer: (d) a and b

Step 1: ➵ Conductor cross-sectional area, liquidtight flexible metal conduit [Table 1, Note 4 and Table 4 of Chapter 9].

Raceway = 0.872 × 0.4 = 0.3488 square inch
Nipple = 0.872 × 0.6 = 0.5232 square inch

Step 2: ➵ Cross-sectional area of existing conductors.

No. 12 THW = 0.0260 square inch × 2 wires = 0.0520 square inch
No. 10 THW = 0.0333 square inch × 2 wires = 0.0666 square inch
No. 12 bare = 0.006 square inch × 1 wire = 0.0060 square inch
Total cross-sectional area of existing conductors = 0.1246 square inch

Note: Ground wires must be counted for raceway fill – Table 1, Note 3 of Chapter 9.

Step 3: ➵ Subtract the area of the existing conductors from the permitted area of conductor fill.

Raceway (more than 24 inches long): .3488 square inch – .1246 square inch = .2242 square inch
Nipple (less than 24 inches long): .5232 square inch – .1246 square inch = .3986 square inch

❑ Conductors in Spare Space Area

An existing 1-inch EMT contains two No. 12 THHN conductors, two No. 10 THHN conductors, and one No. 12 bare (stranded) conductor. How many additional No. 8 THHN conductors can be added to this raceway (Figure 5–13)?

(a) 7 conductors if raceway is more than 24 inches long
(b) 12 conductors if raceway is less than 24 inches long
(c) 15 conductors regardless of the raceway length
(d) a and b

- Answer: (d) a and b

Step 1: ➸ Cross-sectional area permitted for conductor fill [Table 1, Note 4 and Table 4 of Chapter 9].
Raceway: = .864 × 0.4 = .3456 square inch
Nipple: = .864 × 0.6 = .5184 square inch

Step 2: ➸ Cross-sectional area of existing conductors.
No. 10 THHN
.0211 square inch × 2 = .0422 square inch
No. 12 THHN
.0133 square inch × 2 = .0266 square inch
No. 12 bare
.0060 square inch × 1 = .0060 square inch

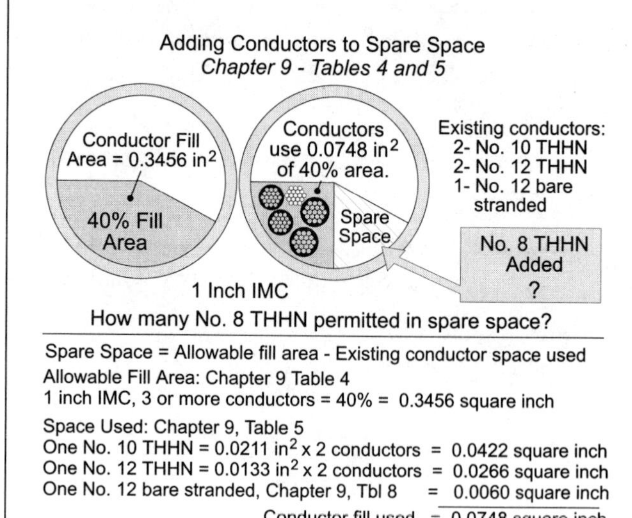

Figure 5–13
Adding Conductors to Spare Space

Total cross-sectional area of existing conductors = .0748 square inch

Note: Ground wires must be counted for raceway fill see Table 1, Note 3 of Chapter 9.

Step 3: ➸ Subtract the area of the existing conductors from the permitted area of conductor fill.
Raceway more than 24 inches long: .3456 square inch – .0748 square inch = .2708 square inch
Nipple (less than 24 inches long): .5184 square inch – .0748 square inch = .4436 square inch

Step 4: ➸ Cross-sectional area of the conductors to be installed [Table 5 of Chapter 9].

No. 10 THHN = 0.0211 square inch.

Step 5: ➸ Divide the spare space area (Step 3) by the conductor area.
Raceway: = 0.2708 square inches/.0366 square inch= 7.4 or 7 conductors
Nipple: = 0.4436 square inch/.0366 square inch = 12.1 or 12 conductors.

We must round down to 18 conductors because Note 7 to Table 1 only applies if all of the conductors are the same size and same insulation.

5–4 TIPS FOR RACEWAY CALCULATIONS

Tip 1: ➸ Take your time.

Tip 2: ➸ Use a ruler or straight-edge when using tables.

Tip 3: ➸ Watch out for the different types of raceways and conductor insulation, particularly RHH/RHW with or without outer cover.

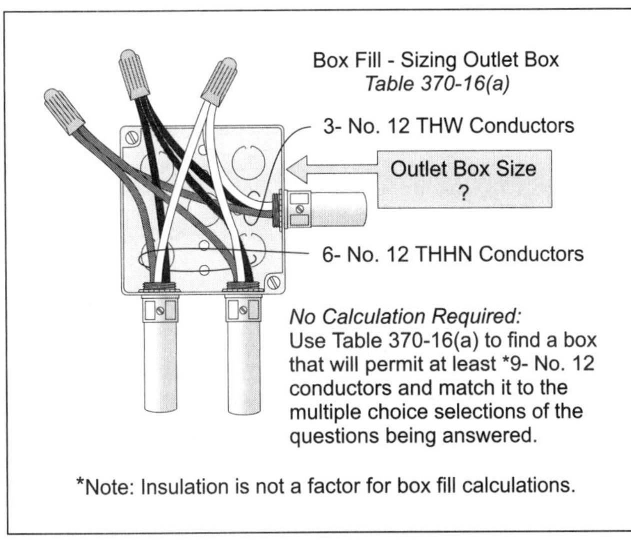

Figure 5–14
Box Fill – Sizing Outlet Box

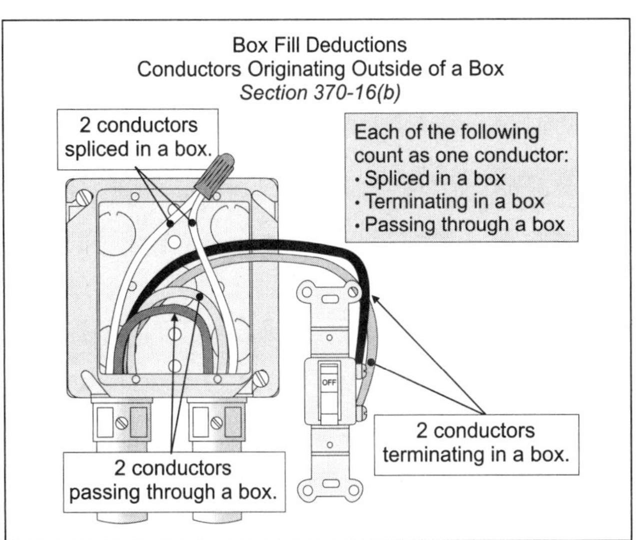

Figure 5–15
Box Fill Deductions

PART B – OUTLET BOX FILL CALCULATIONS

INTRODUCTION [*SECTION 370-16*]

Boxes shall be of sufficient size to provide free space for all conductors. An outlet box is generally used for the attachment of devices and fixtures and has a specific amount of space (volume) for conductors, devices, and fittings. The volume taken up by conductors, devices, and fittings in a box must not exceed the box fill capacity. The volume of a box is the total volume of its assembled parts, including plaster rings, industrial raised covers, and extension rings. The total volume includes only those fittings that are marked with their volume in cubic inches [*Section 370-16(a)*].

5–5 SIZING BOX – CONDUCTORS ALL THE SAME SIZE [Table 370-16(a)]

When all of the conductors in an outlet box are the same size (insulation doesn't matter), Table 370-16(a) of the *National Electrical Code®* can be used to:

(1) Determine the number of conductors permitted in the outlet box, or

(2) Determine the size outlet box required for the given number of conductors.

Note: Table 370-16(a) applies only if the outlet box contains no switches, receptacles, fixture studs, fixture hickeys, manufactured cable clamps, or grounding conductors (not likely).

❑ **Outlet Box Size**

What size outlet box is required for six No. 12 THHN conductors, and three No. 12 THW conductors (Figure 5–14)?

(a) 4 × 1$^{1}/_{4}$ square (b) 4 × 1$^{1}/_{2}$ square (c) 4 × 1$^{1}/_{4}$ round (d) 4 × 1$^{1}/_{2}$ round

• Answer: (b) 4 × 1$^{1}/_{2}$ square

Table 370-16(a) permits nine No. 12 conductors, insulation is not a factor.

❑ **Number of Conductors in Outlet Box**

Using Table 370-16(a), how many No. 14 THHN conductors are permitted in a 4 × 1$^{1}/_{2}$ round box?

(a) 7 conductors (b) 9 conductors (c) 10 conductors (d) 11 conductors

• Answer: (a) 7 conductors

5–6 CONDUCTOR EQUIVALENTS [*SECTION 370-16(b)*]

Table 370-16(a) does not take into consideration the fill requirements of clamps, support fittings, devices, or equipment grounding conductors within the outlet box. In no case can the volume of the box and its assembled sections be less than the fill calculation as listed below:

(1a) Conductor Terminating in the Box. Each conductor that originates outside the box and terminates or is spliced within the box is considered as one conductor (Figure 5–15).

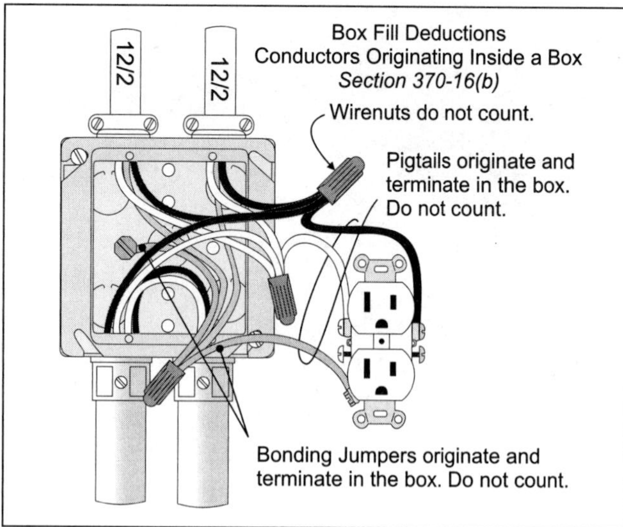

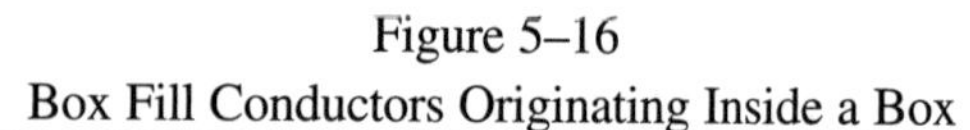
Figure 5–16
Box Fill Conductors Originating Inside a Box

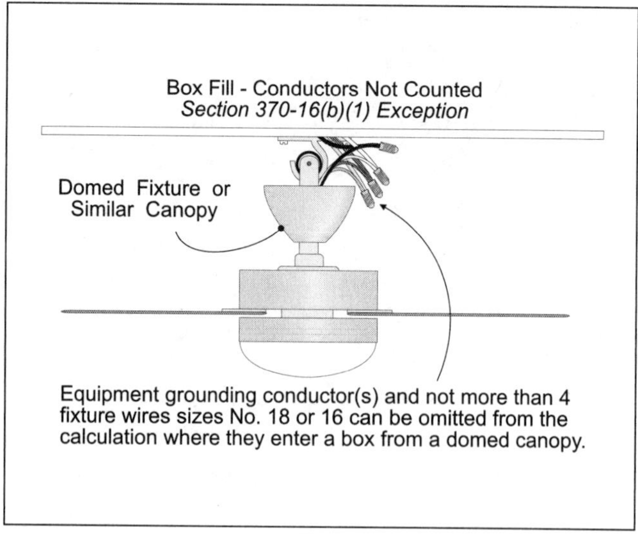

Figure 5–17
Box Fill–Conductors Not Counted

(1b) Conductor Running through the Box. Each conductor that runs through the box is considered as one conductor (Figure 5–15). Conductors, no part of which leaves the box, shall not be counted, this includes equipment bonding jumpers and pigtails (Figure 5–16).

Exception. Fixture wires smaller than No. 14 from a domed fixture or similar canopy are not counted (Figure 5–17).

(2) Cable Fill. One or more internal cable clamps in the box are considered as one conductor volume in accordance with the volume listed in Table 370-16(b), based on the largest conductor that enters the outlet box (Figure 5–18).

Note: Small fittings such as locknuts and bushings are not counted [*Section 370-16(b)*].

(3) Support Fittings Fill. One or more fixture studs or hickeys within the box are considered as one conductor volume, based on the largest conductor that enters the outlet box (Figure 5–19).

(4) Device or Equipment Fill. Each yoke or strap containing one or more devices or equipment is considered as two conductors, based on the largest conductor that terminates on the yoke (Figure 5–20).

(5) Grounding Conductors. One or more grounding conductors are considered as one conductor volume in accordance with the volume based on the largest grounding conductor that enters the outlet box (Figure 5–21).

Note. Fixture ground wires smaller than No. 14 from a domed fixture or similar canopy are not counted [*Section 370-16(b)(1)* Exception].

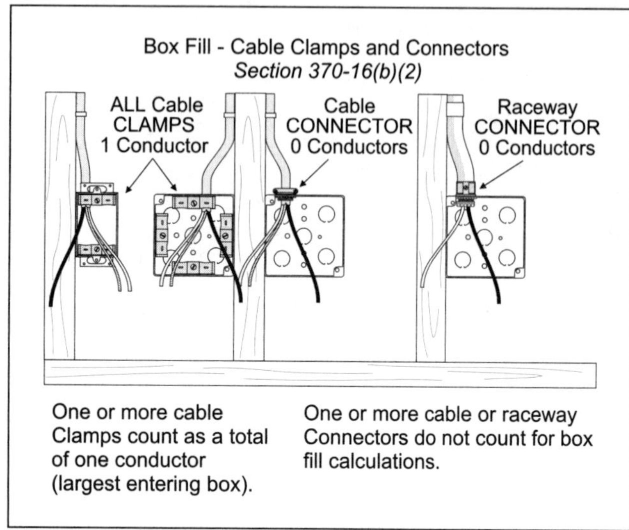

Figure 5–18
Box Fill – Cable Clamps and Connectors

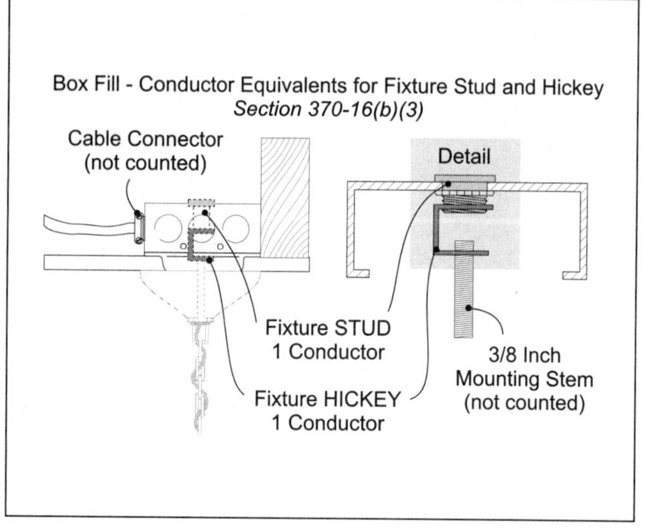

Figure 5–19
Box Fill – Conductor Equivalents for Fixture Stud and Hickey

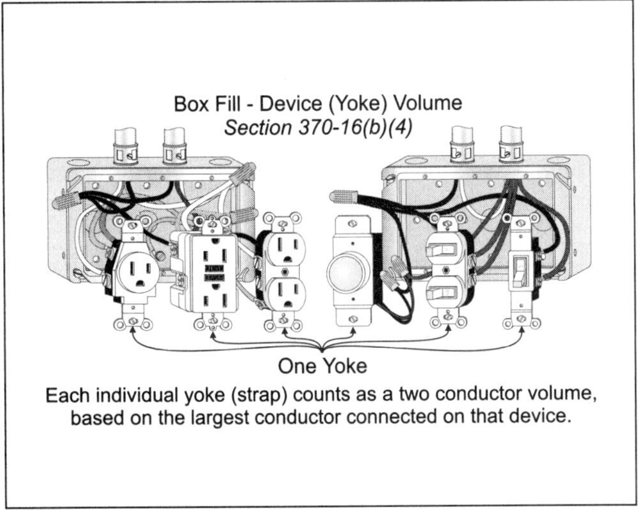

Figure 5–20
Box Fill – Device Volume

Figure 5–21
Box Fill – Grounding Conductor

What's Not Counted

Wirenuts, cable connectors, raceway fittings, and conductors that originate and terminate within the outlet box (such as equipment bonding jumpers and pigtails) are not counted for box fill calculations [*Section 370-16(a)*].

❑ **Number of Conductors**

What is the total number of conductors used for box fill calculations in Figure 5–22?

(a) 5 conductors (b) 7 conductors (c) 9 conductors (d) 11 conductors

• Answer: (d) 11 conductors

Switch –	Five No. 14 conductors, two conductors for device and three conductors terminating
Receptacle –	Four No. 12 conductors, two conductors for the device and two conductors terminating
Ground wire –	One conductor
Cable clamps –	One conductor

5–7 SIZING BOX – DIFFERENT SIZE CONDUCTORS [*SECTION 370-16(b)*]

To determine the size of the outlet box when the conductors are of different sizes (insulation is not a factor), the following steps can be used:

Step 1: ➛ Determine the number and size of conductors equivalents in the box.

Step 2: ➛ Determine the volume of the conductors equivalents from Table 370-16(b).

Step 3: ➛ Size the box by using Table 370-16(a).

Outlet Box Sizing

What size outlet box is required for 14/3 Type NM cable (with ground) that terminates on a switch, with 14/2 NM that terminates on a receptacle, if the box has internal cable clamps factory installed (Figure 5–22)?

(a) $4 \times 1^1/_4$ square (b) $4 \times 1^1/_2$ square (c) $4 \times 2^1/_8$ square (d) any of these

• Answer: (c) $4 \times 2^1/_8$ square

Step 1: ➛ Determine the number and size of conductors.

14/3 NM	3 – No. 14
14/2 NM	2 – No. 14
Cable clamps	1 – No. 14
Switch	2 – No. 14
Receptacles	2 – No. 14
Ground wires	1 – No. 14
Total	11– No. 14

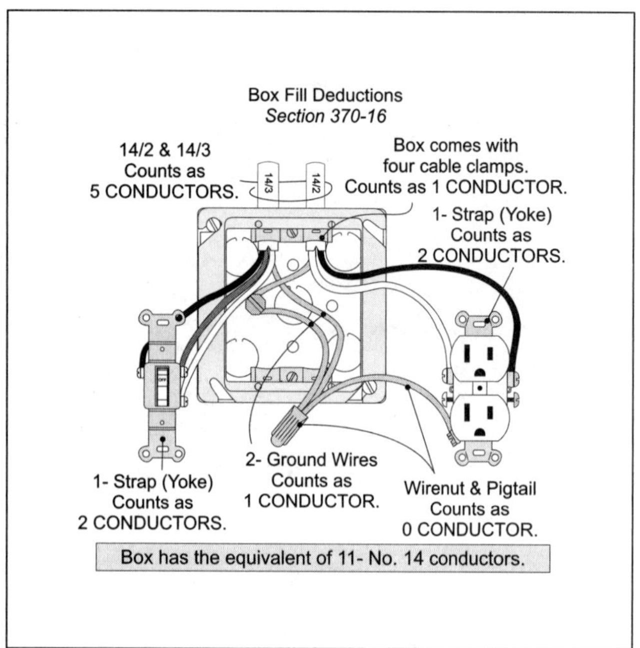

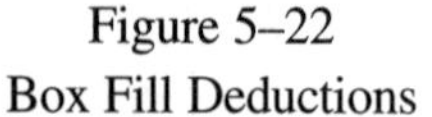

Figure 5–22
Box Fill Deductions

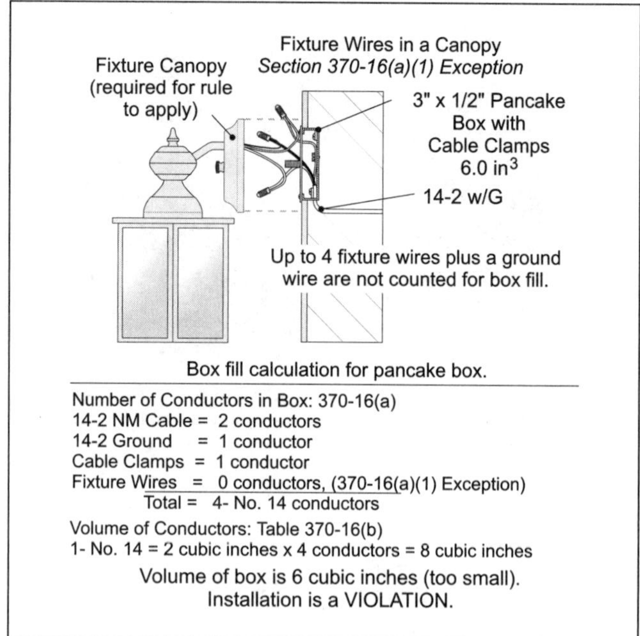

Figure 5–23
Fixture Wires in a Canopy

Step 2: ➻ Determine the volume of the conductors, [Table 370-16(b)].
No. 14 = 2 cubic inches
No. 14 conductors volume = 11 wires × 2 cubic inches = 22 cubic inches

Step 3: ➻ Select the outlet box from Table 370-16(a).
4 × $1^1/_2$ square, 21 cubic inches, too small
4 × $2^1/_8$ square, 30.3 cubic inches, just right

❑ **Domed Fixture Canopy [*Section 370-16(a),* Exception].**

A round 3 × $^1/_2$ box has a total volume of 6 cubic inches and has factory internal cable clamps. Can this pancake box be used with a lighting fixture that has a domed canopy? The branch circuit wiring is 14/2 nonmetallic sheath cable and the paddle fan has three fixture wires and one ground wire all smaller than No. 14 (Figure 5–23).

(a) Yes (b) No

• Answer: (b) No

The box is limited to 6 cubic inches, and the conductors total 8 cubic inches [*Section 370-16(a)*].

Step 1: ➻ Determine the number and size of conductors within the box.

14/2 NM	2 – No. 14
Cable clamps	1 – No. 14
Ground wire	1 – No. 14
Total	4 – No. 14 conductors

Step 2: ➻ Determine the volume of the conductors [Table 370–16(b)].
No. 14 = 2 cubic inches
Four No. 14 conductors = 4 wires × 2 cubic inches = 8 cubic inches

❑ **Conductors Added to Existing Box**

How many No. 14 THHN conductors can be pulled through a 4 × $2^1/_8$ square box that has a plaster ring of 3.6 cubic inches? The box already contains two receptacles, five No. 12 THHN conductors, and one No. 12 bare grounding conductor (Figure 5–24).

(a) 4 conductors (b) 5 conductors (c) 6 conductors (d) 7 conductors

• Answer: (b) 5 conductors

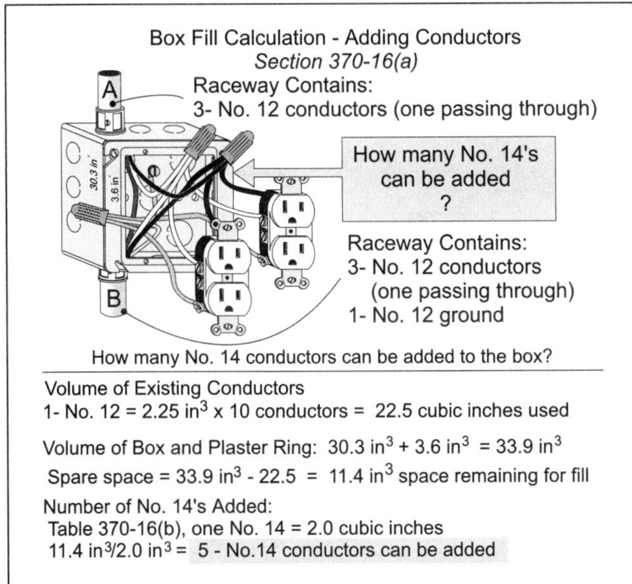

Figure 5–24
Box Fill Calculation – Adding Conductors

Pull And Junction Boxes - No. 4 and Larger
Section 370-28
Straight Pulls
Section 370-28 is used to size pull boxes, junction boxes, and conduit bodies when conductor sizes No. 4 and larger are used.
Angle Pulls
U Pulls

Figure 5–25
Pull and Junction Boxes

Step 1: ➛ Determine the number and size of the existing conductors.

Two receptacles	4 – No. 12 conductors (2 yokes × 2 conductors)
Five No. 12's	5 – No. 12 conductors
One ground	1 – No. 12 conductor
Total	10 - No. 12 conductors

Step 2: ➛ Determine the volume of the existing conductors [Table 370-16(b)].
No. 12 conductor = 2.25 cubic inches, 10 wires × 2.25 cubic inches = 22.5 cubic inches

Step 3: ➛ Determine the space remaining for the additional No. 14 conductors.
Remaining space = Total space less existing conductors
Total space = 30.3 cubic inches (box) [Table 370-16(a)] + 3.6 cubic inches (ring) = 33.9 cubic inches
Remaining space = 33.9 cubic inches – 22.5 cubic inches (ten No. 12 conductors)
Remaining space = 11.4 cubic inches

Step 4: ➛ Determine the number of No. 14 conductors permitted in the spare space.
Conductors added = Remaining space/added conductors volume
Conductors added = 11.4 cubic inches/2 cubic inches [Table 370-16(b)]
Conductors added = Five No. 14 conductors

PART C – PULL, JUNCTION BOXES, AND CONDUIT BODIES

INTRODUCTION

Pull boxes, junction boxes, and *conduit bodies* must be sized to permit conductors to be installed so that the conductor insulation will not be damaged. For conductors No. 4 and larger, we must size pull boxes, junction boxes, and conduit bodies according to the requirements of Section 370-28 (Figure 5–25).

5–8 PULL AND JUNCTION BOX SIZE CALCULATIONS

Straight Pull Calculation [*Section 370-28(a)(1)*]

A straight pull calculation applies when conductors enter one side of a box and leave through the opposite wall of the box. The minimum distance from where the raceway enters to the opposite wall must not be less than eight times the trade size of the largest raceway (Figure 5–26).

Sizing Junction/Pull Boxes for Straight Conductor Pulls
Section 370-28(a)(1)

Straight Pull

3" 3"

8 Times Largest Raceway

8 x 3 Inches = 24 inches

The length of a pull box containing a straight pull must be not less than 8 times the diameter of the largest raceway.

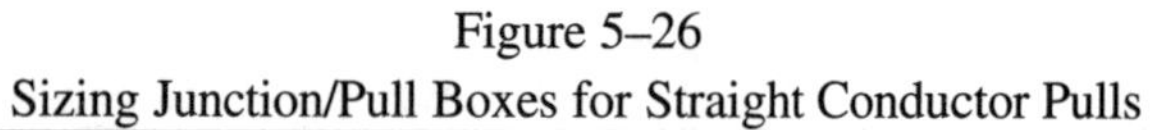

Figure 5–26
Sizing Junction/Pull Boxes for Straight Conductor Pulls

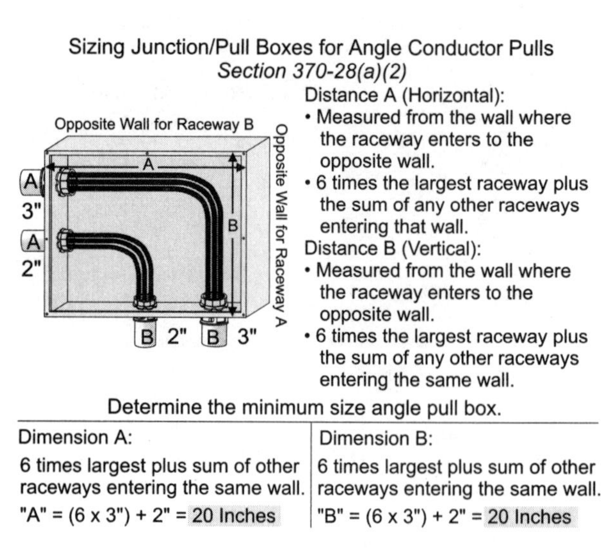

Figure 5–27
Sizing Junction/Pull Boxes for Angle Conductor Pulls

Angle Pull Calculation [*Section 370-28(a)(2)*]

An angle pull calculation applies when conductors enter one wall and leave the enclosure not opposite the wall of the conductor entry. The distance for angle pull calculations from where the raceway enters to the opposite wall must not be less than six times the trade diameter of the largest raceway, plus the sum of the diameters of the remaining raceways on the same wall and *row* (Figure 5–27). When there is more than one row, each row shall be calculated separately, and the row with the largest calculation shall be considered the minimum angle pull dimension.

U-Pull Calculations [*Section 370-28(a)(2)*]

A U-pull calculation applies when the conductors enter and leave from the same wall. The distance from where the raceways enter to the opposite wall must not be less than six times the trade diameter of the largest raceway, plus the sum of the diameters of the remaining raceways on the same wall (Figure 5–28).

Distance between Raceways Containing the Same Conductor Calculation [*Section 370-28(a)(2)*]

After sizing the pull box, the raceways must be installed so that the distance between raceways enclosing the same conductors shall not be less than six times the trade diameter of the largest raceway. This distance is measured from the nearest edge of one raceway to the nearest edge of the other raceway (Figures 5–28 and 5–29).

5–9 DEPTH OF BOX AND CONDUIT BODY SIZING [*SECTION 370-28(a)(2),* Exception]

When conductors enter an enclosure opposite a *removable cover*, such as the back of a pull box or conduit body, the distance from where the conductors enter to the removable cover shall not be less than the distances listed in Table 373-6(a); one wire per terminal (Figure 5–30).

❑ Depth of Pull Or Junction Box

A 24" × 24" pull box has two 2-inch conduits that enter the back of the box with No. 4/0 conductors. What is the minimum depth of the box?

(a) 4 inches (b) 6 inches
(c) 8 inches (d) 10 inches

• Answer: (a) 4 inches, Table 373-6(a)

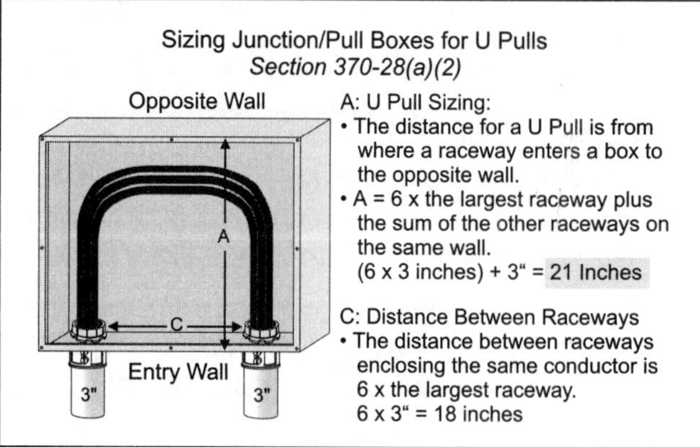

Figure 5–28
Sizing Junction/Pull Boxes for U-Pulls

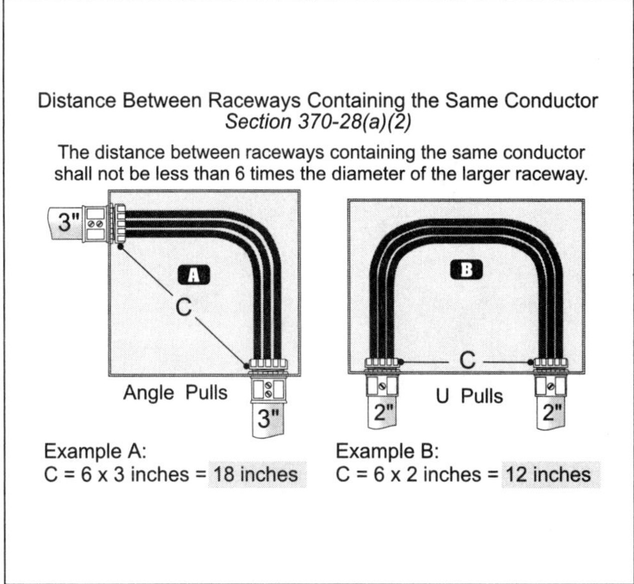

Figure 5–29
Distance between Raceways Containing the Same Conductor

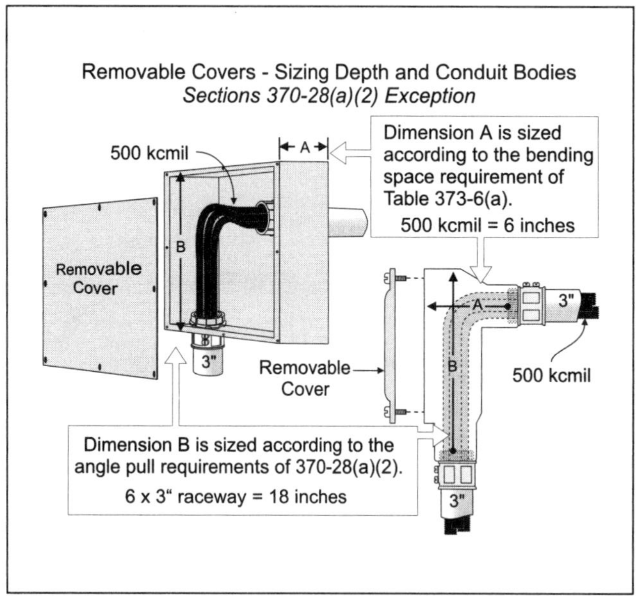

Figure 5–30
Removable Covers – Sizing Depth and Conduit Bodies

5–10 JUNCTION AND PULL BOX SIZING TIPS

When sizing pull and junction boxes, the following suggestions should be helpful.

Step 1: ➻ Always draw out the problem.

Step 2: ➻ Calculate the HORIZONTAL distance(s):

- Left to right straight calculation
- Left to right angle or U-pull calculation
- Right to left straight calculation
- Right to left angle or U-pull calculation

Step 3: ➻ Calculate the VERTICAL distance(s):

- Top to bottom straight calculation
- Top to bottom angle or U-pull calculation
- Bottom to top straight calculation
- Bottom to top angle or U-pull calculation

5–11 PULL BOX EXAMPLES

❑ **Pull Box Sizing**

A junction box contains two 3 inch raceways on the left side and one 3 inch raceway on the right side. The conductors from one of the 3 inch raceways on the left wall are pulled through the 3 inch raceway on the right wall. The other 3 inch raceway's conductors are pulled through a raceway at the bottom of the pull box (Figure 5–31).

➪ **Horizontal Dimension**

What is the horizontal dimension of this box?

(a) 18 inches (b) 21 inches (c) 24 inches (d) none of these

- Answer: (c) 24 inches, Section 370-28 (Figure 5–31, Part A).

Left wall to the right wall angle pull = (6 × 3 inches) + 3 inches = 21 inches

Left wall to the right wall straight pull = 8 × 3 inches = 24 inches

Right wall to left wall angle pull = No calculation

Right wall to the left wall straight pull = 8 × 3 inches = 24 inches

➪ Vertical Dimension

What is the vertical dimension of this box?

(a) 18 inches (b) 21 inches
(c) 24 inches (d) none of these

• Answer: (a) 18 inches, Section 370-28 (Figure 5–31, Part B).

Top to bottom angle = No calculation
Top to bottom straight = No calculation
Bottom to top angle = 6 × 3 inches = 18 inches
Bottom to top straight = No calculation

➪ Distance between Raceways

What is the minimum distance between the two 3 inch raceways that contain the same conductors?

(a) 18 inches (b) 21 inches
(c) 24 inches (d) none of these

• Answer: (a) 18 inches, 6 × 3 inches, Section 370-28 (Figure 5–31, Part C)

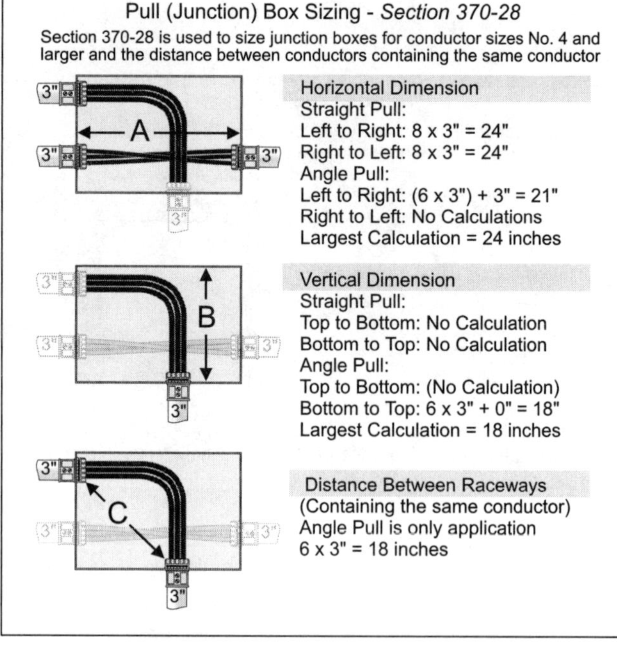

Figure 5–31
Pull (Junction) Box Sizing

❑ Pull Box Sizing

A pull box contains two 4 inch raceways on the left side and two 2 inch raceways on the top.

➪ Horizontal Dimension

What is the horizontal dimension of the box?

(a) 28 inches (b) 21 inches (c) 24 inches (d) none of these

• Answer: (a) 28 inches, Section 370-28(a)(2)

Left wall to the right wall angle pull = (6 × 4 inches) + 4 inches = 28 inches
Left wall to the right wall straight pull = No calculation
Right wall to left wall angle pull = No calculation
Right wall to the left wall straight pull = No calculation

➪ Vertical Dimension

What is the vertical dimension of the box?

(a) 18 inches (b) 21 inches (c) 24 inches (d) 14 inches

• Answer: (d) 14 inches, Section 370-28(a)(2)

Top to bottom wall angle pull = (6 × 2 inches) + 2 inches = 14 inches
Top to bottom wall straight pull = No calculation
Bottom to top wall angle pull = No calculation
Bottom to top wall straight pull = No calculation

➪ Distance between Raceways

What is the minimum distance between the two 4 inch raceways that contain the same conductors?

(a) 18 inches (b) 21 inches (c) 24 inches (d) None of these

• Answer: (c) 24 inches, Section 370-28(a)(2)

6 × 4 inches = 24 inches

Unit 5 – Raceway, Outlet Box, and Junction Boxes Calculations Summary Questions

Part A – Raceway Fill Calculations

5–1 Understanding Chapter 9 Tables

1. When all the conductors are the same size and insulation, the number of conductors permitted in a raceway can be determined simply by looking at the Tables listed in _____ .
(a) Chapter 9 (b) Appendix B (c) Appendix C (d) Appendix D

2. When equipment grounding conductors are installed in a raceway, the actual area of the conductor must be used when calculating raceway fill.
(a) True (b) False

3. When a raceway does not exceed 24 inches, the raceway is permitted to be filled to _____ of its cross–section area.
(a) 53% (b) 31% (c) 40% (d) 60%

4. How many No. 16 TFFN conductors can be installed in a $^{3}/_{8}$ inch electrical metallic tubing?
(a) 40 (b) 26 (c) 30 (d) 29

5. How many No. 6 RHH (without outer cover) can be installed in a 1 inch nonmetallic tubing?
(a) 25 (b) 16 (c) 13 (d) 7

6. How many No. 1/0 XHHW can be installed in a 2 inch flexible metal conduit?
(a) 7 (b) 6 (c) 16 (d) 13

7. How many No. 12 RHH (with outer cover) can be installed in a 1 inch IMC raceway?
(a) 7 (b) 11 (c) 5 (d) 4

8. If we have a 2 inch rigid metal conduit and we want to install three THHN aluminum compact conductors, what is the largest aluminum compact conductor permitted to be installed?
(a) No. 4/0 (b) 250 kcmil (c) 300 kcmil (d) 500 kcmil

9. The actual area of conductor fill is dependent on the raceway size and the number of conductors installed. If there are three or more conductors installed in a raceway the total area of conductor fill is limited to _____ percent.
(a) 53 (b) 31 (c) 40 (d) 60

10. • What is the area in square inches for a No. 10 THW?
(a) 0.0333 (b) 0.0172 (c) 0.0252 (d) 0.0278

11. What is the area in square inches for a No. 14 RHW (without cover)?
(a) 0.0209 (b) 0.0172 (c) 0.0252 (d) 0.0278

12. What is the area in square inches for a No. 10 THHN?
(a) 0.0117 (b) 0.0172 (c) 0.0252 (d) 0.0211

13. What is the area in square inches for a No. 12 RHH (with outer cover)?
(a) 0.0117 (b) 0.0353 (c) 0.0252 (d) 0.0327

14. • What is the area in square inches of a No. 8 bare solid?
(a) 0.013 (b) 0.027 (c) 0.038 (d) 0.045

5–2 Raceway and Nipple Calculations

15. The number of conductors permitted in a raceway is dependent on _____ .
(a) the area of the raceway
(b) the percent area fill as listed in Chapter 9, Table 1
(c) the area of the conductors as listed in Chapter 9, Tables 5 and 8
(d) all of these

16. A 200 ampere feeder installed in rigid nonmetallic conduit schedule 80, has three No. 3/0 THHN, one No. 2 THHN, and one No. 6 THHN. What size raceway is required?
(a) 2 inch (b) $2^1/_2$ inch (c) 3 inch (d) $3^1/_2$ inch

17. What size rigid metal nipple is required for three No. 4/0 THHN, one No. 1/0 THHN and No. 4 THHN?
(a) $1^1/_2$ inch (b) 2 inch (c) $2^1/_2$ inch (d) none of these

5–3 Existing Raceway Calculations

18. • An existing rigid metal nipple contains four No. 10 THHN and one No. 10 (bare stranded) ground wire in a 3/8 inch rigid metal conduit. How many additional No. 10 THHN can be installed?
(a) 5 (b) 7 (c) 9 (d) 11

Part B – Outlet Box Fill Calculations

5–5 Sizing Box–Conductors All the Same Size [Table 370–16]

19. What size box is required for 6 No. 14 THHN and 3 No. 14 THW?
(a) 4 × $1^1/_4$ square (b) 4 × $1^1/_2$ round (c) 4 × $1^1/_4$ round (d) none of these

20. How many No. 10 THHN are permitted in a 4 × $1^1/_2$ square box?
(a) 8 conductors (b) 9 conductors (c) 10 conductors (d) 11 conductors

5–6 Conductor Equivalents [*Section 370-16*]

21. Table 370-16 does not take into consideration the volume of _____ .
(a) switches and receptacles (b) fixture studs and hickeys
(c) manufactured cable clamps (d) all of these

22. When determining the number of conductors for box fill calculations, which of the following statements are true?
(a) A fixture stud or hickey is considered as one conductor for each type, based on the largest conductor that enters the outlet box.
(b) Internal factory cable clamps are considered as one conductor for one or more cable clamps, based on the largest conductor that enters outlet box.
(c) The device yoke is considered as two conductors, based on the largest conductor that terminates on the strap (device mounting fitting).
(d) all of these

23. • When determining the number of conductors for box fill calculations, which of the following statements are true?
(a) Each conductor that runs through the box without loop (without splice) is considered as one conductor.
(b) Each conductor that originates outside the box and terminates in the box is considered as one conductor.
(c) Wirenuts, cable connectors, raceway fittings, and conductors that originate and terminate within the outlet box (equipment bonding jumpers and pigtails) are not counted for box fill calculations.
(d) all of these

24. It is permitted to omit one equipment grounding conductor and not more than _____ that enter a box from a fixture canopy.
(a) four fixture wires (b) four No. 16 fixture wires
(c) four No. 18 fixture wires (d) b and c

25. Can a round 4 × $^1/_2$ box marked as 8 cubic inches with manufactured cable clamps supplied with 14/2 NM be used with a fixture that has two No. 18 TFN and a canopy cover?
(a) Yes (b) No

5–7 Sizing Box–Different Size Conductors [*Section 370-16(b)*]

26. What size outlet box is required for: one 12/2 NM cable that terminates on a switch, one 12/3 NM cable that terminates on a receptacle, and the box has manufactured cable clamps.
(a) 4 × $1^1/_4$ square (b) 4 × $1^1/_2$ square (c) 4 × $2^1/_8$ square (d) none of these

27. • How many No. 14 THHN conductors can be pulled through a 4 × $1^1/_2$ square box with a plaster ring of 3.6 cubic inches? The box contains a two duplex receptacles, five No. 14 THHN and two grounding conductors.
(a) 1 (b) 2 (c) 3 (d) 4

Part C – Pull, Junction Boxes, and Conduit Bodies

5–8 Pull and Junction Box Size Calculations

28. When conductors No. 4 and larger are installed in boxes and conduit bodies, we must size the enclosure according to which of the following requirements?
(a) The minimum distance for straight pull calculations from where the conductors enter to the opposite wall must not be less than eight times the trade size of the largest raceway.
(b) The distance for angle pull calculations from the raceway entry to the opposite wall must not be less than six times the trade diameter of the largest raceway, plus the sum of the diameters of the remaining raceways on the same wall and row.
(c) The distance between raceways enclosing the same conductor(s) shall not be less than six times the trade diameter of the largest raceway.
(d) all of the above are correct.

29. When conductors enter an enclosure opposite a removable cover, the distance from where the conductors enter to the removable cover shall not be less than _____ .
(a) six times the largest raceway (b) eight times the largest raceway
(c) a or b (d) none of these

The following information applies to the next three questions.

A junction box contains two $2^1/_2$ inch raceways on the left side and one $2^1/_2$ inch raceway on the right side. The conductors from one $2^1/_2$ inch raceway (left wall) are pulled through the raceway on the right wall. The other $2^1/_2$ inch raceway conductors (on the side) are pulled through a $2^1/_2$ inch raceway at the bottom of the pull box.

30. What is the distance from the left wall to the right wall?
(a) 18 inches (b) 21 inches (c) 24 inches (d) 20 inches

31. • What is the distance from the bottom wall to the top wall?
(a) 18 inches (b) 21 inches (c) 24 inches (d) 15 inches

32. What is the distance from between the raceways that contain the same conductors?
(a) 18 inches (b) 21 inches (c) 24 inches (d) 15 inches

The following information applies to the next three questions.

A junction box contains two 2 inch raceways on the left side and two 2 inch raceways on the top.

33. What is the distance from the left wall to the right wall?
(a) 28 inches (b) 21 inches (c) 24 inches (d) 14 inches

34. What is the distance from the bottom wall to the top wall?
(a) 18 inches (b) 21 inches (c) 24 inches (d) 14 inches

35. What is the distance between the 2 inch raceways that contain the same conductors?
(a) 18 inches (b) 21 inches (c) 24 inches (d) 12 inches

✯ Challenge Questions

Part A – Raceway Fill Calculations

36. • A 3 inch schedule 40 PVC raceway contains seven No. 1 RHW conductors without outer cover. How many No. 2 THW conductors may be installed in this raceway with the existing conductors?
(a) 11 (b) 15 (c) 20 (d) 25

Part B – Outlet Box Fill Calculations

37. • Determine the minimum cubic inches required for two No. 10 TW passing through a box, four No. 14 THHN terminating, two No. 12 TW terminating to a receptacle, and one No. 12 equipment bonding jumper from the receptacle to the box.
(a) 18.5 cubic inches (b) 22 cubic inches (c) 20 cubic inches (d) 21.75 cubic inches

38. • When determining the number of conductors in a box fill two No. 18 fixture wires, one 14/3 nonmetallic-sheathed cable with ground, one duplex switch, and two cable clamps would count as _____ .
(a) 9 conductors (b) 8 conductors (c) 10 conductors (d) 6 conductors

Part C – Pull, Junction Boxes, and Conduit Bodies

The Figure 5–32 applies to the next four questions.

39. The minimum horizontal dimension for the junction box shown in the diagram is _____ (Figure 5–32).
(a) 21 inches (b) 18 inches
(c) 24 inches (d) 20 inches

40. • The minimum vertical dimension for the junction box is _____ (Figure 5–32).
(a) 16 inches (b) 18 inches
(c) 20 inches (d) 24 inches

41. The minimum distance between the two 2 inch raceways that contain the same conductor "C" would be _____ (Figure 5–32).
(a) 12 inches (b) 18 inches
(c) 24 inches (d) 30 inches

42. • If a 3 inch raceway entry (250 kcmil) is in the wall opposite to a removable cover, the distance from that wall to the cover must not be less than _____ (Figure 5–32)?
(a) 4 inches (b) $4^1/_2$ inches
(c) 5 inches (d) 6 inches

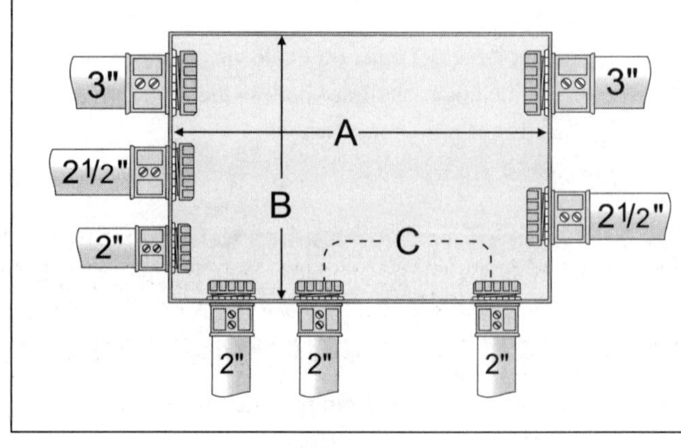

Figure 5–32

Unit 6

Conductor Sizing and Protection Calculations

OBJECTIVES

After reading this unit, the student should be able to briefly explain the following concepts:

- Conductor allowable ampacity
- Conductor ampacity
- Conductor bundling derating factor
- Conductor sizing summary
- Conductor insulation property
- Conductor size – voltage drop
- Conductors in parallel
- Current-carrying conductors
- Equipment conductors size and protection examples
- Minimum conductor size
- Overcurrent protection of equipment conductors
- Overcurrent protection
- Overcurrent protection of conductors
- Terminal ratings

After reading this unit, the student should be able to briefly explain the following terms:

- Ambient temperature
- American Wire Gauge
- Ampacity
- Ampacity derating factors
- Conductor bundling
- Conductor properties
- Continuous load
- Current-carrying conductors
- Fault current
- Interrupting rating
- Overcurrent protection device
- Overcurrent protection
- Parallel conductors
- Temperature correction factors
- Terminal ratings
- Voltage drop

PART A – GENERAL CONDUCTOR REQUIREMENTS

6–1 CONDUCTOR INSULATION PROPERTY [Table 310-13]

Table 310-13 of the *NEC®* provides information on conductor properties such as permitted use, maximum operating temperature, and other insulation details (Figure 6–1).

The following abbreviations and explanations should be helpful in understanding Table 310-13 as well as Table 310-16.

-2	Conductor is permitted to be used at a continuous 90ºC operation temperature
F	Fixture wire (solid or 7 strand) [Table 402-3]
FF	Flexible fixture wire (19 strands) [Table 402-3]
H	75ºC insulation rating
HH	90ºC insulation
N	Nylon outer cover
T	Thermoplastic insulation
W	Wet or damp

Fixture wires, see Article 402, Tables 402-3 and 402-5.

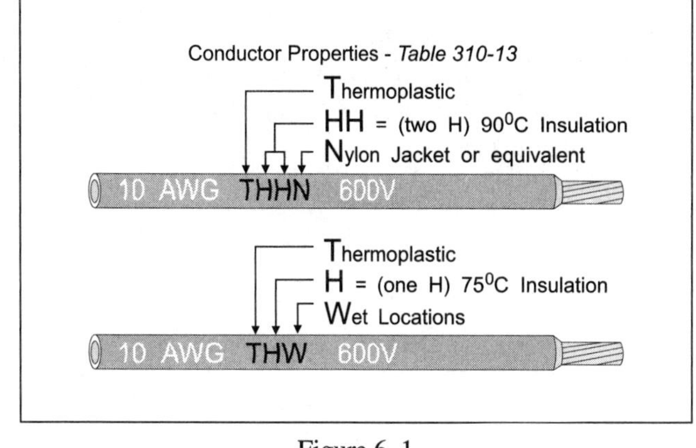

Figure 6–1
Conductor Properties

TABLE 310-13 CONDUCTOR INFORMATION					
	Column 2	**Column 3**	**Column 4**	**Column 5**	**Column 6**
Type Letter	Trade Name	Maximum Operating Temperature	Applications Provisions	Sizes Available	Outer Covering
THHN	Heat resistant thermoplastic	90°C	Dry and damp locations	14 – 1000	Nylon jacket or equivalent
THHW	Moisture- & heat-resistant thermoplastic	75°C 90°C	Wet locations Dry and damp locations	14 – 1000	None
THW	Moisture- & heat-resistant thermoplastic	75°C 90°C	Dry, damp, and wet locations Within electrical discharge lighting equipment *See Section 410–31*	14 – 2000	None
THWN	Moisture- & heat-resistant thermoplastic	75°C	Dry and wet locations	14 – 1000	Nylon jacket or equivalent
TW	Moisture-resistant thermoplastic	60°C	Dry and wet locations	14 – 2000	None
XHHW	Moisture- & heat-resistant cross-linked synthetic polymer	90°C 75°C	Dry and damp locations Wet locations	14 – 2000	None

❑ **Table 310-13**

TW can be described as _____ (Figure 6–2).

(a) thermoplastic insulation

(b) suitable for dry or wet locations

(c) maximum operating temperature of 60ºC

d) all of the above

• Answer: (d) all of the above

❑ **Table 402-3**

TFFN can be described as _____ .

(a) stranded fixture wire

(b) thermoplastic insulation with a nylon outer cover

(c) suitable for dry and wet locations

(d) all of the above except (c)

• Answer: (d) all of the above except (c)

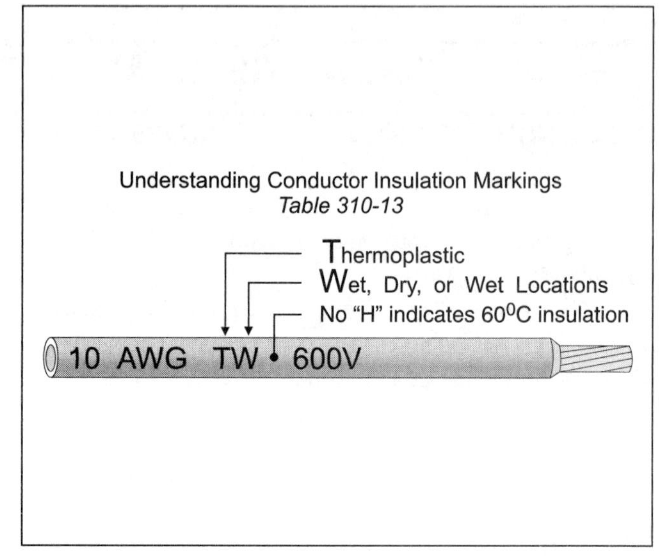

Figure 6–2
Understanding Conductor Insulation Markings

6–2 CONDUCTOR ALLOWABLE AMPACITY [*Section 310-15*]

The ampacity of a conductor is the current the conductor can carry continuously under specific conditions of use [Article 100 definition]. The ampacity of a conductor is listed in Table 310-16 under the condition of no more than three current-carrying conductors bundled together in an ambient temperature of 86ºF (Figure 6–3). The ampacity of a conductor changes if the ambient temperature is not 86ºF or if more than three current-carrying conductors are bundled together for more than two feet.

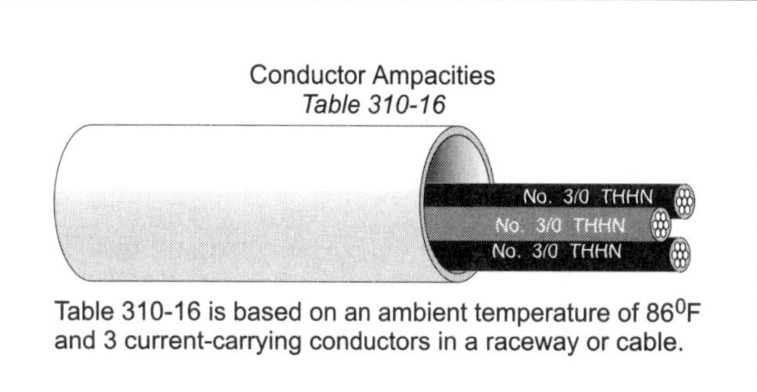

Figure 6–3
Conductor Ampacities

Section 310-15 lists two ways of determining conductor ampacity:

(a) General Requirements

(1) Tables or Engineering Supervision. There are two ways to determine conductor ampacity:

- Tables 310-16 [*Section 310-15(b)*]
- Engineering formula

Note: For all practical purposes, use the ampacities listed in Table 310-16.

TABLE 310–16. ALLOWABLE AMPACITIES OF INSULATED CONDUCTORS
Based On Not More Than Three Current-Carrying Conductors and Ambient Temperature of 30°C (86°F)

Size	Temperature Rating of Conductor, See Table 310-13						Size
	60°C (40°F)	75°C (167°F)	90°C (194°F)	60°C (40°F)	75°C (167°F)	90°C (194°F)	
AWG kcmil	TW	THHN THW THWN XHHW Wet Location	THHN THHW XHHW Dry Location	TW	THHN THW THWN XHHW Wet Location	THHN THHW WHHN Dry Location	AWG kcmil
	COPPER			ALUMINUM/COPPER-CLAD ALUMINUM			
14*	20	20	25				12
12*	25	25	30	20	20	25	10
10*	30	35	40	25	30	35	8
8	40	50	55	30	40	45	8
6	55	65	75	40	50	60	6
4	70	85	95	55	65	75	4
3	85	100	110	65	75	85	3
2	95	115	130	75	90	100	2
1	110	130	150	85	100	115	1
1/0	125	150	170	100	120	135	1/0
2/0	145	175	195	115	135	150	2/0
3/0	165	200	225	130	155	175	3/0
4/0	195	230	260	150	180	205	4/0
250	215	255	290	170	205	230	250
300	240	285	320	190	230	255	300
350	260	310	350	210	250	280	350
400	280	335	380	225	270	305	400
500	320	380	430	260	310	350	500

*See 240-3(d)

FPN: The ampacities listed in Table 310-16 are based on temperature alone and do not take voltage drop into consideration. Voltage drop considerations are for efficiency of operation and not safety; therefore, sizing conductors for voltage drop is not a *Code* requirement; see Sections 215-2(d) FPN No. 2 for more details.

Ampacity Definition. The ampacity of a conductor is the current in ampere that a conductor can carry continuously without exceeding its temperature rating [Article 100].

(b) Table Ampacity

The allowable ampacities of a conductor is listed in Table 310-16. These ampacities are based on the condition where no more than 3 current-carrying conductors are bundled together at an ambient temperature of 86°F.

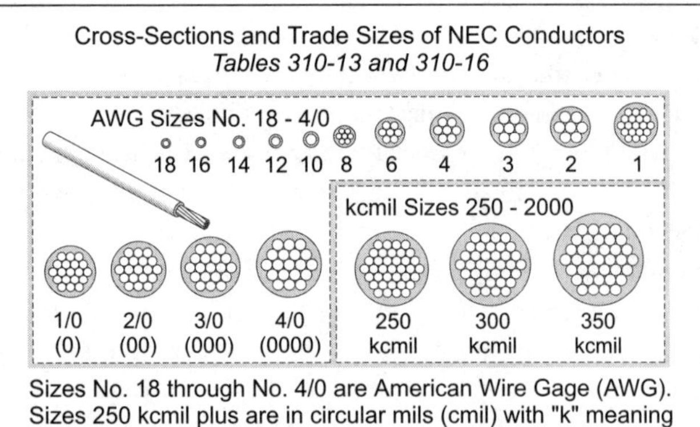

Figure 6–4
Cross-Section and Trade Sizes of *NEC*® Conductors

CAUTION: Section 240-3(d) specifies that the maximum overcurrent protection device permitted for copper No. 14 is 15 amperes, No. 12 is 20 amperes, and No. 10 is 30 amperes. The maximum overcurrent protection device permitted for aluminum No. 12 is 15 amperes and No. 10 is 25 amperes. The overcurrent protection device limitations of Section 240–3(d) does not apply to the following:

Air-conditioning conductor........Section 440–22
Capacitor conductors........Article 460
Class I remote control conductors........Article 725
Cooking equipment taps........Article 725
Feeder taps........Section 240–21
Fixture wires and taps........210–19(c) Exception
Motor branch circuit conductors........Section 430–52
Motor feeder conductor........Section 430–62
Motor control conductors........Section 430–72
Motor taps branch circuits........Section 430–53(d)
Motor taps feeder conductors........Section 430-28
Power loss hazard conductors........Section 240–3(a)
Two-wire transformers........Section 240–3(i)
Transformer tap conductor........Section 240–21(b)(d) and (m)
Welders........Article 630

6–3 CONDUCTOR SIZING [*SECTION 110-6*]

Conductors are sized according to the American Wire Gage (AWG) from Number 40 (No. 40) through Number 0000 (No. 4/0). The smaller the AWG size, the larger the conductor. Conductors larger than No. 4/0 are identified according to their circular mil area, such as 250,000, 300,000, 500,000. The circular mil size is often expressed in kcmil, such as 250 kcmil, 300 kcmil, 500 kcmil etc. (Figure 6–4).

Smallest Conductor Size

The smallest size conductor permitted by the *National Electrical Code*® for branch circuits, feeders, or services is No. 14 copper or No. 12 aluminum [Table 310-5]. Some local codes require a minimum No. 12 for commercial and industrial installations. Conductors smaller than No. 14 are permitted for:

Class I remote control circuits [*Section 402-11, Exception, and 725-27*]
Fixture wire [*Section 402-5 and 410-24*]
Flexible cords [*Section 400-12*]
Motor control circuits [*Section 430-72*]
Nonpower-limited fire alarm circuits [*Section 760-16*]
Power-limited fire alarm circuits [*Section 760-71(a)*].

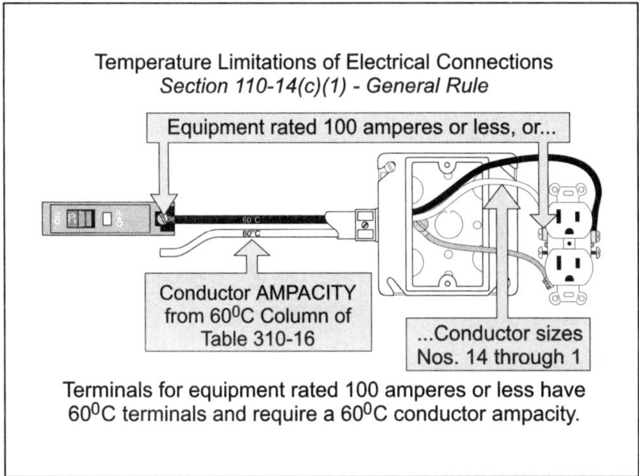

Figure 6–5
Temperature Limitations of Electrical Connections

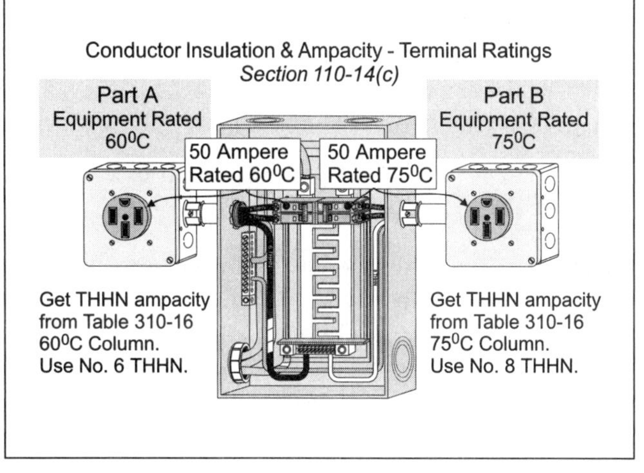

Figure 6–6
Conductor Insulation & Ampacity – Terminal Ratings

6–4 TERMINAL RATINGS [*SECTION 110-14(c)*]

When selecting a conductor for a circuit, always select the conductors not smaller than the terminal ratings of the equipment.

Circuits Rated 100 Ampere and Less [*Section 110-14(c)(1)*]

Equipment terminals rated 100 ampere or less (and pressure connector terminals for No. 14 through No. 1 conductors), shall have the conductor sized no smaller than the 60ºC temperature rating listed in Table 310-16, unless the terminals are marked otherwise (Figure 6–5).

❑ **Terminal Rated 60ºC [*Section 110-14(c)(1)(b)*]**

What size THHN conductor is required for a 50 ampere circuit, listed for use at 60ºC (Figure 6–6, Part A)?

(a) No. 10 (b) No. 8 (c) No. 6 (d) any of these

• Answer: (c) No. 6

Conductors must be sized to the lowest temperature rating of either the equipment or the conductor. THHN insulation can be used, but the conductor size must be selected based on the 60ºC terminal rating of the equipment, not the 90ºC rating of the insulation. Using the 60ºC column of Table 310-16, this 50 ampere circuit requires a No. 6 THHN conductor (rated 55 ampere at 60ºC).

❑ **Terminal Rated 75ºC [*Section 110-14(c)(1)(c)*]**

What size THHN conductor is required for a 50 ampere circuit, listed for use at 75ºC (Figure 6–6, Part B)?

(a) No. 10 (b) No. 8 (c) No. 6 (d) any of these

• Answer: (b) No. 8

Conductors must be sized according to the lowest temperature rating of either the equipment or the conductor. THHN conductors can be used, but the conductor size must be selected according to the 75ºC terminal rating of the equipment, not the 90ºC rating of the insulation. Using the 75ºC column of Table 310-16, this installation would permit No. 8 THHN (rated 50 ampere 75ºC) to supply the 50 ampere load.

Circuits over 100 Ampere [*Section 110-14(c)(2)*]

Terminals for equipment rated over 100 ampere and pressure connector terminals for conductors larger than No. 1 shall have the conductor sized according to the 75ºC temperature rating listed in Table 310-16 (Figure 6–7).

❑ **Over 100 Ampere [*Section 110-14(c)(2)(b)*]**

What size THHN conductor is required to supply a 225 ampere feeder?

(a) No. 1/0 (b) No. 2/0 (c) No. 3/0 (d) No. 4/0

• Answer: (d) No. 4/0

The conductors in this example must be sized to the lowest temperature rating of either the equipment or the conductor. THHN conductors can be used, but the conductor size must be selected according to the 75ºC terminal rating of the equipment.

Using the 75°C column of Table 310-16, this would require a No. 4/0 THHN (rated 230 ampere at 75°C) to supply the 225 ampere load. No. 3/0 THHN is rated 225 ampere at 90°C, but we must size the conductor to the terminal rating at 75°C.

Minimum Conductor Size Table

When sizing conductors, the following table must always be used to determine the minimum size conductor:

Terminal Size and Matching Cooper Conductor		
Terminal ampacity	**60°C Terminals Wire Size**	**75° Terminals Wire Size**
15	14	14
20	12	12
30	10	10
40	8	8
50	6	8
60	4	6
70	4	4
100	1	3
125	1/0	1
150	–	1/0
200	–	3/0
225	–	4/0
250	–	250 kcmil
300	–	350 kcmil
400	–	2 – 3/0
500	–	2 – 250 kcmil

CAUTION: When sizing conductors, we must consider conductor voltage drop, ambient temperature correction, and conductor bundle adjustment factors. These subjects are covered later in this book.

What is the purpose of THHN if we can not use its higher ampacity?

In general, 90°C rated conductor ampacities cannot be used for sizing circuit conductors. However, THHN offers the opportunity of having a greater conductor ampacity for conductor ampacity derating. The higher ampacity of THHN can permit a conductor to be used without having to increase its size because of conductor ampacity derating. Remember, the advantage of THHN is not to permit a smaller conductor, but it might prevent you from having to install a larger conductor because of ampacity adjustments (Figure 6–8).

Note: This is explained in detail in Part B of this unit.

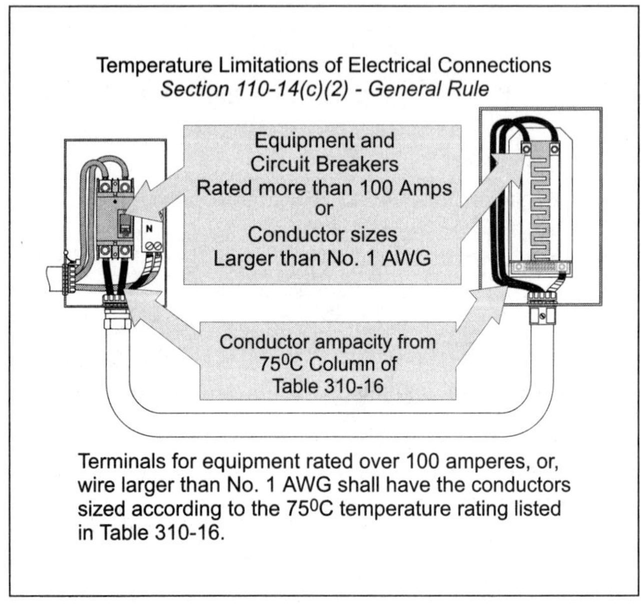

Figure 6–7
Temperature Limitations of Electrical Connections

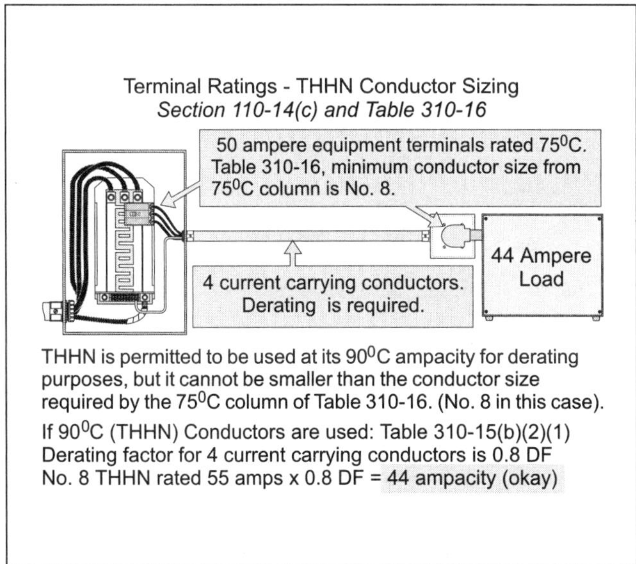

Figure 6–8
Terminal Ratings – THHN Conductor Sizing

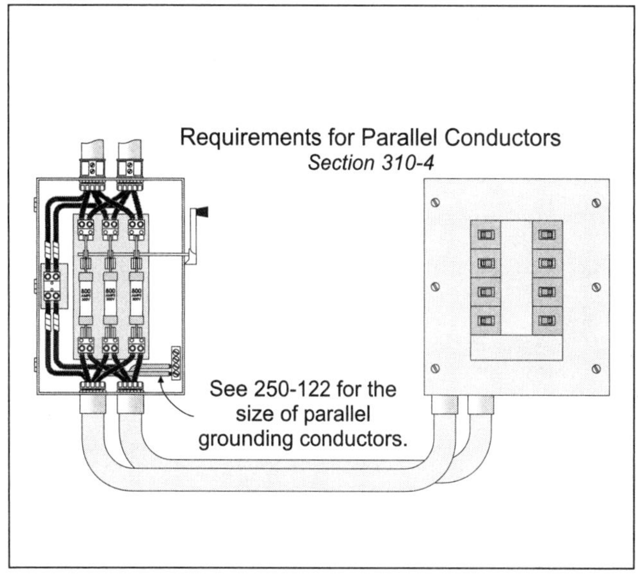

Figure 6–9
Requirements for Parallel Conductors

6–5 CONDUCTORS IN PARALLEL [*SECTION 310-4*]

Parallel conductors (electrically joined at both ends) permit a smaller cross-sectional area per ampere. This can result in significant cost saving for circuits over 300 ampere (Figure 6–9). The following table demonstrates the increase circular mil area required per ampere with larger conductors.

Conductor Size	Circular Mils Chapter 9, Table 8	Ampacity 75°C	Circular Mils Per Ampere
No. 1/0	105,600 cm	150 ampere	704 cm/per amp
No. 3/0	167,600 cm	200 ampere	838 cm/per amp
250 kcmil	250,000 cm	255 ampere	980 cm/per amp
500 kcmil	500,000 cm	380 ampere	1,316 cm/per amp
750 kcmil	750,000 cm	475 ampere	1,579 cm/per amp

❏ **Sizing Parallel Conductors**

What size 75°C conductors are required for a 600-ampere service that has a calculated demand load of 550 ampere (Figure 6–10)?

(a) 500 kcmil (b) 750 kcmil (c) 1,000 kcmil (d) 1,250 kcmil

• Answer: (d) One 1,250 kcmil conductor, rated 590 ampere [Table 310-16]

❏ **Sizing Parallel Conductors**

What size conductors would be required in one raceway (nipple) for a 600-ampere service? The calculated demand load is 550 ampere (Figure 6–11)?

(a) two – 300 kcmil (b) two – 250 kcmil (c) two – 500 kcmil (d) two – 750 kcmil

• Answer: (a) Two–300 kcmil conductors, each rated 285 ampere

285 ampere × 2 conductors = 570 ampere [Table 310-16]

If we parallel the conductors, we can use two 300 kcmil conductors (total 600,000) instead of one 1,250 kcmil conductor.

Grounding Conductors in Parallel [*Section 310-4*]

When equipment grounding conductors are installed with circuit conductors that are run in parallel, each raceway must have an equipment grounding conductor sized according to the overcurrent protection device rating that protects the circuit [*Section 250-122*] (Figure 6–12).

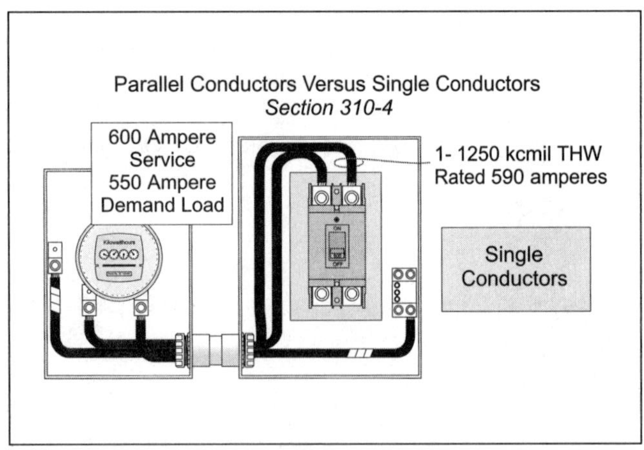

Figure 6–10
Parallel Conductors versus Single Conductors

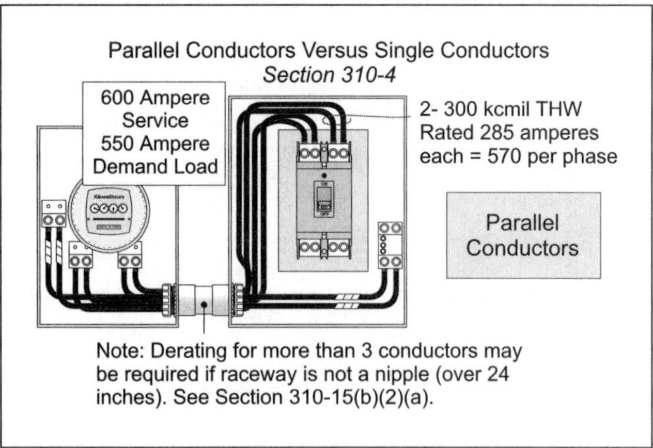

Figure 6–11
Parallel Conductors versus Single Conductors

❑ **Sizing Grounding Conductors in Parallel [*Section 250-122(f)*]**

What size equipment grounding conductor is required in each of two raceways for a 600-ampere feeder (Figure 6–13)?

(a) No. 3 (b) No. 2 (c) No. 1 (d) No. 1/0

• Answer: (c) No. 1

There must be a No. 1 equipment grounding conductor in each of the two raceways.

6–6 CONDUCTOR SIZE – VOLTAGE DROP [*SECTION 210-19(a) FPN No. 4, and 215-2(d) FPN No. 2*]

Conductor voltage drop is the result of the conductor's resistance (R) opposing the current (I) flow, $E_{vd} = R \times I$.

The *NEC®* generally does not require that voltage drop be limited on conductors, but the FPN suggest that we consider its effects. Voltage drop is covered in detail in Unit 8 of this book.

6–7 OVERCURRENT PROTECTION [Article 240]

Overcurrent protection devices are intended to open the circuit to prevent damage to persons or property due to excessive or dangerous heat. Overcurrent protection devices have two ratings, overcurrent and ampere interrupting current (AIC).

Note: Overcurrent protection devices are also used to open the circuit to clear ground-faults.

Overcurrent Rating [*Section 240-1 FPN*]. This is the actual ampere rating of the protection device, such as 15, 20, or 30 ampere. If the current flowing through the protection device exceeds the device setting for a significant period of time, the protection device will open to remove dangerous heat (Figure 6–14).

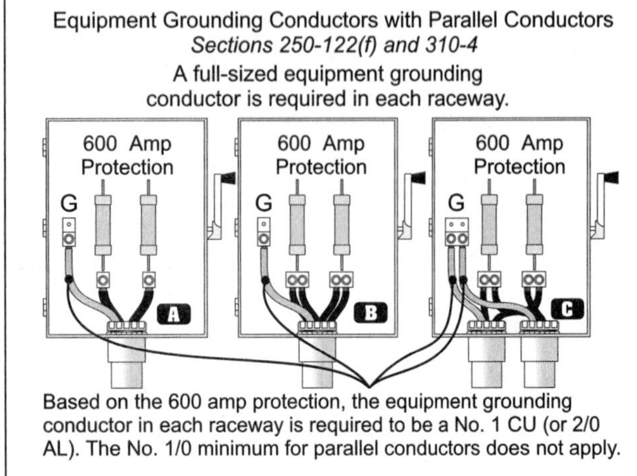

Figure 6–12
Equipment Grounding Conductors with Parallel Conductors

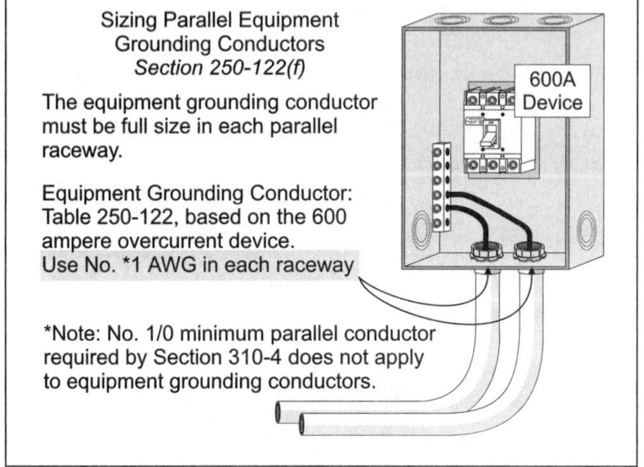

Figure 6–13
Sizing Parallel Equipment Grounding Conductors

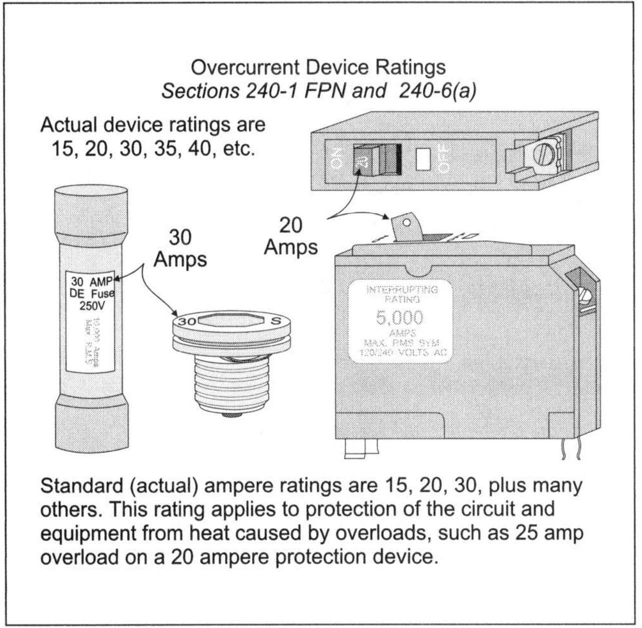

Figure 6–14
Overcurrent Device Ratings

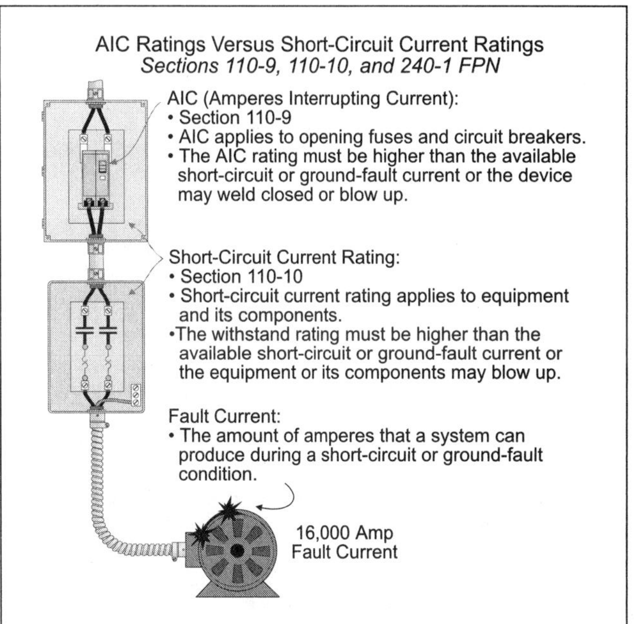

Figure 6–15
A/C Ratings versus Short-Circuit Current Ratings

Interrupting Rating (Short-circuit) [*Section 110-9*]. Overcurrent protection devices, such as circuit breakers and fuses, are intended to open the circuit at fault levels. Accordingly, they are required to have an interrupting rating sufficient for the maximum possible fault current available on the line side terminals of the equipment.

If the overcurrent protection device is not rated for the available fault current, it could explode while attempting to clear the fault, and/or the downstream equipment could suffer serious damage causing possible hazards to people. UL, ANSI, IEEE, NEMA, manufacturers, and other organizations have considerable literature on how to properly size overcurrent protection and methods for calculating available short-circuit current (Figure 6–15).

Note: The minimum interruption rating for a circuit breaker is 5,000 ampere [Section 240-83(c)] and 10,000 ampere for fuses [Section 240-60(c)] Figure 6–16).

Standard Size Protection Devices [*Section 240-6(a)*]. The *National Electrical Code®* list standard sized overcurrent protection devices: 15, 20, 25, 30, 35, 40, 45, 50, 60, 70, 80, 90, 100, 110, 125, 150, 175, 200, 225, 250, 300, 350, 400, 450, 500, 600, 700, 800, 1,000, 1,200, 1,600, 2,000, 2,500, 3,000, 4,000, 5,000, and 6,000 ampere. Exception: Additional standard ratings for fuses include : 1, 3, 6, 10, and 601 ampere.

Continuous Load. Overcurrent protection devices are sized no less than 125 percent of the continuous load, plus 100 percent of the noncontinuous load. [*Section 210-20(a), 215-3,* and *384-16(d)*] (Figure 6–17).

❑ **Continuous Load**

What size protection device is required for a 100 ampere continuous load (Figure 6–18)?

(a) 150 ampere (b) 100 ampere (c) 125 ampere (d) any of these

• Answer: (c) 125 ampere, 100 ampere × 1.25 = 125 ampere [*Section 240-6(a)*]

6–8 OVERCURRENT PROTECTION OF CONDUCTORS – GENERAL REQUIREMENTS [*SECTION 240-3*]

There are many different rules for sizing and protecting conductors and equipment. It is not simply No. 12 wire and a 20 ampere breaker. The general rule is that conductors must be protected according to their ampacity as listed in Table 310-16. Other methods of protection are permitted or required as listed in subsections (b) through (g) of this Section.

(b) Next Higher Overcurrent Device Rating. The next higher protection device is permitted if all of the following conditions are met (Figure 6–19):

(1) Conductors do not supply multioutlet receptacle branch circuits for portable cord- and plug-connected loads.

(2) The ampacity of a conductor does not correspond with the standard ampere rating of a fuse or circuit breaker as listed in Section 240-6(a).

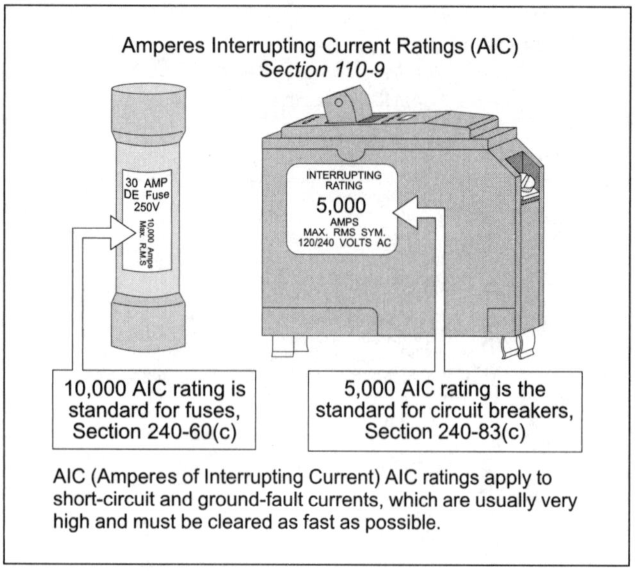

Amperes Interrupting Current Ratings (AIC)
Section 110-9

30 AMP DE Fuse 250V
10,000 Amps Max. R.M.S

INTERRUPTING RATING
5,000
AMPS
MAX. RMS SYM.
120/240 VOLTS AC

10,000 AIC rating is standard for fuses, Section 240-60(c)

5,000 AIC rating is the standard for circuit breakers, Section 240-83(c)

AIC (Amperes of Interrupting Current) AIC ratings apply to short-circuit and ground-fault currents, which are usually very high and must be cleared as fast as possible.

Figure 6–16
Amperes Interrupting Current Ratings (A/C)

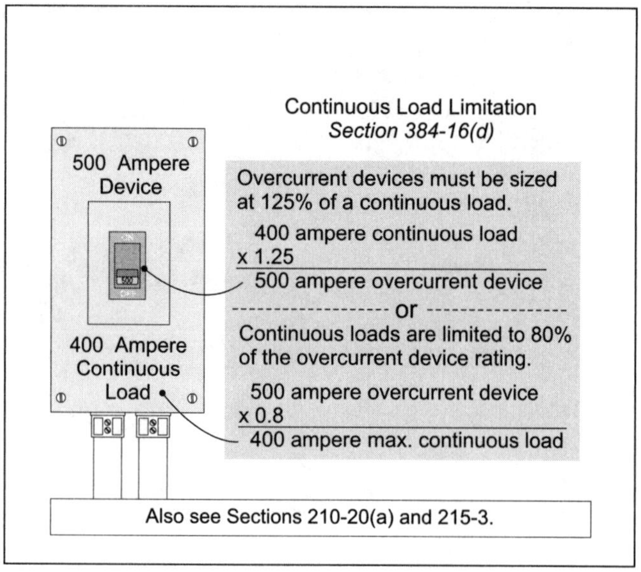

Continuous Load Limitation
Section 384-16(d)

500 Ampere Device

400 Ampere Continuous Load

Overcurrent devices must be sized at 125% of a continuous load.
400 ampere continuous load
x 1.25
500 ampere overcurrent device
------------------ or ----------------
Continuous loads are limited to 80% of the overcurrent device rating.
500 ampere overcurrent device
x 0.8
400 ampere max. continuous load

Also see Sections 210-20(a) and 215-3.

Figure 6–17
Continuous Load Limitation

(3) The next size up breaker or fuse does not exceed 800 ampere.

❑ **Overcurrent Protection of Conductors**

What size conductor is required for a 104-ampere continuous load that is protected with a 150-ampere breaker (Figure 6–20)?

(a) No. 1/0 (b) No. 1 (c) No. 2 (d) any of these

• Answer: (b) No. 1

The conductor must be sized no less than 125% of the continuous load: 104 amperes × 1.25 = 130 amperes

No. 1 THHN is rated 30 amperes at 75°C [*Section 110-14(c)(2)(b)*] and can be protected by a 150-ampere protection device.

(c) Circuits with Overcurrent Protection over 800 Ampere. If the circuit overcurrent protection device exceeds 800 ampere, the circuit conductor ampacity must not be less than the rating of the overcurrent protection device as listed in Section 240-6(a) (Figure 6–21).

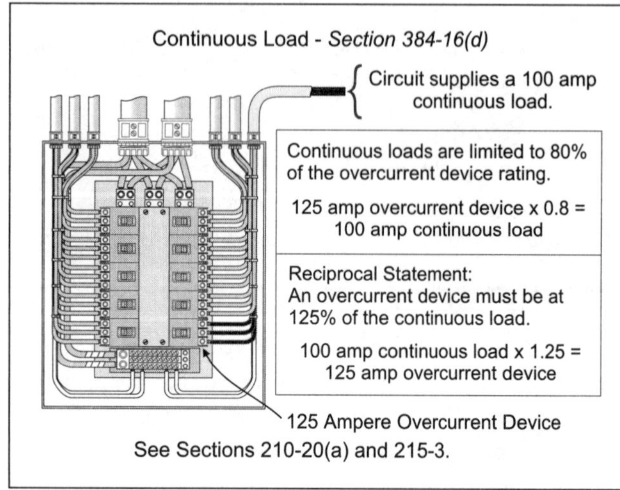

Figure 6–18
Continuous Loads

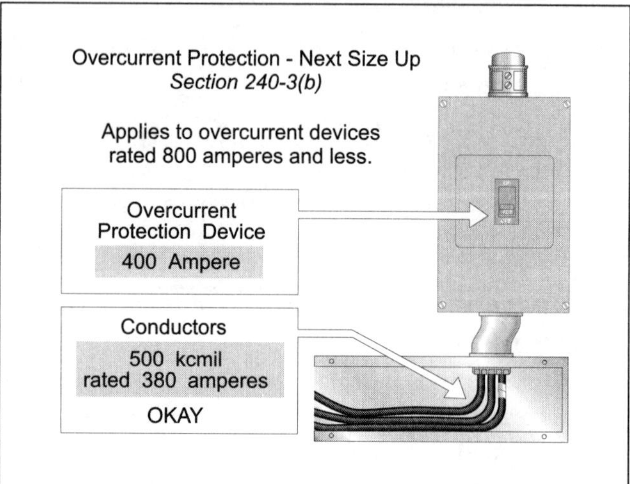

Figure 6–19
Overcurrent Protection – Next Size Up

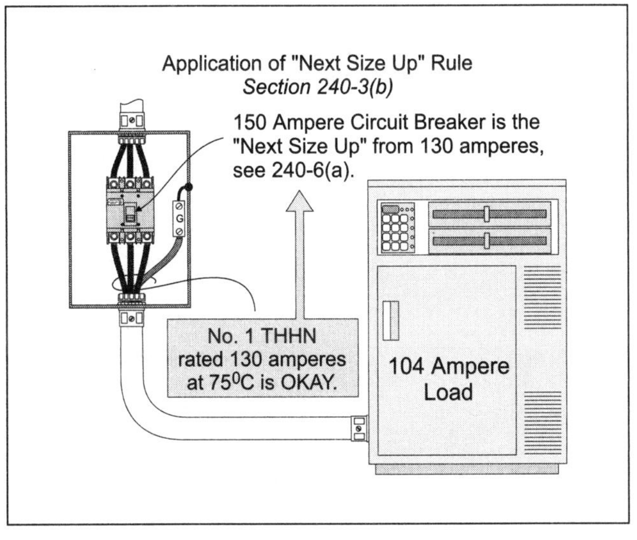

Figure 6–20
Application of "Next Size Up" Rule

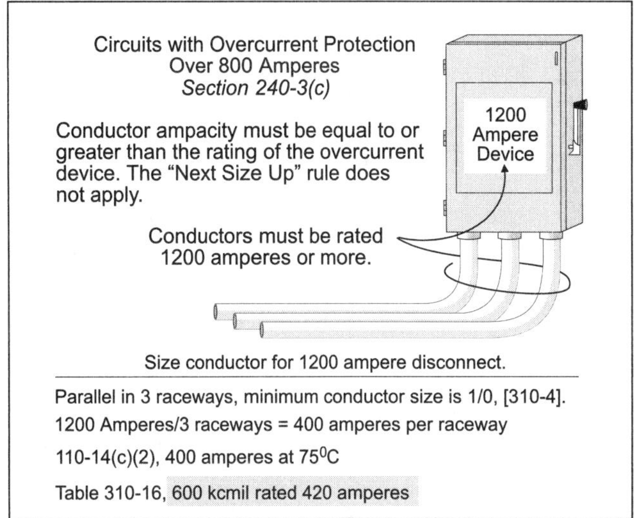

Figure 6–21
Circuits with Overcurrent Protection over 800 Amperes

(d) Small Conductors. Unless specifically permitted in 240-3(e) through 240-3(g), overcurrent protection shall not exceed 15 ampere for No. 14, 20 ampere for No. 12, and 30 ampere for No. 10 copper, or 15 ampere for No. 12, and 25 ampere for No. 10 aluminum and copper-clad aluminum after ampacity correction (Figure 6–22).

6–9 OVERCURRENT PROTECTION OF CONDUCTORS – SPECIFIC REQUIREMENTS

When sizing and protecting conductors for equipment, be sure to apply the specific *NEC®* requirement.

Equipment

Air-Conditioning [*Section 440-22,* and *440-32*]
Appliances [*Section 422-10* and *422-11*]
Cooking Appliances [*Section 210-19(c), 210-21(b)(4),* Note 4 of Table 220-19]
Electric Heating Equipment [*Section 424-3(b)*]
Fire Protective Signaling Circuits [*Section 760-23*]
Motors:
Branch Circuits [*Section 430-22(a),* and *430-52*]
Feeders [*Section 430-24,* and *430-62*]
Remote Control [*Section 430-72*]
Panelboard [*Section 384-16(a)*]
Transformers [*Section 240-21* and *450-3*]

Feeders and Services

Dwelling Unit Feeders and Neutral [*Section 215-2, 310-15(b)(6)*]
Feeder Conductor [*Section 215-2* and *215-3*]
Service Conductors [*Section 230-42* and *230-90(a)*]
Temporary Conductors [*Section 305-4*]

Grounded (neutral) Conductor

Neural Calculations [*Section 220-22*]
Grounded Service Size [*Section 230-24(b)*]

Tap Conductors

Tap - Ten foot [*Section 240-21(b)(1)*]
Tap - Twenty-five foot [*Section 240-21(b)(2)*]
Tap - One hundred foot [*Section 240-21(b)(4)*]
Tap - Outside Feeder [*Section 240-21(b)(5)*]

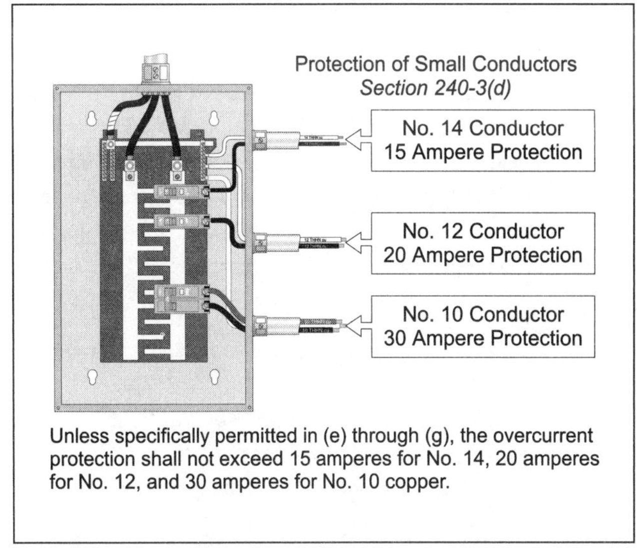

Figure 6–22
Protection of Small Conductors

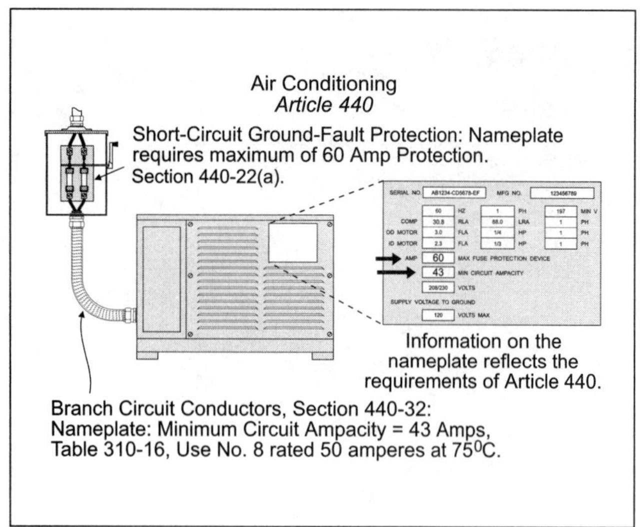

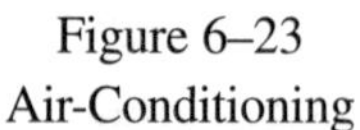

Figure 6–23
Air-Conditioning

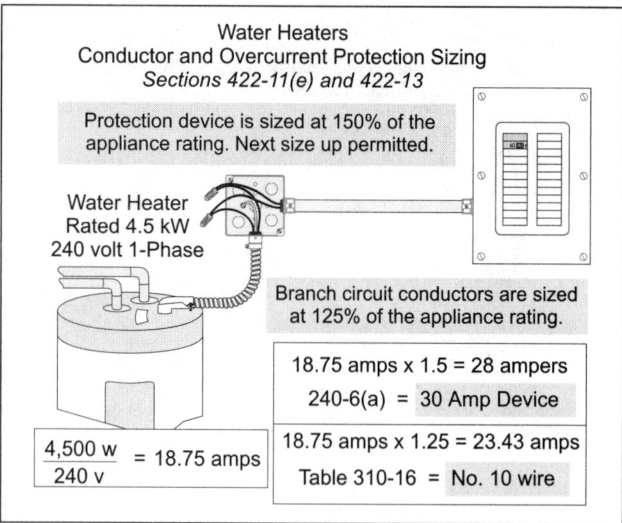

Figure 6–24
Water Heaters

6–10 EQUIPMENT CONDUCTORS SIZE AND PROTECTION EXAMPLES

Air-Conditioning

An air-conditioner nameplate indicates the minimum circuit ampacity of 43 ampere and maximum fuse size of 60 ampere. What is the minimum size branch circuit conductor and the maximum size overcurrent protection device (Figure 6–23)?

(a) No. 6 with a 60 ampere fuse
(b) No. 8 with a 60 ampere fuse
(c) No. 8 with a 50 ampere fuse
(d) No. 10 with a 30 ampere fuse

• Answer: (b) No. 8 with a 60 ampere fuse

Conductor: The conductors must be sized based on the 75°C column of Table 310-16: No. 8 rated 50 amperes.

Overcurrent Protection: The protection device must not be greater than a 60-ampere fuse, either one-time or dual-element.

❑ Water Heater [*Section 422-11(e) and 422-13*]

What size conductor and protection device is required for a 4,500 VA, 240 volt water heater (Figure 6–24)?

(a) No. 10 wire with 20-ampere protection
(b) No. 10 wire with 25-ampere protection
(c) No. 10 wire with 30-ampere protection
(d) b or c

• Answer: (d) b or c

I = VA/E = 4,500 VA/240 volts = 18.75 ampere

Conductor Size: The conductor is sized at 125 percent of the water heater rating [*Section 422-13*].

Minimum conductor = 18.75 ampere × 1.25 = 23.4 ampere

Conductor is sized according to the 60°C column of Table 310-16 = No. 10 rated 30 ampere

Overcurrent Protection: Overcurrent protection device sized no more than 150 percent of appliance rating, [*Section 422-11(e)*].

8.75 ampere × 1.50 = 28.1 ampere, next size up = 30 ampere

❑ Motor

What size branch circuit conductor and short-circuit protection (circuit breaker) is required for a 2 horsepower (12 ampere) motor rated 230 volts (Figure 6–25)?

(a) No. 14 with a 15-ampere breaker
(b) No. 12 with a 20-ampere breaker
(c) No. 12 with a 30-ampere breaker
(d) No. 14 with a 30-ampere breaker

• Answer (d) No. 14 with a 30-ampere protection device

Conductors: Conductors are sized no less than 125 percent of the motor full-load current [*Section 430-6(a)* and *430-22(a)*].

12 ampere × 1.25 = 15 ampere, Table 310-16, No. 14 is rated 20 ampere.

Overcurrent Protection: The short-circuit protection (circuit breaker) is sized at 250 percent of motor full-load current.

12 ampere × 2.5 = 30 ampere [*Section 240-6(a)* and *430-52(c)(1)* Exception No. 1]

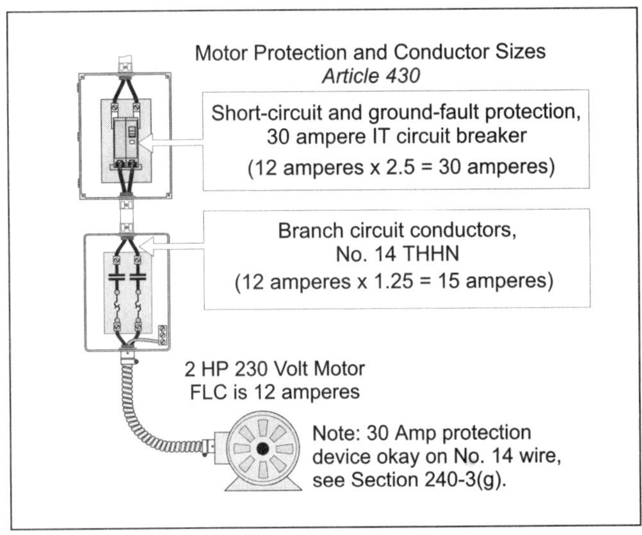

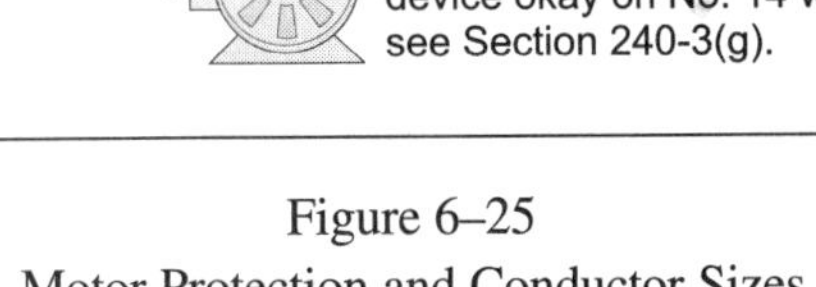

Figure 6–25
Motor Protection and Conductor Sizes

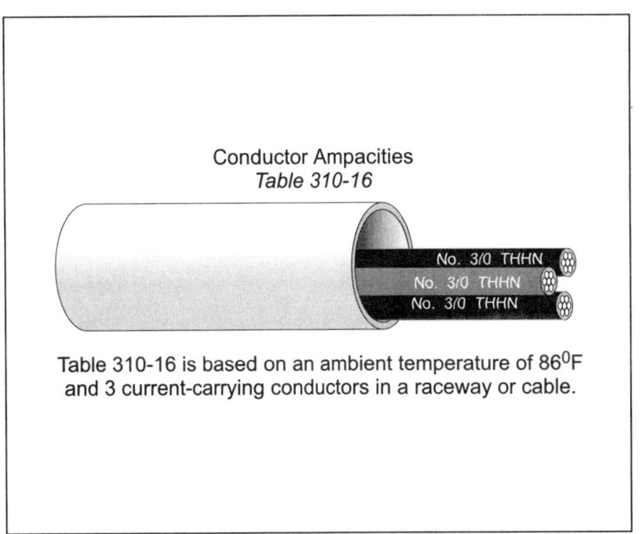

Figure 6–26
Conductor Ampacities

PART B – CONDUCTOR AMPACITY CALCULATIONS

6–11 CONDUCTOR AMPACITY [*SECTION 310-10*]

The insulation temperature rating of a conductor is limited to an operating temperature that prevents serious heat damage to the conductor's insulation. If the conductor carries excessive current, the I^2R heating within the conductor can destroy the conductor insulation. To limit elevated conductor operation temperatures, the current flow (ampacity) in the conductors must be limited.

Allowable Ampacities

The ampacity of a conductor is the current the conductors can carry continuously under the specific condition of use [Article 100 definition]. The ampacity of a conductor is listed in Table 310-16 under the condition of no more than three current-carrying conductors bundled together in an ambient temperature of 86°C. The ampacity of a conductor changes if the ambient temperature is not 86°F or if more than three current-carrying conductors are bundled together in any way (Figure 6–26).

6–12 AMBIENT TEMPERATURE DERATING FACTOR [Table 310-16]

The ampacity of a conductor as listed on Table 310-16 is based on the conductor operating at an ambient temperature of 86°F (30°C). When the ambient temperature is different than 86°F (30°C) for a prolonged period of time, the conductor ampacity listed in Table 310-16 must be adjusted to a new ampacity.

In general, 90°C rated conductor ampacities cannot be used for sizing circuit conductors. However, higher insulation temperature rating offers the opportunity of having a greater conductor ampacity for conductor ampacity derating and a reduced temperature correction factor. The temperature correction factors used to determine the new conductor ampacity are listed at the bottom of Table 310-16. The following formula can be used to determine the conductors new ampacity when the ambient temperature is not 86°F (30°C) (Figure 6–27):

Ampacity = Allowable Ampacity × Temperature Correction Factor

Note: Conductor ampacity adjustment does not apply if the different ambient is 10 feet or less and does not exceed 10 percent of the total length of the conductor [*Section 310-15(a)(2)* Exception].

❑ **Ambient Temperature below 86°F (30°C)**

The ampacity of No. 12 THHN when installed in a walk-in cooler that has an ambient temperature of 50°F (Figure 6–28)?

(a) 31 ampere (b) 35 ampere (c) 30 ampere (d) 20 ampere

- Answer: (a) 31 ampere

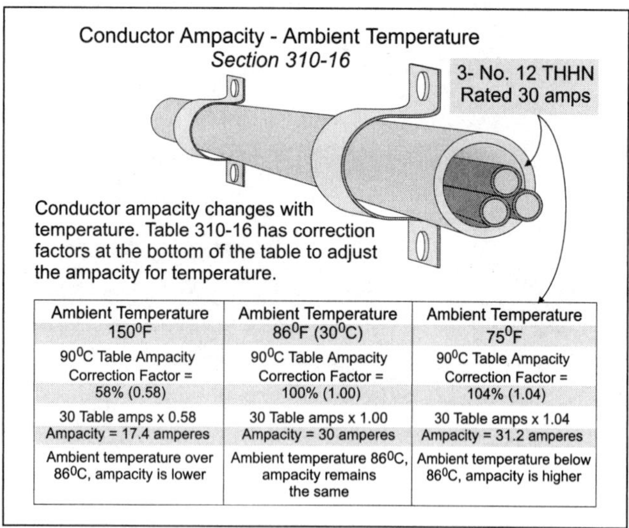

Figure 6–27
Conductor Ampacity – Ambient Temperature

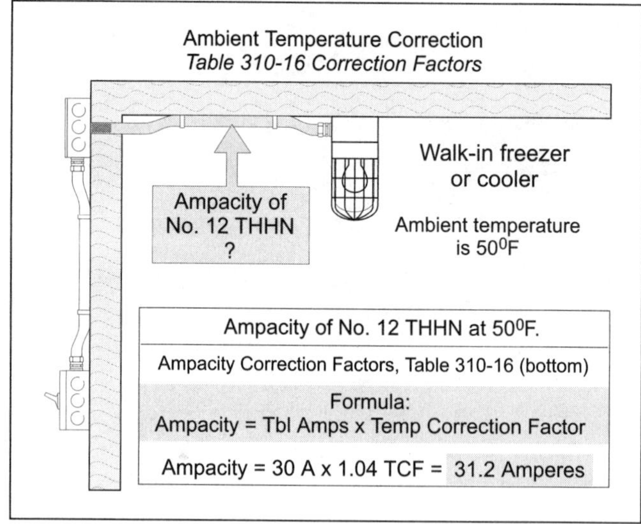

Figure 6–28
Ambient Temperature Correction

New Ampacity = Table 310-16 Ampacity × Temperature Correction Factor

Table 310-16 ampacity for No. 12 THHN is 30 ampere at 90°C.

Temperature Correction Factor for 90°C conductor installed at 50°C is 1.04.

New Ampacity = 30 ampere × 1.04 = 31.2 ampere

Note: Ampacity increases when the ambient temperature is less than 86°F (30°C).

❑ Ambient Temperature above 86°F (30°C)

What is the ampacity of No. 6 THHN when installed on a roof that has an ambient temperature of 60°C (Figure 6–29)?

(a) 53 ampere (b) 35 ampere (c) 75 ampere (d) 60 ampere

• Answer: (a) 53 ampere

New Ampacity = Table 310-16 Ampacity × Ambient Temperature Correction Factor

Table 310-16 ampacity for No. 6 THHN is 75 ampere at 90°C.

Temperature Correction Factor for 90°C conductor rating installed at 60°C is 0.71.

New Ampacity = 75 ampere × 0.71 = 53.25 ampere

Note: Ampacity decreases when the ambient temperature is more than 86°F (30°C).

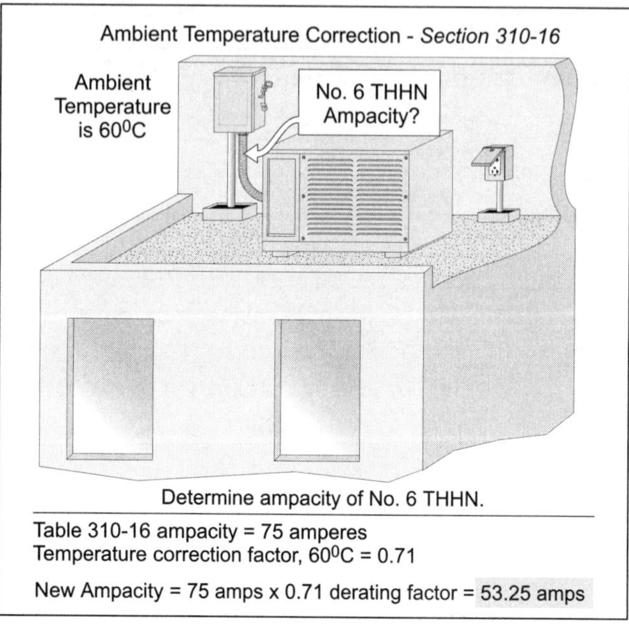

Figure 6–29
Ambient Temperature Correction

❑ Conductor Size

What size conductor is required to supply a 40-ampere load? The conductors pass through a room where the ambient temperature is 100°F (Figure 6–30).

(a) No. 10 THHN (b) No. 8 THHN
(c) No. 6 THHN (d) any of these

• Answer: (b) No. 8 THHN

The conductor to the load must have an ampacity of 40 ampere after applying the ambient temperature derating factor.

New Ampacity = Table 310-16 Ampere × Ambient Temperature Correction

No. 10 THHN = 40 ampere × 0.91 = 36.4 ampere

No. 8 THHN = 55 ampere × 0.91 = 50.0 ampere

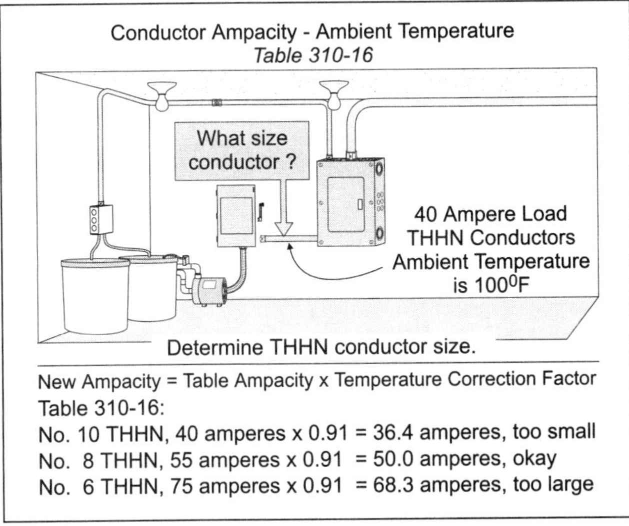

Figure 6–30
Conductor Ampacity – Ambient Temperature

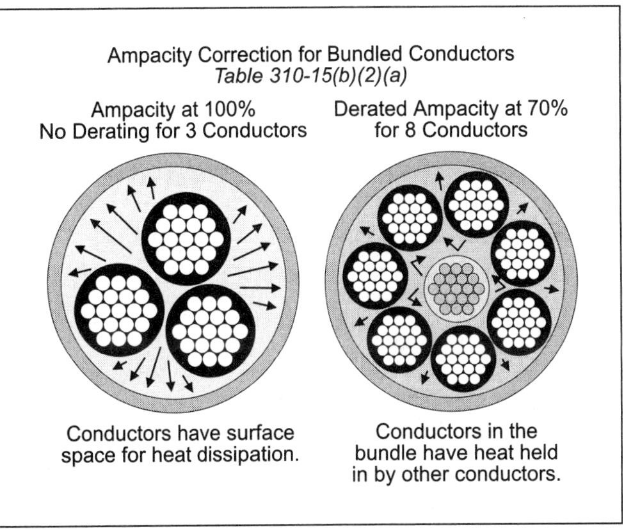

Figure 6–31
Ampacity Correction for Bundled Conductors

6–13 CONDUCTOR BUNDLING DERATING FACTOR [Table 310-15(b)(2)]

When many conductors are bundled together, the ability of the conductors to dissipate heat is reduced. The *National Electrical Code®* requires that the ampacity of a conductor be reduced whenever four or more current-carrying conductors are bundled together (Figure 6–31). In general, 90ºC rated conductor ampacities cannot be used for sizing a circuit conductor. However, higher insulation temperature rating offers the opportunity of having a greater conductor ampacity for conductor ampacity derating.

The ampacity derating factors used to determine the new ampacity is listed in Table 310-15(b)(2)(a). The following formula can be used to determine the new conductor ampacity when more than three current-carrying conductors are bundled together:

New Ampacity = Table 310-16 Ampacity × Correction Factor

Note: Conductor bundle ampacity adjustment factors do not apply to conductors in a nipple that does not exceeding 24 inches, see Exception 3 to Section 310-15(b)(2)(a) (Figure 6–32).

❑ **Conductor Ampacity**

What is the ampacity of four current-carrying No. 10 THHN conductors installed in a raceway or cable (Figure 6–33)?

(a) 20 ampere (b) 24 ampere (c) 32 ampere (d) none of these

- Answer: (c) 32 ampere

  Ampacity = Table 310-16 Ampacity × Adjustment

  Table 310-16 ampacity for No. 10 THHN is 40 ampere at 90ºC.

  Bundle adjustment factor for four current-carrying conductors is 0.8.

  New Ampacity = 40 ampere × 0.8 = 32 ampere

❑ **Conductor Size**

A raceway contains four current-carrying conductors. What size conductor is required to supply a 40-ampere non-continuous load (Figure 6–34)?

(a) No. 10 THHN (b) No. 8 THHN
(c) No. 6 THHN (d) none of these

- Answer: (b) No. 8 THHN

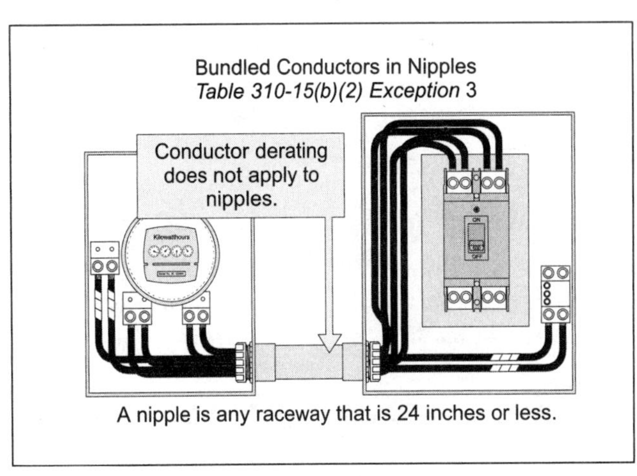

Figure 6–32
Bundled Conductors in Nipples

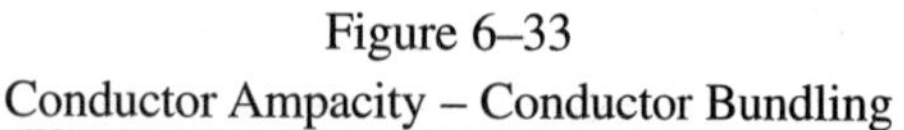

Figure 6–33
Conductor Ampacity – Conductor Bundling

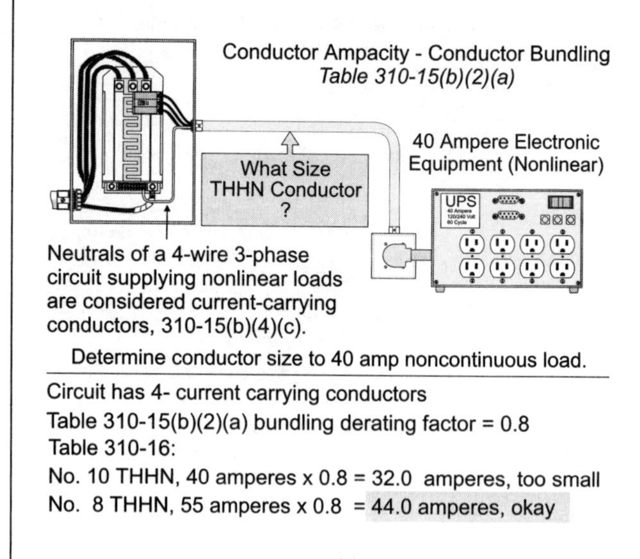

Figure 6–34
Ampacity Correction for Bundled Conductors

The conductor must have an ampacity of 40 ampere after applying bundle adjustment factor.

Ampacity = 310-16 Ampacity × Bundle Adjustment Factor

No. 10 THHN = 40 ampere × 0.8 = 32 ampere, too small

No. 8 THHN = 55 ampere × 0.8 = 44 ampere, just right

No. 6 THHN = 75 ampere × 0.8 = 60 ampere, larger than required by the *NEC®*

6–14 AMBIENT TEMPERATURE AND CONDUCTOR BUNDLING DERATING FACTORS

If the ambient temperature is different than 86°F (30°C) and there are more than three current-carrying conductors bundled together, then the ampacity listed in Table 310-16 must be adjusted for both conditions (Figure 6–35).

The following formula can be used to determine the new conductor ampacity when both ambient temperature and bundle adjustment factors apply:

Ampacity = Table 310-16 Ampacity × Temperature Correction Factor × Bundle Adjustment Factor

Note: Ampacity adjustment does not apply if the length of conductors or ambient temperature is 10 feet or less and does not exceed 10 percent of the conductor length [*Section 310-15(a)(2)* Exception].

❑ **Conductor Ampacity**

What is the ampacity of four current-carrying No. 8 THHN conductors installed in ambient temperature of 100°F (Figure 6–36)?

(a) 25 ampere (b) 40 ampere (c) 55 ampere (d) 60 ampere

• Answer: (b) 40 ampere

New Ampacity = Table 310-16 Ampacity × Temperature Factor × Bundle Factor

Table 310-16 ampacity of No. 8 THHN is 55 ampere at 90°C.

Temperature Correction factor for 90°C conductor insulation at 100°F is 0.91.

Bundle adjustment factor for four conductors is 0.8.

New Ampacity = 55 ampere × 0.91 × 0.8 = 40 ampere

6–15 CURRENT-CARRYING CONDUCTORS

Table 310-15(b)(2)(a) adjustment factors only apply when there is more than three current-carrying conductors bundled. Naturally, all phase conductors are considered current-carrying, and the following should be helpful in determining which other conductors are considered current-carrying:

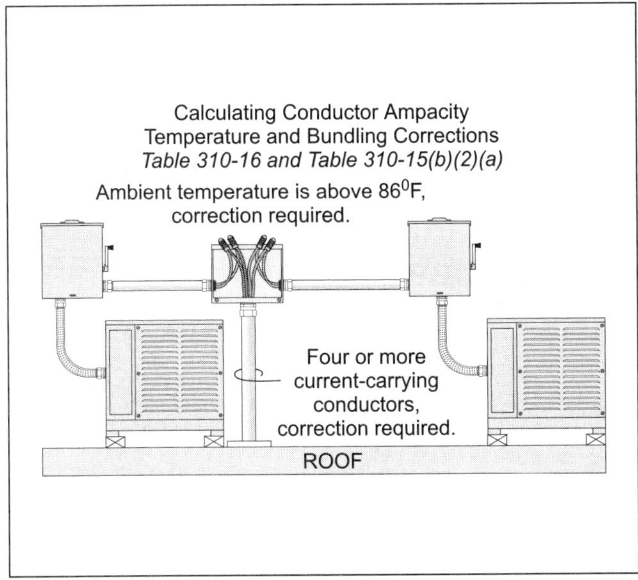

Figure 6–35
Calculating Conductor Ampacity with Temperature and Bunching

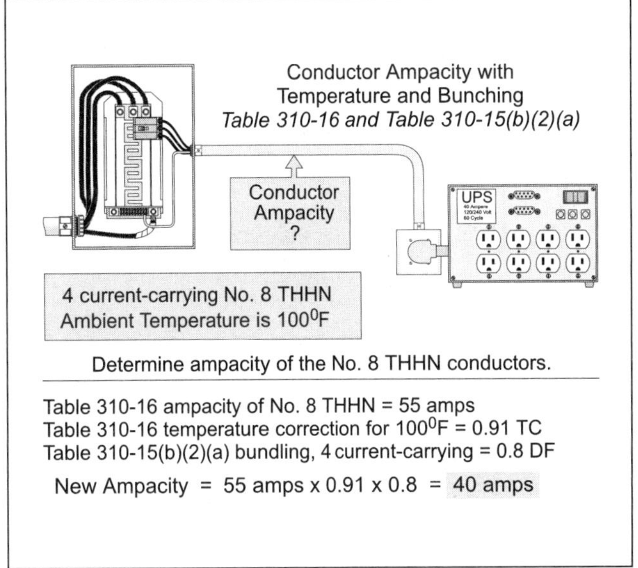

Figure 6–36
Conductor Ampacity with Temperature and Bunching

Grounded (neutral) Conductor – Balanced Circuits, Section 310-15(b)(4)(a)

The grounded (neutral) conductor of a balanced 3-wire circuit, or a balanced 4-wire wye circuit is not considered a current-carrying conductor (Figure 6–37).

Grounded (neutral) Conductor – Unbalanced 3-wire Wye Circuit, Section 310-15(b)(4)(b)

The grounded (neutral) conductor of balanced 3-wire wye circuit is considered a current-carrying conductor (Figure 6–38).

This can be proven with the following formula:

$$I_{Neutral} = \sqrt{(L_1^2 + L_2^2)-(L_1 \times L_2)} \quad L_1 = \text{Current of one phase} \quad L_2 = \text{Current of the other phase}$$

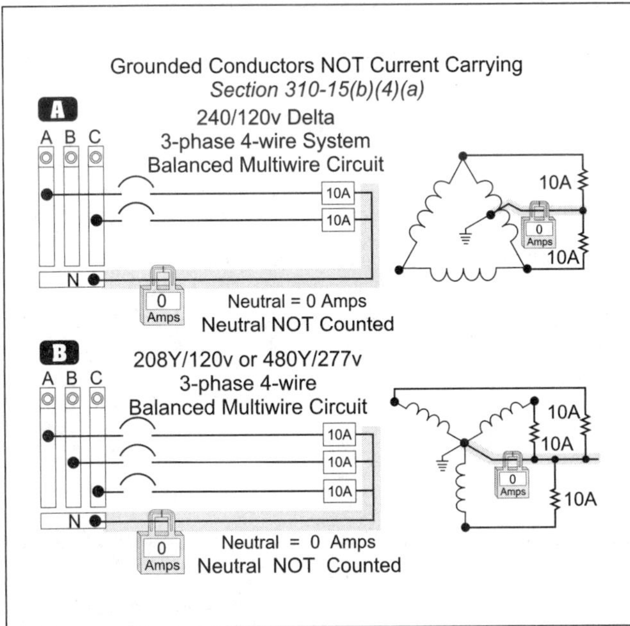

Figure 6–37
Grounded Conductors Not Current Carrying

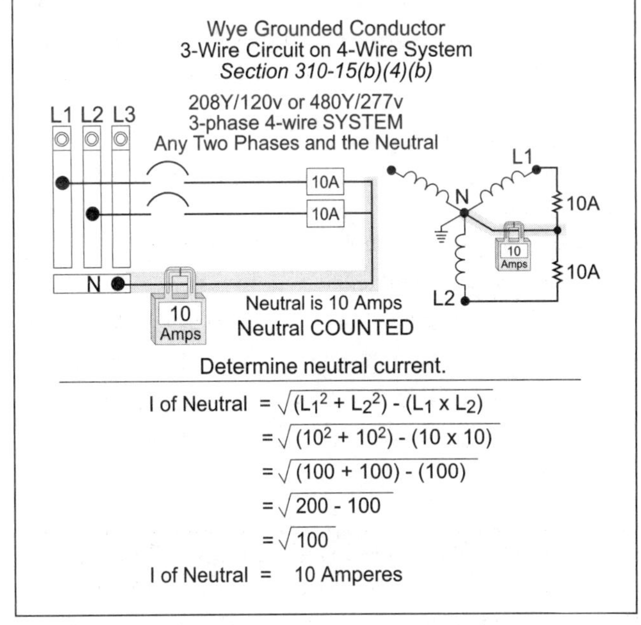

Figure 6–38
Wye Grounded Conductor

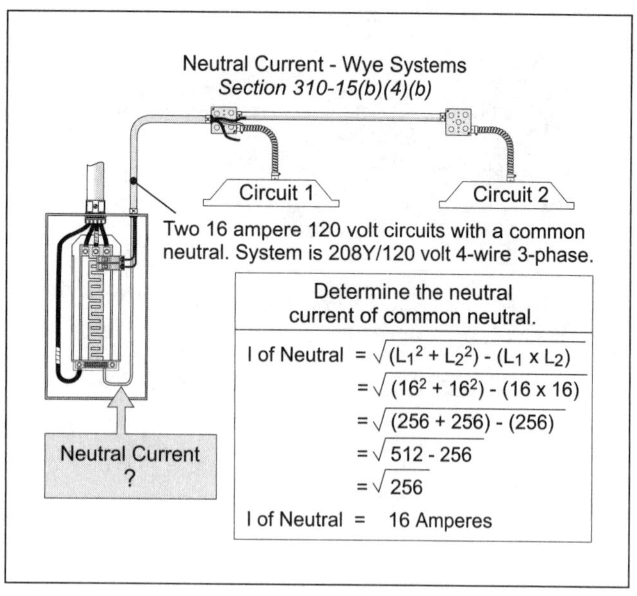

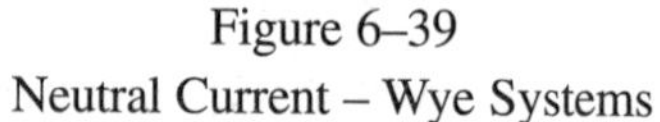
Figure 6–39
Neutral Current – Wye Systems

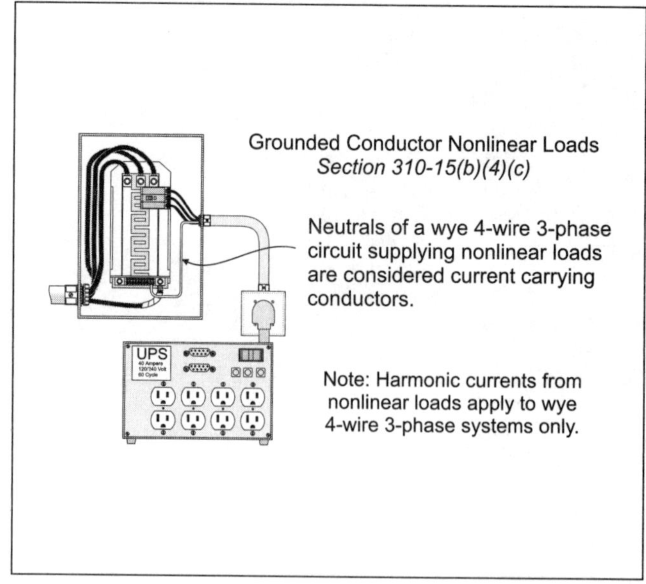

Figure 6–40
Grounded Conductor Nonlinear Loads of a 4-Wire, 3-Phase System

❑ Grounded (neutral) Conductor

What is the neutral current for a balanced 16 ampere, 3-wire, 208Y/120-volt branch circuit of a 4-wire, 3-phase wye system that supplies fluorescent lighting (Figure 6–39)?

(a) 8 ampere (b) 16 ampere (c) 32 ampere (d) none of these

- Answer: (b) 16 ampere

$I_{Neutral} = \sqrt{(L_1^2 + L_2^2) - (L_1 \times L_2)}$, $= \sqrt{(16^2 + 16^2) - (16 \times 16)}$, $= \sqrt{512 - 256} = \sqrt{256} = 16$ ampere

Grounded (neutral) Conductor – Nonlinear Loads, Section 310-15(b)(4)(c)

The grounded (neutral) conductor of a balanced 4-wire wye circuit that is at least 50 percent loaded with nonlinear loads (computers, electric discharge lighting, etc.) is considered a current-carrying conductor (Figure 6–40).

CAUTION: Nonlinear loads produce harmonic currents that add on the neutral conductor, and the current on the neutral can be doubled (Figure 6–41).

Two-wire Circuits

Both the grounded and ungrounded conductor of a two-wire circuit carries current and both are considered current-carrying (Figure 6–42).

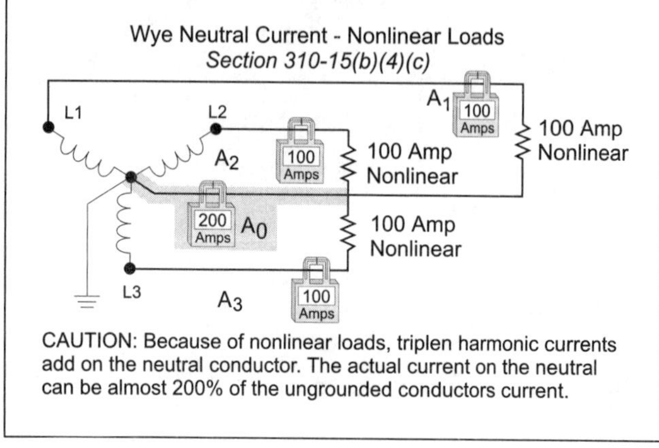

Figure 6–41
Wye-Neutral Current – Nonlinear Loads

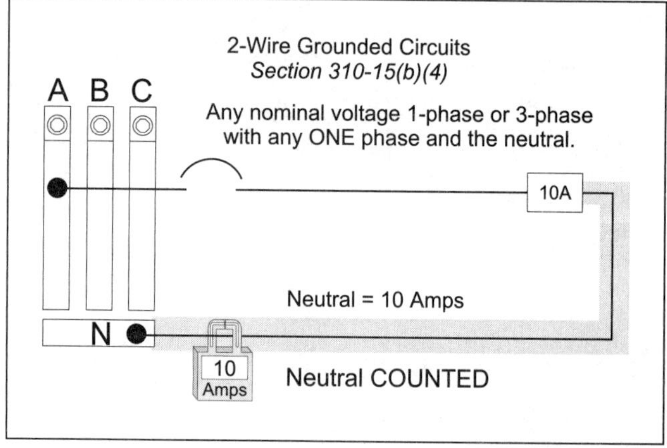

Figure 6–42
2-Wire Grounded Circuits

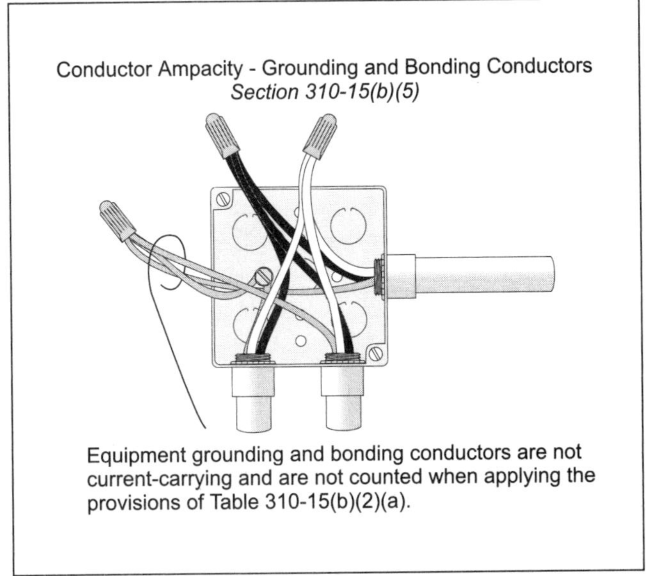

Figure 6–43
Conductor Ampacity – Grounding and Bonding Conductors

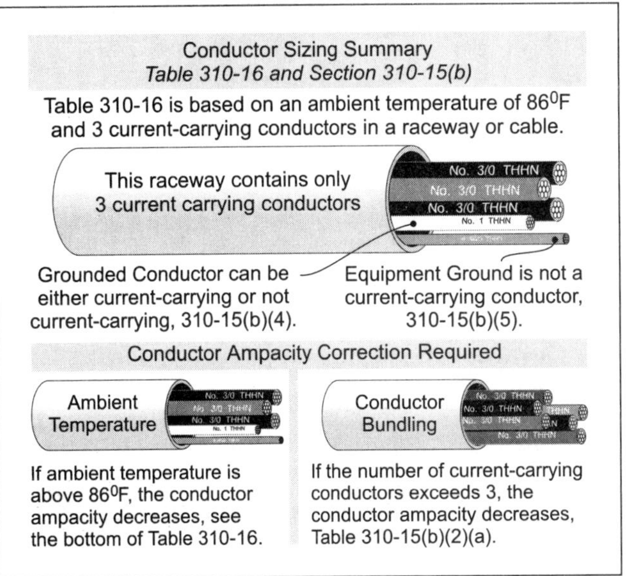

Figure 6–44
Conductor Sizing Summary

Grounding and Bonding Conductors, Section 310-15(b)(b)

Grounding and bonding conductors do not normally carry current and are not considered current-carrying (Figure 6–43).

Note: Grounding and bonding conductors are not counted when adjusting conductor ampacity for the effects of conductor bunching.

6–16 CONDUCTOR SIZING SUMMARY

The ampacity of a conductor changes with changing conditions. The factors that affect conductor ampacity are (Figure 6–44):

1. The allowable ampacity as listed in Table 310-16.
2. The ambient temperature correction factors if the ambient temperature is not 86°F.
3. Conductor ampacity derating factors apply if three or more current-carrying conductors are bundled together.

Terminal Ratings, Section 110-14(c)

Equipment rated 100 ampere or less must have the conductor sized no smaller than the 60°C column of Table 310-16. Equipment rated over 100 ampere must have the conductors sized no smaller than for the 75°C column of Table 310-16. However, higher insulation temperature rating offers the opportunity of having a greater conductor ampacity for conductor ampacity derating.

Unit 6 – Conductor Sizing and Protection Summary Questions

6–1 Conductor Insulation Property [Table 310-13]

1. THHN can be described as _____ .
   (a) thermoplastic insulation with a nylon outer cover
   (b) suitable for dry and wet locations
   (c) having a maximum operating temperature of 90ºC
   (d) a and c

6–2 Conductor Allowable Ampacity [*Section 310-15*]

2. • The maximum overcurrent protection device size for No. 14 is 15 ampere, No. 12 is 20 ampere, and No. 10 is 30 ampere. This is a general rule, but it does not apply to motors or air-conditioners according to Section 240-3.
   (a) True (b) False

6–3 Conductor Sizing

3. • Conductor sizes are expressed in American Wire Gauge (AWG) from No. 40 through No. 4/0. Conductors larger than _____ are express in circular mils.
   (a) No. 1/0 (b) No. 1 (c) No. 3/0 (d) No. 4/0

4. The smallest size conductor permitted for branch circuits, feeders, and services for residential, commercial, and industrial locations is _____ .
   (a) No. 14 copper (b) No. 12 aluminum (c) No. 12 copper (d) a and b

6–4 Terminal Rating [*Section 110-14(c)*]

5. Equipment terminals rated 100 ampere or less (circuit breakers, fuses, etc.) and pressure connector terminals for No. 14 through No. 1 conductors shall have the conductor sized according to 60ºC temperature rating as listed in Table 310-16.
   (a) True (b) False

6. • What is the minimum size THHN conductor that is permitted to terminate on a 70-ampere circuit breaker of fuse? Be sure to comply with the requirements of Section 110-14(c)(1).
   (a) No. 8 (b) No. 6 (c) No. 4 (d) none of these

7. • What size THHN conductor is required for a 70-ampere branch circuit if the circuit breaker and equipment is listed for 75ºC terminals and the load does not exceed 65 ampere?
   (a) No. 10 (b) No. 8 (c) No. 6 (d) No. 4

8. Terminals for equipment rated over 100 ampere and pressure connector terminals for conductors larger than No. 1 shall have the conductor sized according to 75ºC temperature rating as listed in Table 310-16.
   (a) True (b) False

9. • What size THHN conductor is required for an air-conditioning unit if the nameplate requires a conductor ampacity of 34 ampere? Terminals of all the equipment and circuit breakers are rated 75ºC.
   (a) No. 12 (b) No. 10 (c) No. 8 (d) No. 14

10. What is the minimum size THHN conductor required for a 150-ampere circuit breaker or fuse? Be sure to comply with the requirements of Section 110-14(c)(2).
(a) No. 1/0 (b) No. 2/0 (c) No. 3/0 (d) No. 4/0

11. In general, THHN (90°C) conductor ampacities cannot be used when sizing conductors; but when more than three current-conductors conductors are bundled together, or if the ambient temperature is greater than 86°F, the allowable conductor ampacity must be decreased. THHN offers the opportunity of having a greater ampacity for conductor derating purposes, thereby permitting the same conductor to be used without having to increase the conductor size.
(a) True (b) False

12. • What size conductor is required to supply a 190-ampere load in a dry location? Terminals are rated 75°C.
(a) 300 kcmil (b) No. 4/0 (c) No. 3/0 (d) none of these

6–5 Conductors In Parallel [*Section 310-4*]

13. • Phase and grounded (neutral) conductors sized No. 1 AWG and larger are permitted to be connected in parallel.
(a) True (b) False

14. To ensure that the currents are evenly distributed between the parallel conductors, each conductor within a parallel set must be installed in the same type of raceway (metallic or nonmetallic) and must be the same length, material, circular mils, insulation type, and must terminate in the same method.
(a) True (b) False

15. • Paralleling of conductors is done by sets.
(a) True (b) False

16. When an electric relay (coil) is energized, the initial current can be very high, causing significant voltage drop. The reduced voltage at the coil (because of voltage drop) can cause the coil contacts to chatter (open and close like a buzzer) or not close at all. Paralleling of control wiring conductors is often necessary and permitted by the *NEC®* to reduce the effects of voltage drop for long control runs.
(a) True (b) False

17. When equipment grounding conductors are installed in parallel, each raceway must have a full-size equipment grounding conductor sized according to the overcurrent protection device rating of that circuit.
(a) True (b) False

18. All parallel equipment grounding conductors are required to be a minimum No. 1/0.
(a) True (b) False

19. What size equipment grounding conductor is required in each raceway for an 800 ampere, 500 kcmil feeder parallel in two raceways?
(a) No. 3 (b) No. 2 (c) No. 1 (d) No. 1/0

20. If we have an 800 ampere service with a calculated demand load of 750 ampere, what size 75°C conductors would be required if parallel in two raceways?
(a) No. 4/0 (b) 250 kcmil (c) 500 kcmil (d) 750 kcmil

21. • What are the conductors required for a 250-ampere feeder paralleled in two raceways?
(a) No. 3 (b) No. 2 (c) No. 2/0 (d) No. 1/0

6–6 Conductor Size – Voltage Drop [*Section 210–196)* FPN No. 4, and *215–2(b)* FPN No. 2]

22. There is no mandatory rule in the *NEC®* limiting the voltage drop on conductors, but the *Code* recommends that we consider its effect.
(a) True (b) False

6–7 Overcurrent Protection [Article 240]

23. One of the purposes of conductor overcurrent protection is to protect the conductors against excessive or dangerous heat.
(a) True (b) False

24. Overcurrent devices must be designed and rated to clear fault current and must have a short-circuit interrupting rating sufficient for the available fault levels. The minimum interruption rating for circuit breakers is _____ ampere and _____ ampere for fuses.
(a) 10,000; 10,000 (b) 5,000; 5,000 (c) 5,000; 10,000 (d) 10,000; 5,000

25. The following are the standard sized circuit breakers and fuses: _____ .
(a) 25 ampere (b) 90 ampere (c) 350 ampere (d) any of these

26. The maximum continuous load permitted on a overcurrent protection device is limited to _____ of the device rating.
(a) 80% (b) 100% (c) 125% (d) 150%

6–8 Overcurrent Protection of Conductors – General Requirements [*Section 240-3*]

27. If the ampacity of a conductor does not correspond with the standard ampere rating of a fuse or circuit breaker, the next size up protection device is permitted. This applies only if the conductors supply multioutlet receptacles for portable cord- and plug-connected loads.
(a) True (b) False

28. What size conductor (75°C) is required for a 70-ampere breaker that supplies a 70-ampere load?
(a) No. 8 (b) No. 6 (c) No. 4 (d) any of these

6–11 Conductor Ampacity [*Section 310-10* and *310-15*]

29. The temperature rating of a conductor is the maximum operating temperature the conductor insulation can withstand (without serious damage) over a prolonged period of time. The _____ provide guidance for adjusting the conductor's ampacities for the different conditions.
(a) conductor allowable ampacities (b) ambient temperature correction factors
(c) correction factors (d) all of these

6–12 Ambient Temperature Derating Factor [Table 310-16]

30. The ampacities listed in Table 310-16 apply only when the ambient temperature is 40°C and there are no more than two current-carrying conductors bundled together. If the ambient temperature is not 40°C, or there are more than two current-carrying conductors in a raceway, the allowable ampacities must be adjusted to reflect the ampacity under the condition of use.
(a) True (b) False

31. • What is the ampacity of No. 8 THHN conductors when installed in a walk-in cooler if the ambient temperature is 50°F?
(a) 40 ampere (b) 50 ampere (c) 55 ampere (d) 57 ampere

32. • What size conductor is required to feed a 16-ampere load when the conductors are in an ambient temperature of 100°F? The circuit is protected with a 20-ampere overcurrent protection device.
(a) No. 14 THHN (b) No. 12 THHN (c) No.10 THHN (d) No. 8 THHN

6–13 Conductor Bundle Derating Factor [Table 310-15(b)(2)(a)]

33. When four or more current-carrying conductors are bundled together for more than _____ , the conductor allowable ampacity must be reduced according to the factors listed in Table 310-15(b)(2)(a).
(a) 12 inches (b) 24 inches (c) 36 inches (d) 48 inches

34. What is the ampacity of four No. 1/0 THHN conductors?
(a) 111 ampere (b) 136 ampere (c) 153 ampere (d) 171 ampere

35. • A raceway contains eight current-carrying conductors. What size conductor is required to feed a 21 ampere noncontinuous lighting load? The overcurrent protection device is rated 30 ampere.
(a) No. 14 THHN (b) No. 12 THHN (c) No. 10 THHN (d) any of these

6–14 Ambient Temperature and Conductor Bundling Derating Factors

36. What is the ampacity of eight current-carrying No. 10 THHN conductors installed in ambient temperature of 100ºF?
(a) 21 ampere (b) 26 ampere (c) 32 ampere (d) 40 ampere

6–15 Current-Carrying Conductors

37. • The neutral conductor of a balanced 3-wire, delta circuit, or 4-wire wye circuit is considered a current-carrying conductor for the purpose of applying the derating factors of Table 310-15(b)(2)(a).
(a) True (b) False

38. The neutral conductor of a balanced 4-wire wye circuit that is at least 50 percent loaded with nonlinear loads (electric-discharge lighting, electronic ballast, dimmers, controls, computers, laboratory test equipment, medical test equipment, recording studio equipment, etc.) is not considered a current-carrying conductor for the purpose of applying bundle adjustment factors.
(a) True (b) False

39. • The neutral conductor of a balanced 3-wire wye circuit is not considered a current-carrying conductor for the purpose of applying bundle adjustment factors.
(a) True (b) False

6–16 Conductor Sizing Summary

40. • The ampacity of a conductor can be different along the length of the conductor. The higher calculated ampacity can be used if the length of the lower ampacity is no more than 10 feet or no more than 10 percent of the length of the circuit conductors.
(a) True (b) False

41. Most terminals are rated 60ºC for equipment 100 ampere and less and 75ºC for equipment terminals rated over 100 ampere. Regardless of the conductor ampacity, conductors must be sized no smaller than the terminal temperature rating.
(a) True (b) False

✯ Challenge Questions

6–7 Overcurrent Protection [Article 240]

42. • A continuous load of 27 ampere requires the circuit overcurrent protection device to be sized at _____ ampere.
(a) 20 (b) 30 (c) 40 (d) 35

43. • What size overcurrent protection device is required for a 45 ampere continuous load? The circuit is in a raceway with 14 current-carrying conductors.
(a) 45 ampere (b) 50 ampere (c) 60 ampere (d) 70 ampere

44. • A 65 ampere continuous load requires a _____ ampere overcurrent protection device.
(a) 60 (b) 70 (c) 75 (d) 90

45. • A department store (continuous load) feeder supplies a lighting load of 103 ampere. The minimum size overcurrent protection device permitted for this feeder is _____ ampere.
(a) 110 (b) 125 (c) 150 (d) 175

6–12 Ambient Temperature Derating Factor [Table 310-16]

46. • A No. 2 TW conductor is installed in a location where the ambient temperature is expected to be 102°F. The temperature correction factor for conductor ampacity in this location is _____ .
(a) .96 (b) .88 (c) .82 (d) .71

47. • If the ambient temperature is 71°C, the minimum insulation that a conductor must have and still have the capacity to carry current is _____ .
(a) 60°C (b) 105°C (c) 90°C (d) any of these

6–13 Conductor Bundling Derating Factor [Table 310-15(b)(2)(a)]

48. • The ampacity of six current-carrying No. 4/0 XHHW aluminum conductors installed in a ground floor slab (wet location) is _____ .
(a) 135 ampere (b) 185 ampere (c) 144 ampere (d) 210 ampere

6–14 Ambient Temperature and Conductor Bundle Derating Factors

49. • The ampacity of 15 current-carrying No. 10 RHW aluminum conductors in an ambient temperature of 75°F would be _____ .
(a) 30 ampere (b) 22 ampere (c) 16 ampere (d) 12 ampere

50. • A No. _____ THHN conductor is required for a 19.7-ampere load if the ambient temperature is 75°F and there are nine current-carrying conductors in the raceway.
(a) 14 (b) 12 (c) 10 (d) 8

51. • The ampacity of the nine current-carrying No. 10 THW conductors installed in a 20 inch-long raceway is _____ .
(a) 25 ampere (b) 30 ampere (c) 35 ampere (d) none of these

52. • The ampacity of 10 current-carrying No. 6 THHW conductors installed in an 18 inch-long raceway with an ambient temperature of 39°C is _____ .
(a) 47 ampere (b) 68 ampere (c) 66 ampere (d) 75 ampere

6–15 Current-Carrying Conductors

53. A raceway contains the following: One 4-wire multiwire branch circuit that supplies a balanced incandescent 120-volt lighting load; one 4-wire multiwire branch circuit that supplies a balanced 120-volt fluorescent lighting load; two conductors that supply a receptacle; and there is one equipment grounding conductor. The system is 3-phase 208Y/120 volts. Taking all of these factors into consideration, how many of these conductors are considered current-carrying?
(a) 7 conductors (b) 8 conductors (c) 9 conductors (d) 11 conductors

54. • There is a total of nine No. 10 THW conductors in a raceway. The system voltage is 3-phase, 208Y/120 volts. One conductor is an equipment grounding conductor; four conductors supply a 4-wire multiwire branch circuit for balanced electric-discharge lighting, and the remaining conductors supply a 4-wire multiwire branch circuit for balanced incandescent lighting. Taking all of these factors into consideration, how many of these conductors are considered current-carrying?
(a) 6 conductors (b) 7 conductors (c) 9 conductors (d) 10 conductors

Unit 7

Motor Calculations

OBJECTIVES

After reading this unit, the student should be able to briefly explain the following concepts:

- Branch circuit short-circuit ground fault protection
- Feeder protection
- Feeder conductor size
- Highest rated motor
- Motor overcurrent protection
- Motor branch circuit conductors
- Motor VA calculations
- Motor calculation steps
- Overload protection

After reading this unit, the student should be able to briefly explain the following terms:

- Dual-element fuse
- Ground-fault
- Inverse time breaker
- Heaters
- Motor full-load current
- Motor nameplate current rating
- Next smaller device size
- One-time fuse
- Overcurrent protection
- Overcurrent
- Overload
- Service factor (SF)
- Short-circuit
- Temperature rise

INTRODUCTION

When sizing conductors and overcurrent protection for motors, we must comply with the requirements of the *National Electrical Code®*, specifically Article 430 [*Section 240-3(g)*].

7–1 MOTOR BRANCH CIRCUIT CONDUCTORS [*SECTION 430-22(a)*]

Branch circuit conductors to a single motor must have an ampacity of not less than 125 percent of the motor's full-load current as listed in Tables 430-147 through 430-150 [*Section 430-6(a)*]. The actual conductor size must be selected from Table 310-16 according to the terminal temperature rating (60° or 75°C) of the equipment [*Section 110-14(c)*] (Figure 7–1).

❑ **Motor Branch Circuit Conductors**

What size THHN conductor is required for a 2 horsepower 230-volt, single-phase motor (Figure 7–2)?

(a) No. 14 (b) No. 12
(c) No. 10 (d) No. 8

• Answer: (a) No. 14

Motor Full-Load Current – Table 430-148:

2 horsepower 230 volts single-phase = 12 ampere

Conductor Sized no more than 125 percent of Motor Full-Load Current:

12 ampere × 1.25 = 15 ampere, Table 310-16,
No. 14 THHN rated 20 ampere at 60°C

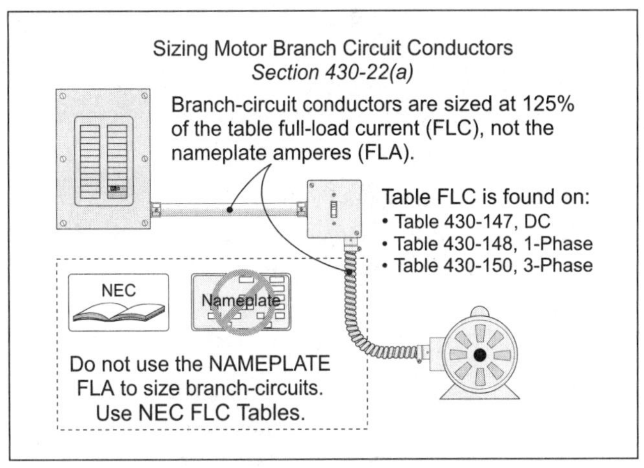

Figure 7–1
Sizing Motor Branch Circuit Conductors

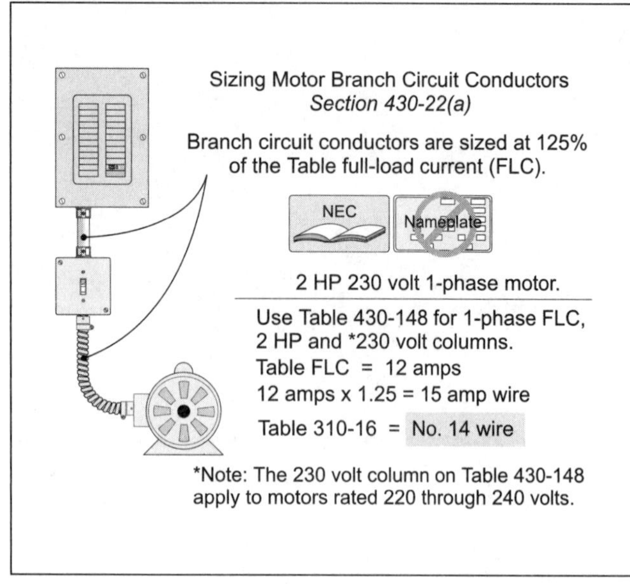

Figure 7–2
Sizing Motor Branch Circuit Conductors

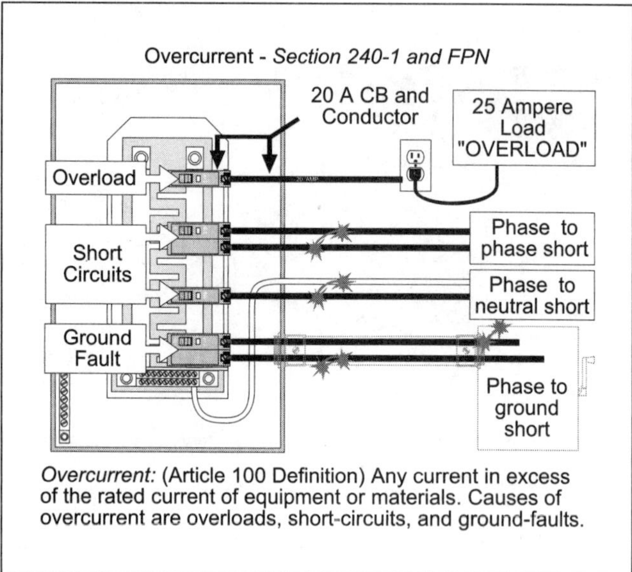

Figure 7–3
Overcurrent

Note: The minimum size conductor permitted for building wiring is No. 14 [*Section 310-5*]; however, local codes and many industrial facilities have requirements that No. 12 be used as the smallest branch circuit wire.

7–2 MOTOR OVERCURRENT PROTECTION

Motors and their associated equipment must be protected against overcurrent (overload, short-circuit, or ground-fault) [Article 100] (Figure 7–3). Due to the special characteristics of induction motors, overcurrent protection is generally accomplished by having the overload protection separated from the short-circuit and ground-fault protection device (Figure 7–4). Article 430, Part C contains the requirements for motor overload protection, and Part D of Article 430 contains the requirements for motor short-circuit and ground-fault protection.

Overload Protection

Overload is the condition in which current exceeds the equipment ampere rating, which can result in equipment damage due to dangerous overheating [Article 100]. Overload protection devices, sometimes called heaters, are intended to protect the motor, the motor-control equipment, and the branch circuit conductors from excessive heating due to motor overload [*Section 430-31*].

Overload protection is not intended to protect against short-circuits or ground-fault currents (Figure 7–5).

If overload protection is to be accomplished by the use of fuses, a fuse must be installed to protect each ungrounded conductor [*Section 430-36* and *430-55*].

Short-Circuit and Ground-Fault Protection

Branch circuit short-circuit and ground-fault protection devices are intended to protect the motor, the motor control apparatus, and the conductors against short-circuit or ground-faults, but they are not intended to protect against an overload [*Section 430-51*] (Figure 7–6).

Note: The ground-fault protection device required for motor circuits is not the type required for personnel [*Section 210-8*], feeders [*Section 215-9* and *240-13*], services [*Section 230-95*], or temporary wiring for receptacles [*Section 305-6*].

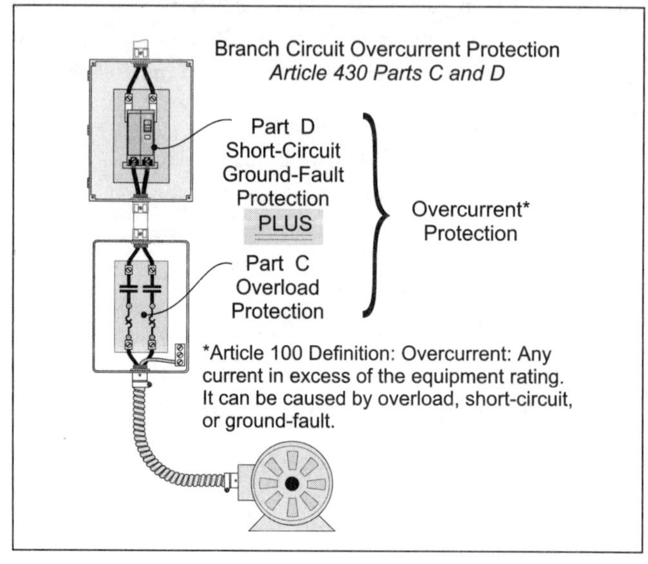

Figure 7–4
Branch Circuit Overcurrent Protection

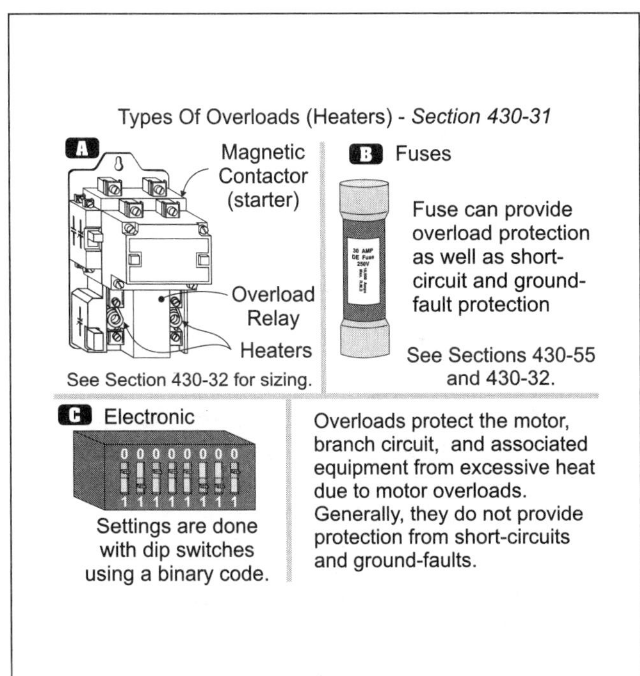

Figure 7–5
Types of Overloads

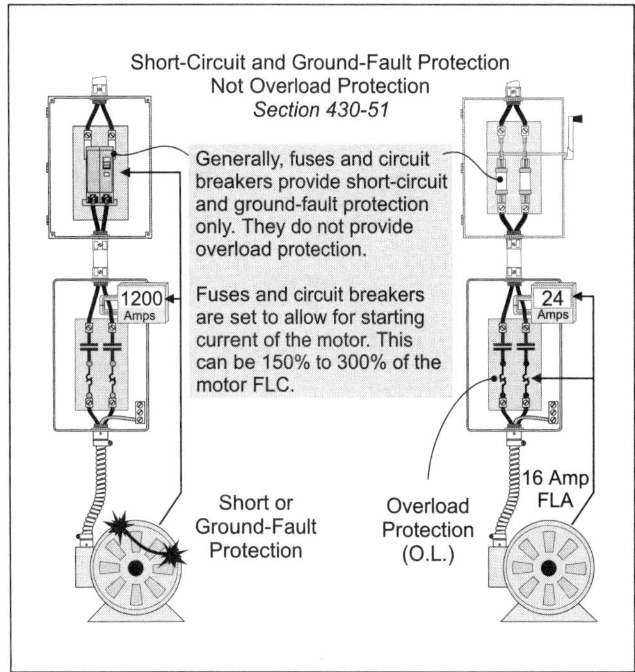

Figure 7–6
Short-Circuit and Ground-Fault Protection

7–3 OVERLOAD PROTECTION [*SECTION 430-32(a)*]

In addition to short-circuit and ground-fault protection, motors must be protected against overload. Generally, the motor overload device is part of the motor starter; however, a separate overload device like a dual-element fuse can be used. Motors rated more than 1 horsepower without integral thermal protection, and motors 1 horsepower or less (automatically started) [*Section 430-32(c)*], must have an overload device sized in response to the motor nameplate current rating [*Section 430-6(a)*]. The overload device must be sized no larger than the requirements of Section 430-32.

Service Factor

Motors with a nameplate service factor (SF) rating of 1.15 or more shall have the overload protection device sized no more than 125 percent of the motor nameplate current rating.

Service Factor

If a dual-element fuse is used for overload protection, what size fuse is required for a 5 horsepower 230-volt, single-phase motor, service factor 1.16, if the motor nameplate current rating is 28 ampere (Figure 7–7)?

(a) 25 ampere (b) 30 ampere (c) 35 ampere (d) 40 ampere

• Answer: (c) 35 ampere

Overload protection is sized to motor nameplate current rating [*Section 430-6(a), 430-32(a)(1),* and *430-55*].

28 ampere × 1.25 = 35 ampere [*Section 240-6(a)*]

Temperature Rise

Motors with a nameplate temperature rise rating not over 40ºC shall have the overload protection device sized no more than 125 percent of motor nameplate current rating.

Temperature Rise

If a dual-element fuse is used for the overload protection, what size fuse is required for a 50 horsepower 460-volt, 3-phase motor, temperature rise 39ºC, motor nameplate current rating of 60 ampere (FLA) (Figure 7–8)?

(a) 40 ampere (b) 50 ampere (c) 60 ampere (d) 70 ampere

• Answer: (d) 70 ampere

Overloads are sized according to the motor nameplate current rating, not the motor full-load current rating.

60 ampere × 1.25 = 75 ampere, 70 ampere [*Section 240-6(a)* and *430-32(a)(1)*]

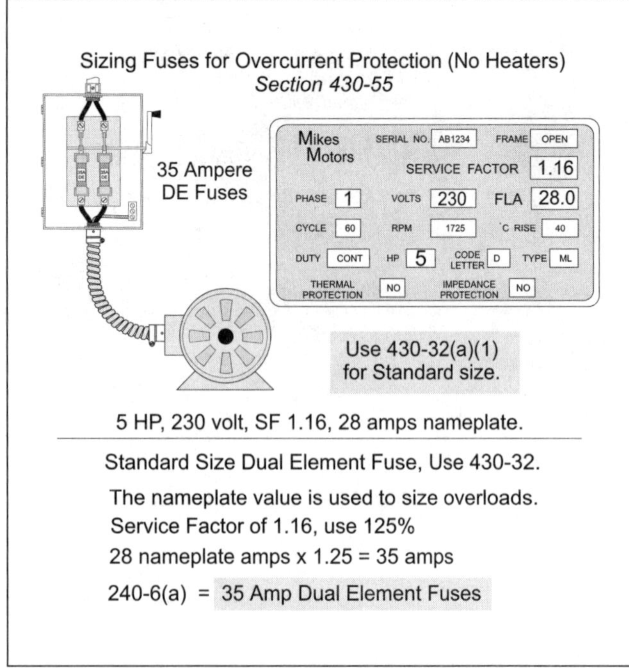

Figure 7–7
Sizing Fuses for Overcurrent Protection

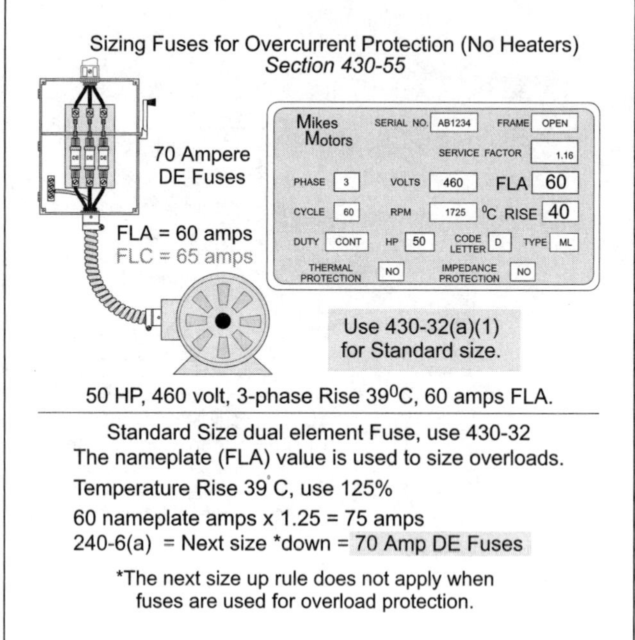

Figure 7–8
Sizing Fuses for Overcurrent Protection

All Other Motors

Motors that do not have a service factor rating of 1.15 and up, or a temperature rise rating of 40ºC and less, must have the overload protection device sized at not more than 115 percent of the motor nameplate ampere rating.

Number of Overloads [*Section 430-37*)

An overload protection device must be installed in each ungrounded conductor according to the requirements of Table 430-37. If a fuse is used for overload protection, a fuse must be installed in each ungrounded conductor [*Section 430-36*].

7–4 BRANCH CIRCUIT SHORT-CIRCUIT GROUND-FAULT PROTECTION [*SECTION 430-52(c)(1)*]

In addition to overload protection, each motor and its accessories requires short-circuit and ground-fault protection. *NEC*® Section 430-52(c) requires the motor branch circuit short-circuit and ground-fault protection (except torque motors) to be sized no greater than the percentages listed in Table 430-152. When the short-circuit ground-fault protection device value determined from Table 430-152 does not correspond with the standard rating or setting of overcurrent protection devices as listed in Section 240-6(a), the next higher protection device size may be used [*Section 430-52(c)(1)* Exception No. 1] (Figure 7–9).

To determine the percentage from Table 430-152 to be used to size the motor branch circuit, short-circuit, and ground-fault protection device, the following steps should be helpful:

Step 1: ➛ Locate the motor type on Table 430-152: Wound rotor, direct-current, or all other motors.

Step 2: ➛ Select the percentage from Table 430-152 according to the type of protection device such as nontime delay (one-time) fuse, dual element fuse, or inverse time circuit breaker.

***NEC*® Table 430–152**			
	Percent of Full-Load Current (FLC) Tables 430–147, 148 and 150		
Type of Motor	**One-Time Fuse**	**Dual-Element Fuse**	**Circuit Breaker**
Direct Current and Wound-Rotor Motors	150	150	150
All Other Motors	300	175	250

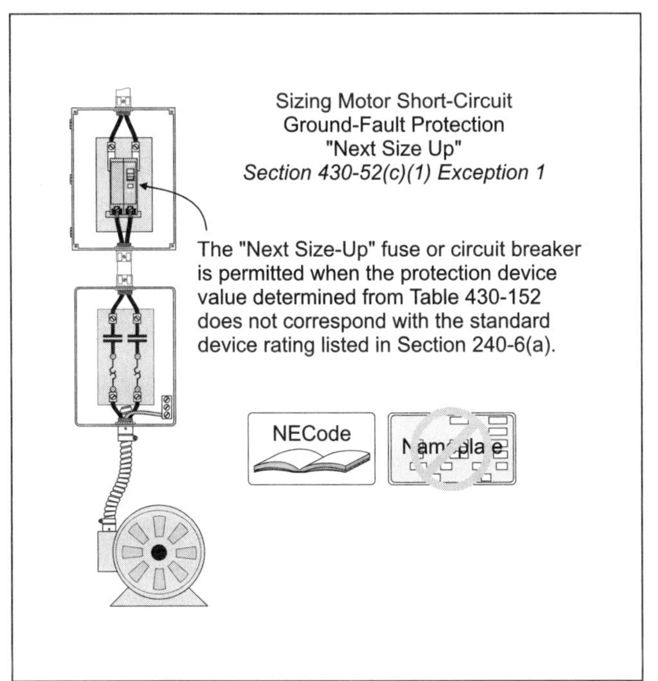

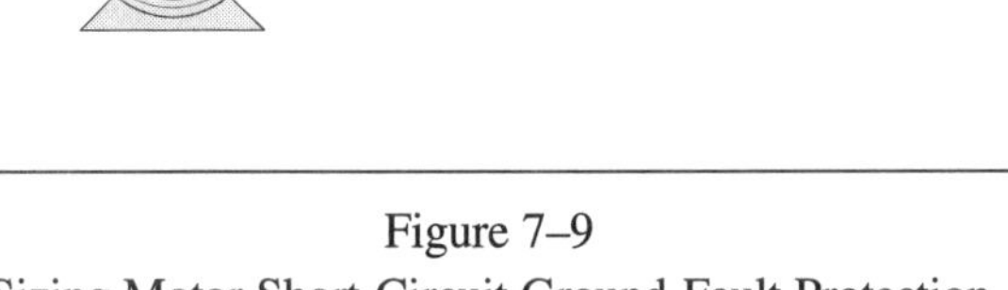

Figure 7–9
Sizing Motor Short-Circuit Ground-Fault Protection

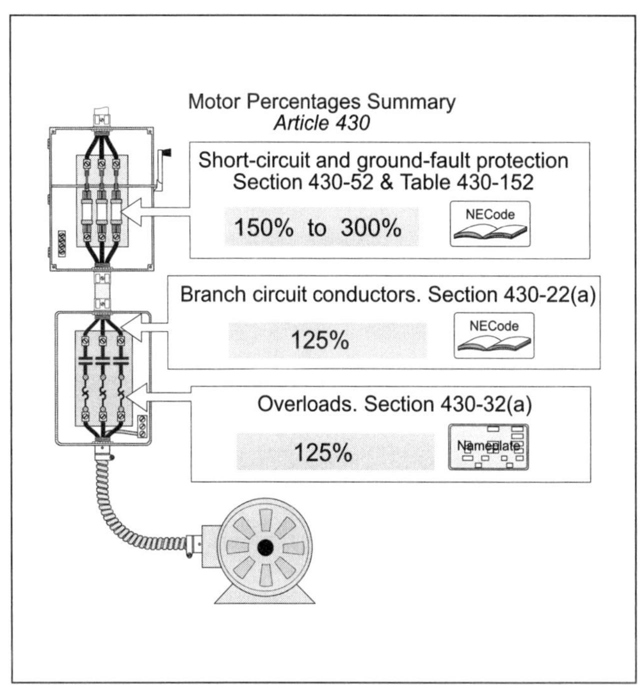

Figure 7–10
Motor Percentages Summary

Note: Where the protection device rating determined by Table 430-152 does not correspond with the standard size or rating of fuses or circuit breakers listed in Section 240-6(a), use of the next higher standard size or rating shall be permitted [*Section 430-52(c)(1)* Exception No. 1].

❑ **Branch Circuit**

Which of the following statements are true?

(a) The branch circuit short-circuit protection (nontime delay fuse) for a 3 horsepower 115-volt motor, single-phase shall not exceed 110 ampere.

(b) The branch circuit short-circuit protection (dual element fuse) for a 5 horsepower Design E, 230-volt motor, single-phase shall not exceed 50 ampere.

(c) The branch circuit short-circuit protection (inverse time breaker) for a 25 horsepower synchronous, 460-volt motor, 3-phase shall not exceed 70 ampere.

(d) all of these are true

• Answer: (d) all of these are true

Short-circuit and ground-fault protection, 430-53(c)(1) Exception No. 1 and Table 430-152:
Table 430-148 – 34 ampere × 3.00 = 102 ampere, next size up permitted, 110 ampere
Table 430-148 – 28 ampere × 1.75 = 49 ampere, next size up permitted, 50 ampere
Table 430-150 – 26 ampere × 2.50 = 65 ampere, next size up permitted, 70 ampere

WARNING: Conductors are sized at 125 percent of the motor full-load current [*Section 430-22(a)*], overloads are sized from 115 percent to 125 percent of the motor nameplate current rating [*Section 430-32(a)(1)*], and the short-circuit ground-fault protection device is sized from 175 percent to 300 percent of the motor full-load current [Table 430-152]. There is no relationship between the branch circuit conductor ampacity (125 percent) and the short-circuit ground-fault protection device (175 percent up to 300 percent) (Figure 7–10).

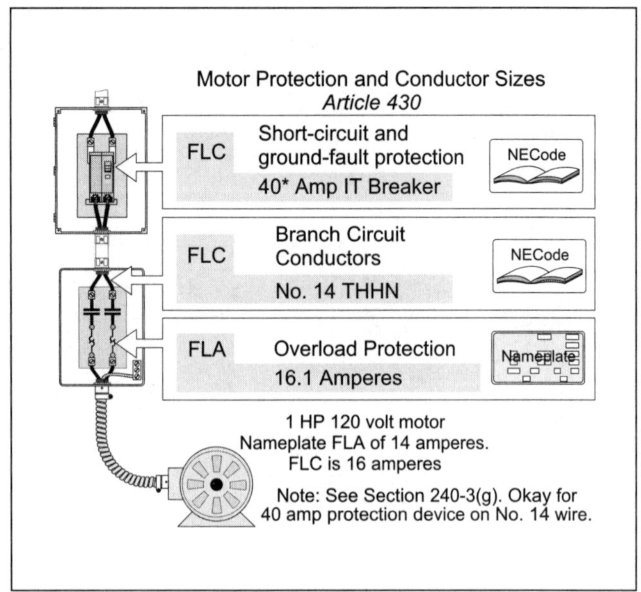

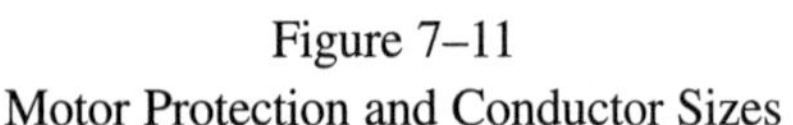

Figure 7–11
Motor Protection and Conductor Sizes

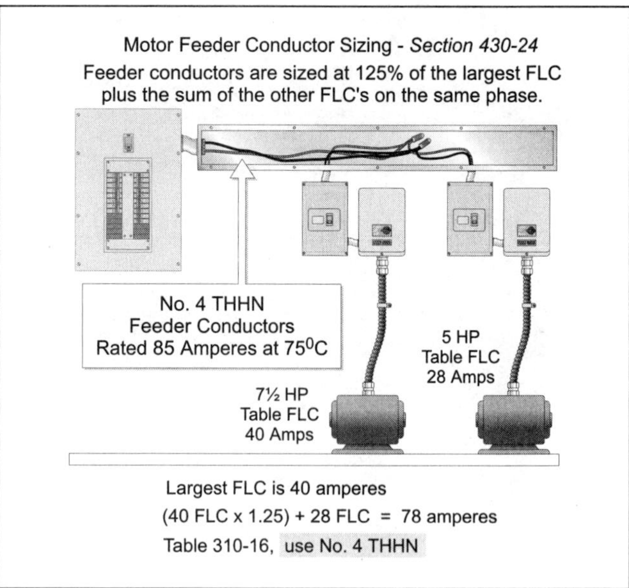

Figure 7–12
Motor Feeder Conductor Sizing

❑ Branch Circuit

Which of the following statements are true for a 1 horsepower 120-volt motor, nameplate current rating of 14 ampere (Figure 7–11)?

(a) The branch circuit conductors can be No. 14 THHN.

(b) Overload protection is from 16.1 ampere to 18.2 ampere.

(c) Short-circuit and ground-fault protection is permitted to be a 40-ampere circuit breaker.

(d) all of these are true

• Answer: (d) all of these are true

Conductor Size [*Section 430-22(a)*]

16 ampere × 1.25 = 20 ampere = No. 14 at 60°C, Table 310-16

Overload Protection Size

Standard [*Section 430-32(a)(1)*] – 14 ampere (nameplate) × 1.15 = 16.1 ampere

Short-Circuit and Ground-Fault Protection [*Section 430-52(c)(1), Tables 430-152* and *240-6*]

16 ampere × 2.50 = 40-ampere circuit breaker

This bothers many electrical people, but the No. 14 THHN conductors and motor are protected against overcurrent by the 16-ampere overload protection device and the 40-ampere short-circuit protection device.

7–5 FEEDER CONDUCTOR SIZE [*SECTION 430-24*]

Conductors that supply several motors must have an ampacity of not less than:

(1) 125 percent of the highest rated motor full-load current (FLC) [*Section 430-17*], plus

(2) The sum of the full-load currents of the other motors (on the same phase) [*Section 430-6(a)*].

❑ Feeder Conductor Size

What size feeder conductor (in ampere) is required for two motors. Motor 1 – 7$^1/_2$ horsepower single-phase, 230 volts (40 ampere) and motor 2–5 horsepower single-phase 230 volts (28 ampere)? Terminals rated for 75°C (Figure 7–12).

(a) 50 ampere (b) 60 ampere (c) 70 ampere (d) 80 ampere

• Answer: (d) 80 ampere, (40 ampere × 1.25) + 28 ampere = 78 ampere

Note: No. 4 conductor at 75°C, Table 310-16 is rated for 85 ampere.

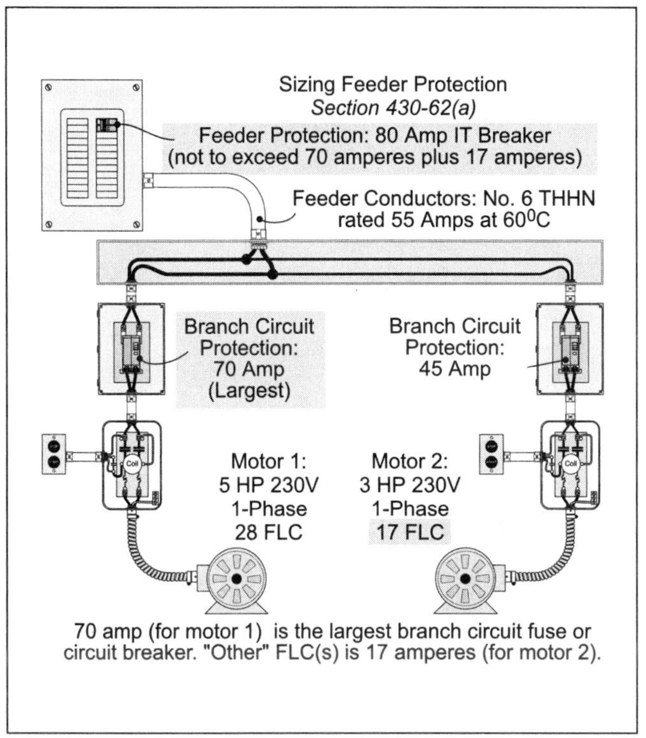

Figure 7–13
Sizing Feeder Protection

Highest Rated Motor (Largest Line)
for Feeder Conductor Size
Section 430-17 and 430-24

L3
L2
L1
N

The highest rated motor in a group is determined by the highest FLC, not the highest horsepower.

Highest Rated Motor

A: 10 HP 208 v 3-Ph. 30.8 FLC
B: 5 HP 208 v 1-Ph. 30.8 FLC
C: 3 HP 120v 1-Ph. 34 FLC

Size feeder using above motors.

Feeder Conductor Size - 430-24 and 430-17.

L1	L2	L3
30.8 -----	30.8 ----	30.8
30.8 -----	30.8	34.0

Highest rated motor = 120v motor at 34 FLC
Other motor(s) in group is the 10 HP 3-phase motor, see shaded area.

Line 3: (34 FLC x 1.25) + 30.8 = 73.3 amps

Table 310-16 = No. 3 wire at 60°C

Figure 7–14
Highest Rated Motor

7–6 FEEDER PROTECTION [*SECTION 430-62(a)*]

Motor feeder conductors must have protection against short-circuits and ground-faults but not overload. The protection device must be sized not greater than the largest branch circuit short-circuit ground-fault protection device [*Section 430-52(c)*] of any motor of the group, plus the sum of the full-load currents of the other motors (on the same time and phase).

❑ **Feeder Protection**

What size feeder protection (inverse time breaker) is required for a 5 horsepower 230-volts, single-phase motor and a 3 horsepower 230-volt, single-phase motor (Figure 7–13)?

(a) 30-ampere breaker (b) 40-ampere breaker (c) 50-ampere breaker (d) 80-ampere breaker

• Answer: (d) 80 ampere breaker

Motor Full-Load Current [Table 430-148]

5 horsepower motor FLC = 28 ampere [Table 430-148]

3 horsepower motor FLC = 17 ampere [Table 430-148]

Branch Circuit Protection [*Section 430-52(c)(1)*, *Table 430-152* and *240-6(a)*]

5 horsepower: 28 ampere × 2.5 = 70 ampere

3 horsepower: 17 ampere × 2.5 = 42.5 ampere: Next size up, 45 ampere

Feeder Conductor [*Section 430-24(a)*]

28 ampere × 1.25 + 17 ampere = 52 ampere, No. 6 rated 55 ampere at 60°C, Table 310-16

Feeder Protection [*Section 430-62*]

Not greater than 70-ampere protection, plus 17 ampere = 87 ampere: Next size down, 80 ampere

7–7 HIGHEST-RATED MOTOR [*SECTION 430-17*]

When selecting the feeder conductors or feeder short-circuit ground-fault protection device, the highest-rated motor shall be the highest-rated motor full-load current, not the highest-rated horsepower [*Section 430-17*].

❑ **Highest-Rated Motor**

Which is the highest-rated motor of the following (Figure 7–14)?

(a) 10 horsepower 3-phase, 208 volt (b) 5 horsepower single-phase, 208 volt

(c) 3 horsepower single-phase, 120 volt (d) any of these

• Answer: (c) 3 horsepower single-phase, 120-volt, full-load current of 34 ampere

10 horsepower = 30.8 ampere [Table 430-150]

5 horsepower = 30.8 ampere [Table 430-148]

3 horsepower = 34.0 ampere [Table 430-148]

7-8 *MOTOR CALCULATIONS STEPS*

Steps & *NEC* Rules	M1	M2	M3
Step 1: Motor FLC Tables 430–147, 148 & 150	_______ FLC	_______ FLC	_______ FLC
Step 2: Overload Protection Standard 430–32(a)(1)	____ × 1. ____ = ____	____ × 1. ____ = ____	____ × 1. ____ = ____
Step 3: Branch Circuits Conductor. 430-22(a), Table 310–16	_____ × 1.25 = _____	_____ × 1.25 = _____	_____ × 1.25 = _____
Step 4: Branch Circuit Protection Table 430–152, 430–52(c) 240–6(a)	_____ × _____ = _____ Next Size Up	_____ × _____ = _____ Next Size Up	_____ × _____ = _____ Next Size Up
Step 5: Feeder Conductor 430–24 and Table 310–16	_____ × 1.25 + _____ + _____ + _____ = _____ Table 310–16, Use No. _____ (60 or 75°C?)		
Step 6: Feeder Protection 430–62, Table 430–152, and 240–6(a)	_____ + _____ + _____ + _____ = _____ Next Size Down		

❑ Motor Calculation Steps Example

Given: One 10-horsepower, 208-volt, 3-phase motor and three 1 horsepower, 120-volt motors. Determine the standard and maximum overload sizes, branch circuit conductor (THHN) branch circuits short-circuit ground-fault protection device (Inverse Time Breaker), for all motors; then determine the feeder conductor and protection size (Fig. 7–15).

	M1	M2
Step 1: Motor FLC Tables 430–147, 148 or 150	Table 430–150 30.8 Full-Load Current	Table 430–148 16 Full-Load Current
Step 2: Overload Protection Nameplate 430–32(a)(1)	No Nameplate, Use FLC 30.8 ampere × 1.15 = 35.4 ampere	No Nameplate, Use FLC 16 ampere × 1.15 = 18.4 ampere
Step 3: Branch Circuit Conductors 125 Percent of FLC 110–14(c), 430–22(a), Table 310-16	30.8 ampere × 1.25 = 38.5 ampere Table 310–16 No. 8 THHN 60°C	16 ampere × 1.25 = 20 ampere Table 310–16 No. 14 THHN 60°C
Step 4: Branch Circuit Protection FLC × Table 430–152 percent 430–52(c), 240–6(a)	30.8 ampere × 2.5 = 77 ampere Next size up 240–6(a) = 80 ampere	16 ampere × 2.5 = 40 ampere 240–6(a) = 40 ampere
Step 5: Feeder Conductor FLC × 125 percent + FLC others 430–24, Table 310–16	(30.8 ampere × 1.25) + 16 ampere = 54.5 ampere Table 310–16, Use No. 6 THHN, rated 55 ampere 60°C	
Step 6: Feeder Protection Largest Branch protection + FLCs of other motors 430–62, Table 430–152, 240–6(a)	Inverse Time Breaker 80 ampere + 16 ampere = 96 ampere, next size down, 90 ampere	

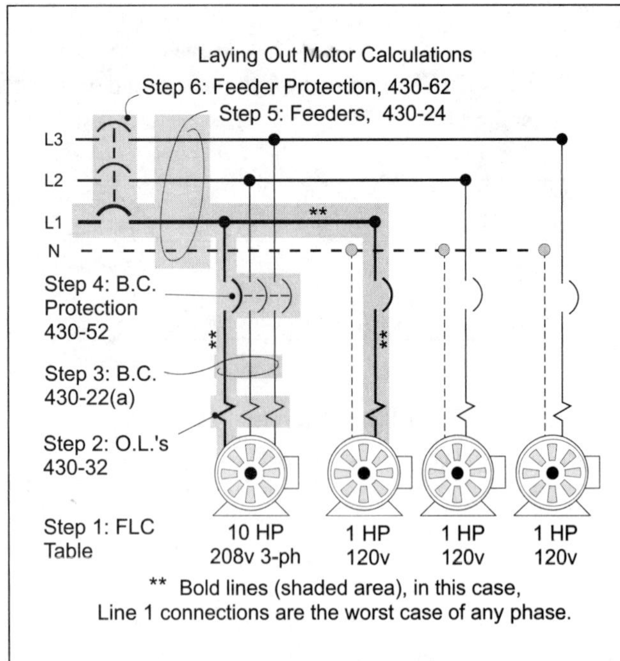

Figure 7–15
Laying Out Motor Calculations

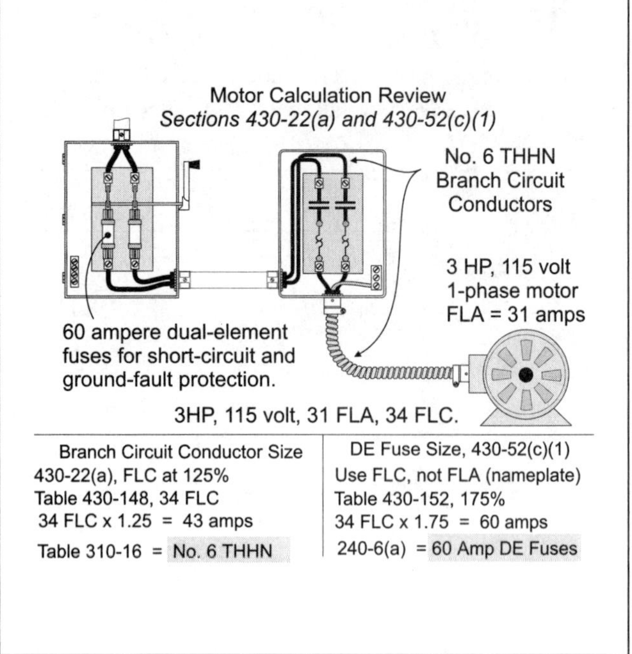

Figure 7–16
Motor Calculation Review

7–9 MOTOR CALCULATION REVIEW

❑ Branch Circuit

Size the branch circuit conductors (THHN) and short-circuit ground-fault protection device for a 3 horsepower 115-volt, single-phase motor. The motor nameplate full-load ampere is 31 ampere and dual-element fuses are to be used for short-circuit and ground-fault protection (Figure 7–16).

Branch Circuit Conductors [*Section 430-22(a)*]

Branch circuit conductors to a single motor must have an ampacity of not less than 125 percent of the motor full-load current as listed in Tables 430-147 through 430-150 [*Section 430-6(a)*].

34 ampere × 125 percent = 43 ampere

Table 310-16, 60°C Terminals: The conductor must be a No. 6 THHN rated 55 ampere [*Section 110-14(c)(1)*].

Branch Circuit Short-Circuit Protection [*Section 430-52(c)(1)*]

The branch circuit short-circuit and ground-fault protection device protects the motor, the motor control apparatus, and the conductors against overcurrent due to short-circuits or ground-faults, but not overload [*Section 430-51*]. The branch circuit short-circuit and ground-faults protection devices sized by considering the type of motor and the type of protection device according to the motor full-load current listed in Table 430-152. When the protection device values determined from Table 430-152 do not correspond with the standard rating of overcurrent protection devices as listed in Section 240-6(a), the next higher overcurrent protection device must be installed.

34 ampere × 175 percent = 60 ampere dual-element fuse [*Section 240-6(a)*]

See Example No. D8 in Appendix D of the *NEC*®.

❑ Feeder

Size the feeder conductor (THHN) and protection device (inverse time breakers 75°C terminal rating) for the following motors: Three 1 horsepower 115-volt single-phase motors; three 5 horsepower 208-volt, single-phase motors; and one 15 horsepower wound-rotor 208-volt, 3-phase motor (Figure 7–17).

Branch circuit short-circuit protection [*Section 240-6(a), 430-52(c)(1),* and *Table 430-152*]

15 horsepower: 46.2 ampere × 150 percent (wound-rotor) = 69 ampere: Next size up = 70 ampere

5 horsepower: 30.8 ampere × 250 percent = 77 ampere: Next size up = 80 ampere

1 horsepower: 16 ampere × 250 percent = 40 ampere

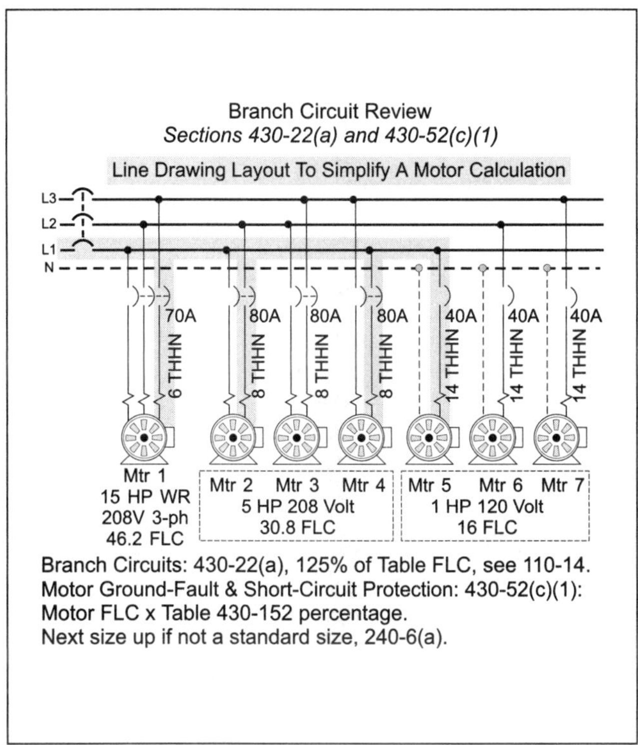

Figure 7–17
Branch Circuit Review

Feeder Conductor Review - *Section 430-24*

No. 1/0 THHN rated 150 amperes at 75⁰C.

L3 L2 L1 N

70A 80A 80A 80A 40A 40A 40A

6 THHN 8 THHN 8 THHN 8 THHN 14 THHN 14 THHN 14 THHN

Mtr 1 15 HP WR 208V 3-ph 46.2 FLC

Mtr 2 Mtr 3 Mtr 4 5 HP 208 Volt 30.8 FLC

Mtr 5 Mtr 6 Mtr 7 1 HP 120 Volt 16 FLC

	Line 1	Line 2	Line 3
Mtr 1, 15 HP	46.2	46.2	46.2
Mtr 2, 5 HP	30.8	30.8	
Mtr 3, 5 HP		30.8	30.8
Mtr 4, 5 HP	30.8		30.8
Mtr 5, 6, & 7, 1 HP	16.0	16.0	16.0
Largest Line	123.8	123.8	123.8

(46.2 x 1.25) + 30.8 + 30.8 + 16.0 = 136 amp for feeder
Table 310-16, No. 1/0 THHN rated 150 amps at 75⁰C

Figure 7–18
Feeder Conductor Review

Feeder Conductor [*Section 430-24*]

Conductors that supply several motors must have an ampacity of not less than 125 percent of the highest rated motor full-load current [*Section 430-17*], plus the sum of the other motor full-load currents [*Section 430-6(a)*].

(46.2 ampere $\times$ 1.25) + 30.8 ampere + 30.8 ampere + 16 ampere = 136 ampere

Table 310-16, No. 1/0 THHN, rated 150 ampere (Figure 7–18)

Note: When sizing the feeder conductor, be sure to only include the motors that are on the same phase. For that reason, only four motors are used for this feeder calculation.

Feeder Protection [*Section 430-62*]

Feeder conductors must be protected against short-circuits and ground-faults sized not to be greater than the maximum branch circuit short-circuit ground-fault protection device [*Section 430-52(c)(1)*] plus the sum of the full-load currents of the other motors (on the same phase).

80 ampere + 30.8 ampere + 46.2 ampere + 16 ampere = 174 ampere

Next size down, 150 ampere circuit breaker [*Section 240-6(a)*]. See Example No. D8 in Appendix D of the *NEC®* (Figure 7–19).

Note: When sizing the feeder protection, be sure to only include the motors that are on the same phase. For that reason, only four motors are used for the feeder calculation.

7–10 MOTOR VA CALCULATIONS

The input VA of a motor is determined by multiplying the motor volts by the motor ampere. To determine the *motor VA rating*, the following formulas can be used:

Motor VA (single-phase) = Volts $\times$ Motor Ampere

Motor VA (three-phase) = Volts $\times$ Motor Ampere $\times \sqrt{3}$

Note: Many people believe that a 230-volt motor consumes less power that a 115-volt motor, but both motors consume the same amount of power.

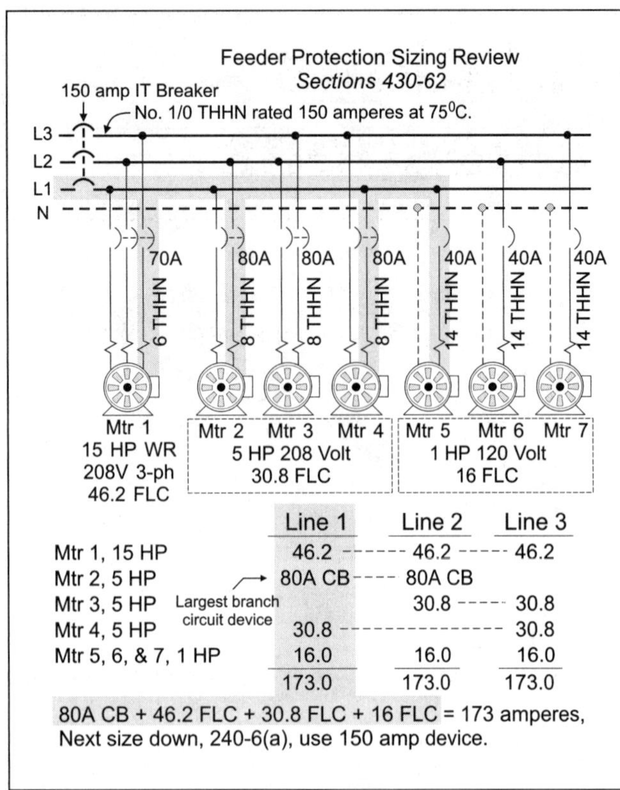

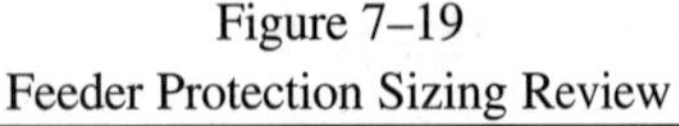
Figure 7–19
Feeder Protection Sizing Review

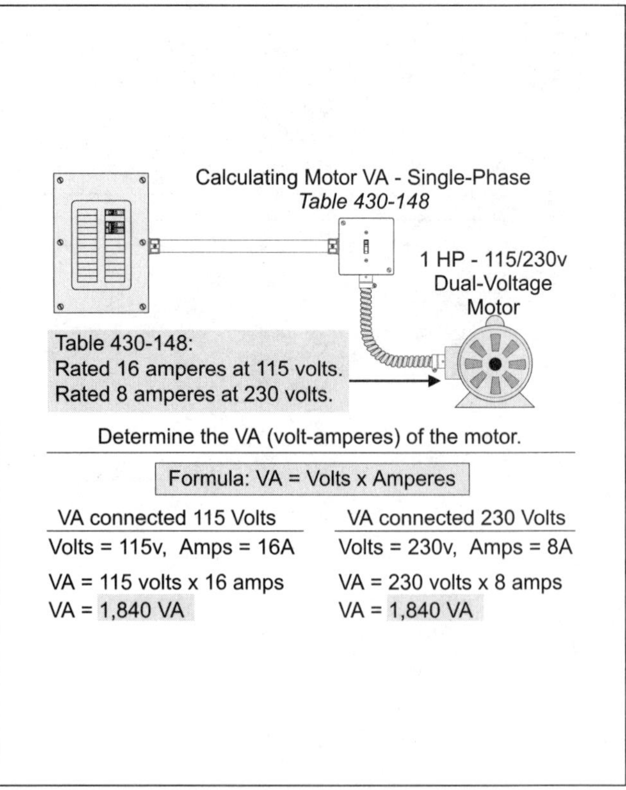

Figure 7–20
Calculating Motor VA – Single Phase

❑ Motor VA – Single-Phase

What is the motor input VA of a 1 horsepower motor rated 115/230 volts (Figure 7–20)?

(a) 1,840 VA at 115 volts (b) 1,840 VA at 230 volts (c) a and b (d) none of these

• Answer: (c) a and b, Table 430-148

Motor VA = Volts × Full-Load Current

VA at 230 volts = 230 volts × 8 ampere = 1,840 VA

VA at 115 volts = 115 volts × 16 ampere = 1,840 VA

❑ Motor VA – Three-Phase

What is the input VA of a 5 horsepower 230-volt, 3-phase motor (Figure 7–21)?

(a) 6,055 VA (b) 3,730 VA
(c) 6,440 VA (d) 8,050 VA

• Answer: (a) 6,055 VA, Table 430-150

Motor VA = Volts × Full-Load Current × $\sqrt{3}$

Table 430-150 Full-Load Current = 15.2 ampere

Motor VA = 230 volts × 15.2 ampere × 1.732

Motor VA = 6,055 VA

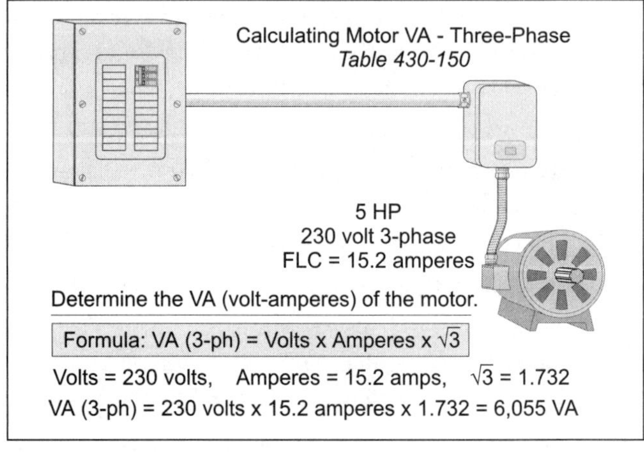

Figure 7–21
Calculating Motor VA – Three-Phase

Unit 7 – Motor Calculations Summary Questions

Introduction

1. When sizing conductors and overcurrent protection for motors, you must comply with the requirements of *NEC*® Article 430, not Article 240.
(a) True (b) False

7–1 Motor Branch Circuit Conductors [*Section 430-22(a)*]

2. What size THHN conductor is required for a 5 horsepower 230-volt, single-phase motor? Terminals are rated 75°C.
(a) No. 14 (b) No. 12 (c) No. 10 (d) No. 8

7–2 Motor Overcurrent Protection

3. Motors and their associated equipment must be protected against overcurrent (overload, short-circuit, or ground-fault), but because of the special characteristics of induction motors, overcurrent protection is generally accomplished by having the overload protection separate from the short-circuit and ground-fault protection.
(a) True (b) False

4. • Which parts of Article 430 contain the requirements for motor overcurrent protection?
(a) Overload protection – Part C (b) Short-circuit ground-fault protection — Part D
(c) a and b (d) none of these

5. • Overload is the condition where current is greater than the equipment ampacity rating resulting in equipment damage due to dangerous overheating [Article 100]. Overload protection devices, sometimes called heaters, are intended to protect the _____ from dangerous overheating.
(a) motor (b) motor-control equipment
(c) branch circuit conductors (d) all of these

6. The branch-circuit short-circuit and ground-fault protection device is intended to protect the motor, the motor control apparatus, and the conductors against overcurrent due to _____ .
(a) short-circuits (b) ground-faults (c) overloads (d) a and b

7–3 Overload Protection [*Section 430-32(a)*]

7. • The *NEC*® requires motor overload protection devices to be sized according to the motor full-load current rating as listed in Tables 430-147, 148, or 150.
(a) True (b) False

8. The standard size overload protection device must be sized according to the requirements of Section 430-32 of the *NEC*®. If the overload protection relay sized according to Section 430-32 is not capable of carrying the motor starting and running current, the next size up overload can be used if sized according to the requirements of Section 430-34.
(a) True (b) False

9. Motors with a nameplate service factor (SF) rating of 1.15 or more must have the overload protection device sized at no more than _____ percent of the motor nameplate current rating.
(a) 100 (b) 115 (c) 125 (d) 135

10. Motors with a nameplate temperature rise rating not over 40°C must have the overload protection device sized at no more than _____ percent of motor nameplate current rating.
(a) 100 (b) 115 (c) 125 (d) 135

11. Motors that have a service factor of 1.10 must have the overload protection device sized at not more than _____ percent of the motor nameplate ampere rating.
(a) 100 (b) 115 (c) 125 (d) 135

12. If a dual-element fuse is used for overload protection, what size fuse is required for a 5 horsepower 208-volt, 3-phase motor, service factor 1.16, motor nameplate current rating of 16 ampere (FLA)?
(a) 20 ampere (b) 25 ampere (c) 30 ampere (d) 35 ampere

13. If a dual-element fuse is used for the overload protection, what size fuse is required for a 30 horsepower 460-volt, 3-phase synchronous motor, temperature rise 39ºC?
(a) 20 ampere (b) 25 ampere (c) 30 ampere (d) 40 ampere

7–4 Branch Circuit Short-Circuit Ground-Fault Protection [*Section 430-52(c)(1)*]

14. In addition to overload protection, each motor and its accessories requires short-circuit and ground-fault protection according to the requirements of Section 430-52. When sizing the branch circuit protection device, we must consider which of the following factors?
(a) the motor type, such as induction, synchronous, wound-rotor, etc.
(b) the motor *Code* letter starting characteristics
(c) the type of protection device to be used, fuse or breaker
(d) all of these

15. The *NEC*® requires motor branch circuit short-circuit and ground-fault protection to be sized not greater than the percentages listed in Table 430-152. When the short-circuit ground-fault protection device value determined from Table 430-152 does not correspond with the standard rating of overcurrent protection devices as listed in Section 240-6(a), the next _____ device size must be used.
(a) smaller (b) larger (c) a or b (d) none of these

16. To determine the percentage of the motor FLC from Table 430-152 that is to be used to size the motor branch circuit short-circuit ground-fault protection device, which of the following steps should be used:
(a) Locate the motor type on Table 430-152, such as direct current, wound rotor, high-reactance, autotransformer start, or all other motors.
(b) Locate the motor starting conditions, such as *Code* letter or no *Code* letter.
(c) Select the percentage from Table 430-152 according to the type of protection device, such as one-time fuse, dual-element fuse, or circuit breaker.
(d) all of these

17. If the branch circuit short-circuit ground-fault protection dual-element fuse selected (sized not greater than the percentages listed in Table 430-152,) is not capable of carrying the load, the next larger size dual-element fuse can be used. The next size dual-element fuse cannot exceed _____ of the motor full-load current rating.
(a) 125% (b) 150% (c) 175% (d) 225%

18. Conductors are sized at _____ percent of the motor full-load currents [*Section 430-6* and *430-22*], overloads from _____ percent, and the motor short-circuit ground-fault protection device (inverse time circuit breaker) is sized up to _____. (There is no relationship between the branch circuit conductor ampacity and the short-circuit ground-fault protection device!)
(a) 125%, 115%, 250% (b) 100%, 125%, 150%
(c) 125%, 125%, 125% (d) 100%, 100%, 100%

19. Which of the following statements are true for a 10 horsepower 208-volt, 3-phase motor, nameplate current 29 ampere?
(a) The branch circuit conductors can be No. 8 THHN
(b) Overload protection is from 33 ampere to 38 ampere
(c) Short-circuit and ground-fault protection is an 80-ampere circuit breaker
(d) all of these

7–5 Feeder Conductor Size [*Section 430-24*]

20. Feeder conductors that supply several motors must have an ampacity of not less than:
(a) 125 percent of the highest rated motor FLC.
(b) the sum of the full-load currents of the other motors on the same phase [*Section 430-6(a)*].
(c) a or b
(d) a and b

21. Motor feeder conductors (sized according to 430-24) must have a feeder protection device to protect against short-circuits and ground-faults (not overloads), sized not greater than:
(a) the largest branch circuit short-circuit ground-fault protection device [*Section 430-52*] of any motor of the group.
(b) the sum of full-load currents of the other motors on the same time phase.
(c) a or b
(d) a and b

7–6 Feeder Protection [*Section 430-62(a)*]

22. • Which of the following statements about a 30 horsepower 460-volt, 3-phase synchronous motor and a 10 horsepower 460-volt, 3-phase motor are true?
(a) The 30 horsepower motor has No. 8 THHN with a 80-ampere breaker.
(b) The 10 horsepower motor has No. 14 THHN with a 35-ampere breaker.
(c) The feeder conductors must be No. 6 THHN with a 90-ampere breaker.
(d) all of these.

7–7 Highest-Rated Motor [*Section 430-17*]

23. When selecting the feeder conductors and short-circuit ground-fault protection device, the highest rated motor shall be the highest rated _____ .
(a) horsepower (b) full-load current (c) nameplate current (d) any of these

24. • Which is the highest rated motor of the following?
(a) 25 horsepower synchronous, 3-phase, 460 volt
(b) 20 horsepower 3-phase, 460 volt
(c) 15 horsepower 3-phase, 460 volt
(d) 3 horsepower 120 volt

7–10 Motor VA Calculations

25. What is the VA input of a dual voltage 5 horsepower 3-phase motor rated 460/230 volts?
(a) 3,027 VA at 460 volts
(b) 6,055 VA at 230 volts
(c) 6,055 VA at 460 volts
(d) b and c

26. What is the input VA of a 3 horsepower 208-volt, single-phase motor?
(a) 3,890 VA (b) 6,440 VA (c) 6,720 VA (d) none of these

☆ Challenge Questions

7–1 Motor Branch Circuit Conductors [*Section 430-22(a)*]

27. • The branch circuit conductors of a 5 horsepower 230-volt motor with a nameplate rating of 25 ampere shall have an ampacity of not less than _____ . Note: The motor is used for intermittent duty and cannot run for more than 5 minutes at any one time due to the nature of the apparatus it drives.
(a) 33 ampere (b) 37 ampere (c) 21 ampere (d) 23 ampere

7–3 Overload Protection [*Section 430-32(a)(1)*]

28. The standard overload protection device for a 2 horsepower 115-volt motor that has a full-load current rating of 24 ampere, and a nameplate rating of 21.5 ampere shall not exceed _____ .
(a) 20.6 ampere (b) 24.7 ampere (c) 29.9 ampere (d) 33.8 ampere

29. • The maximum overload protective device relay for a 2 horsepower 115-volt motor with a nameplate rating of 22 ampere is _____ . The Service factor is 1.2.
(a) 30.8 ampere (b) 33.8 ampere (c) 33.6 ampere (d) 22.6 ampere

Ultimate Trip Setting [*Section 430-32(a)(2)*]

30. The ultimate trip overload device of a thermally protected 1 horsepower 120-volt motor would be rated no more than _____ .
(a) 31.2 ampere (b) 26 ampere (c) 28 ampere (d) 23 ampere

7–4 Branch Circuit Short-Circuit Ground-Fault Protection [*Section 430-52* and *Table 430-152*]

31. A 2 horsepower 120-volt wound rotor motor requires a _____ branch circuit short-circuit protection device.
(a) 15 ampere (b) 20 ampere (c) 25 ampere (d) 40 ampere

32. • The branch circuit short-circuit protection device for a 10 horsepower 230-volt, single-phase wound rotor motor shall not exceed _____ . Note: Use an inverse-time breaker for protection.
(a) 125 ampere (b) 50 ampere (c) 75 ampere (d) 80 ampere

33. The branch circuit protection for a 125 horsepower 240-volt, direct current motor is _____ .
(a) 400 ampere (b) 600 ampere (c) 700 ampere (d) 800 ampere

7–5 Feeder Conductor Size [*Section 430-24*]

34. • The motor controller (below) requires a No. _____ THHN for the feeder, if the motor terminals are rated for 75°C.
(a) 2 (b) 3/0 (c) 4/0 (d) 250 kcmil

7–6 Feeder Protection [*Section 430-62(a)*]

35. • There are three motors: one 5 horsepower 230-volt motor with service factor of 1.2 and two 1 horsepower 120-volt motors. The three motors are fed with a 3-wire cable. Using an inverse time breaker, the feeder conductor protection device after balancing all three motors would be _____ .
(a) 60 ampere (b) 70 ampere (c) 80 ampere (d) 90 ampere

36. • If an inverse-time breaker is used for the feeder short-circuit protection, what size protection is required for the following 3-phase motors?
Motor 1 = 40 horsepower 52 FLC
Motor 2 = 20 horsepower 27 FLC
Motor 3 = 10 horsepower 14 FLC
Motor 4 = 5 horsepower 7.6 FLC
(a) 225 ampere (b) 200 ampere (c) 125 ampere (d) 175 ampere

37. • If dual-element fuses are used to protect a three-wire 115/230-volt feeder conductor for twenty-two $^1/_2$ horsepower single-phase, 115-volt motors, the fuse size selected should not be greater than _____ .
(a) 125 ampere (b) 90 ampere (c) 100 ampere (d) 110 ampere

38. • The feeder protection for a 25 horsepower 208-volt, 3-phase and three 3 horsepower 120-volt motors would be _____ , after balancing. Note: Use Inverse time breakers (ITB).
(a) 225 ampere (b) 200 ampere (c) 300 ampere (d) 250 ampere

Unit 8

Voltage Drop Calculations

OBJECTIVES

After reading this unit, the student should be able to briefly explain the following concepts:

- Alternating current resistance as compared to direct current
- Conductor resistance
- Conductor resistance – alternating current circuits
- Conductor resistance – direct current circuits
- Determining circuit voltage drop
- Extending circuits
- Limiting current to limit voltage drop
- Limiting conductor length to limit voltage drop
- Resistance – alternating current
- Sizing conductors to prevent excessive voltage drop
- Voltage drop considerations
- Voltage drop recommendations

After reading this unit, the student should be able to briefly explain the following terms:

- American wire gauge
- CM = circular mils
- Conductor
- Cross-sectional area
- D = distance
- Eddy currents
- E_{VD} = Conductor voltage drop expressed in volts
- I = load in ampere at 100 percent
- K = direct current constant
- Ohm's Law method
- Q = alternating current adjustment factor
- R = resistance
- Skin effect
- Temperature coefficient
- VD = volts dropped

PART A – CONDUCTOR RESISTANCE CALCULATIONS

8–1 CONDUCTOR RESISTANCE

Metals that carry electric current are called conductors or wires and oppose the flow of electrons. Conductors can be solid, stranded, copper, or aluminum. The conductor's opposition to the flow of current (resistance) is determined by the material type (copper/aluminum), the cross-sectional area (wire size), and the conductor's length and operating temperature. The *resistance* of a conductor is expressed in ohms.

Material

Silver is the best conductor because it has the lowest resistance, but its high cost limits its use to special applications. *Aluminum* is often used when weight or cost are important considerations, but *copp*er is the most common type of metal used for electrical conductors (Figure 8–1).

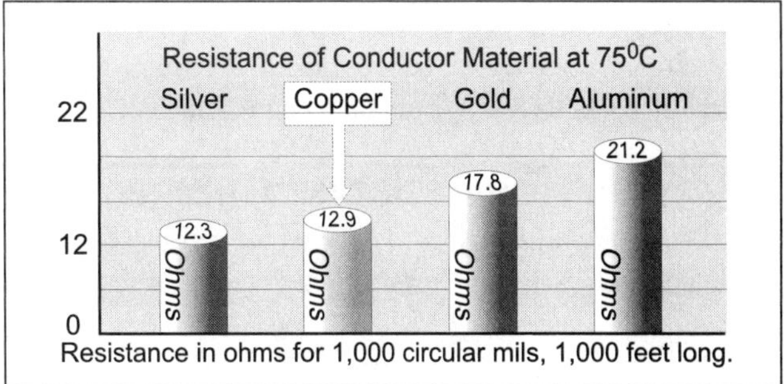

Figure 8–1
Resistance of Conductor Material at 75°C

Cross-Sectional Area

The *cross-sectional area* of a conductor is the conductor's surface area of a cross-section expressed in circular mils. The greater the conductor cross-sectional area, the greater the number of available electron paths and the lower the conductor resistance. Conductors are sized according to the *American Wire Gauge,* which ranges from a No. 40 to the largest, No. 4/0. Conductor resistance varies inversely with the conductor's diameter; that is, the smaller the wire size, the greater the resistance, and the larger the wire size, the lower the resistance (Figure 8–2).

Conductor Length

The resistance of a conductor is directly proportional to its length. The following table provides examples of conductor resistance and *circular mils area* for conductor lengths of 1,000 feet. Naturally, longer or shorter lengths will result in different conductor resistances.

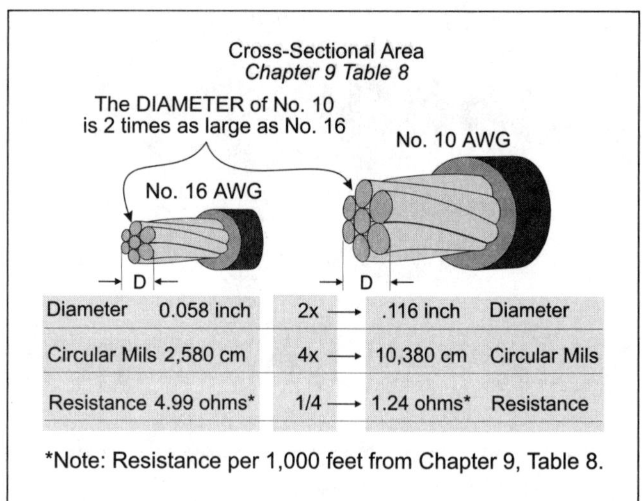

Figure 8–2
Cross-Sectional Area

Conductor Properties – *NEC*® Chapter 9, Table 8			
Conductor Size American Wire Gauge	**Conductor Resistance Per 1,000 Feet at 75°C**	**Conductor Diameter**	**Conductor Area Circular Mils**
No. 14	3.140 ohm (solid)	0.073	4,110
No. 12	1.980 ohm (stranded)	0.092	6,530
No. 10	1.240 ohm (stranded)	0.116	10,380
No. 8	0.778 ohm (stranded)	0.146	16,510
No. 6	0.491 ohm (stranded)	0.184	26,240

Temperature

The resistance of conductors changes with changing temperature; this is called *temperature coefficient.* Temperature coefficient describes the effect that temperature has on the resistance of a conductor. Positive temperature coefficient indicates that as the temperature rises, the conductor resistance will also rise. Examples of conductors that have a positive temperature coefficient are silver, copper, gold, and aluminum conductors. Negative temperate coefficient means that as the temperature increases, the conductor resistance decreases.

The conductor resistances listed in the *National Electrical Code*® Table 8 and 9 of Chapter 9 are based on an operating temperature of 75ºC. A 3-degree change in temperature results in a 1 percent change in conductor resistance for both copper and aluminum conductors. The formula to determine the change in conductor resistance with changing temperature is listed at the bottom of Table 8. For example, the resistance of copper at 90°C is about 5 percent more than at 75°C.

8–2 CONDUCTOR RESISTANCE – DIRECT CURRENT CIRCUITS, [Chapter 9, Table 8]

The *National Electric Code*® lists the resistance and area circular mils for both direct current and alternating current circuit conductors. Direct current circuit conductor resistances are listed in Chapter 9, Table 8, and alternating current circuit conductor resistances are listed in Chapter 9, Table 9.

The *direct current conductor resistances* listed in Chapter 9, Table 8 apply to conductor lengths of 1,000 feet. The following formula can be used to determine the conductor resistance for conductor lengths other than 1,000 feet:

$$\textbf{Direct Current Conductor Resistance} = \frac{\textbf{Conductor Resistance Ohms}}{\textbf{1,000 Feet}} \times \textbf{Conductor Length}$$

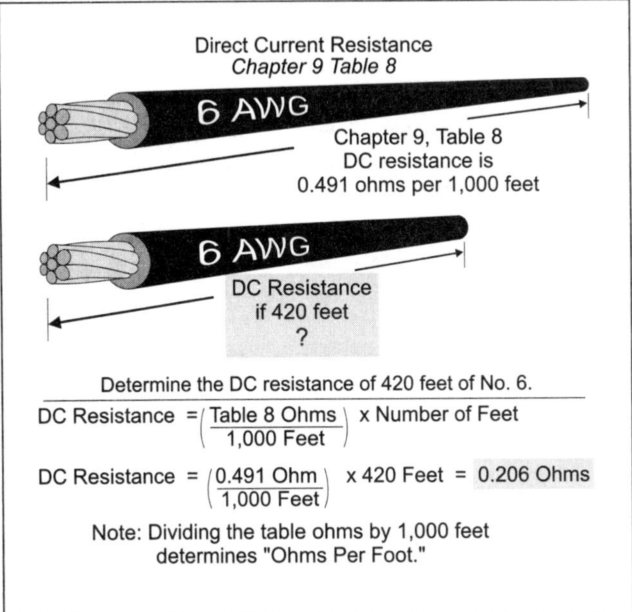

Figure 8–3
Direct Current Resistance

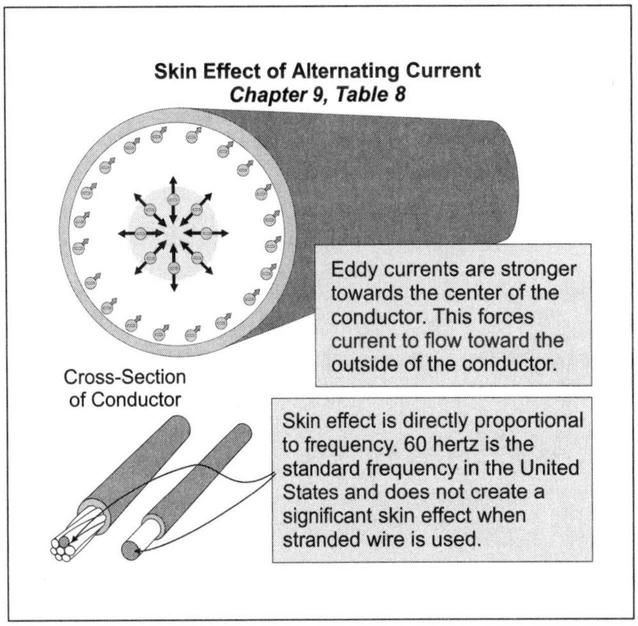

Skin Effect of Alternating Current
Chapter 9, Table 8

Cross-Section
of Conductor

Eddy currents are stronger towards the center of the conductor. This forces current to flow toward the outside of the conductor.

Skin effect is directly proportional to frequency. 60 hertz is the standard frequency in the United States and does not create a significant skin effect when stranded wire is used.

Figure 8–4
Skin Effect of Alternating Current

❑ Conductor Resistance Copper

What is the resistance of 420 feet of No. 6 copper (Figure 8–3)?

(a) 0.49 ohm (b) 0.29 ohm (c) 0.72 ohm (d) 0.21 ohm

• Answer: (d) 0.21 ohm

The resistance of No. 6 copper 1,000 feet long is 0.491 ohm, Chapter 9, Table 8.

The resistance of 420 feet is: (0.491 ohm/1,000 feet) × 420 feet = 0.206 ohm, rounded to 0.21

❑ Conductor Resistance Aluminum

What is the resistance of 1,490 feet of No. 3 aluminum?

(a) 0.60 ohm (b) 0.29 ohm (c) 0.72 ohm (d) 0.21 ohm

• Answer: (a) 0.60 ohm

The resistance of No. 3 aluminum 1,000 feet long is 0.403 ohm, Chapter 9, Table 8.

The resistance of 1,490 feet is: (0.403 ohm/1,000 feet) × 1,490 feet = 0.60 ohm

8–3 CONDUCTOR RESISTANCE – ALTERNATING CURRENT CIRCUITS

In direct current circuits, the only property that opposes the flow of electrons is resistance. In alternating current circuits, the expanding and collapsing magnetic field within the conductor induces an electromotive force that opposes the flow of alternating current. This opposition to the flow of alternating current is called *inductive reactance* which is measured in ohms.

In addition, alternating current flowing through a conductor generates small, erratic, independent currents called *eddy currents*. Eddy currents are greatest in the center of the conductors and repel the flowing electrons toward the conductor surface; this is known as *skin effect* (Figure 8–4).

Because of skin effect, the effective cross-sectional area of an alternating current conductor is reduced, which results in an increase of the conductor resistance. The total opposition to the flow of alternating current (resistance and inductive reactance) is called *impedance* and is measured in ohms.

8–4 ALTERNATING CURRENT RESISTANCE AS COMPARED TO DIRECT CURRENT

The opposition to current flow is greater for alternating current as compared to direct current circuits, because of inductive reactance, eddy currents, and skin effect. The following two tables give examples of the difference between alternating current circuits as compared to direct current circuits (Figure 8–5).

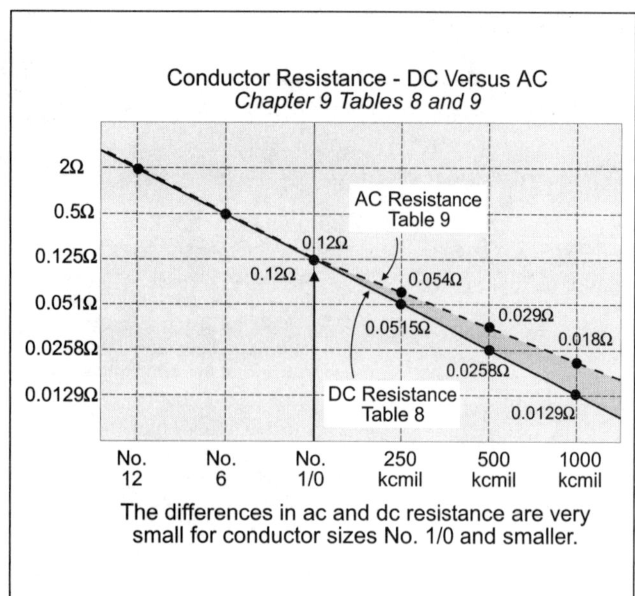

Figure 8–5
Conductor Resistance

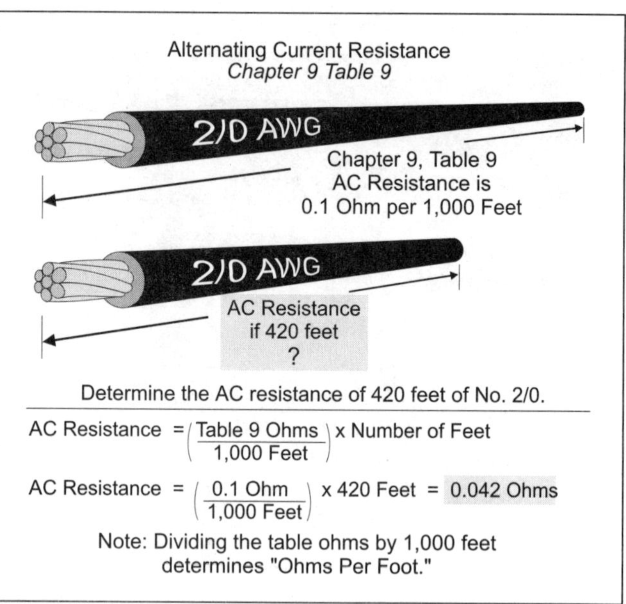

Figure 8–6
Alternating Current Resistance

COPPER – Alternating Current versus Direct Current Resistance at 75°C			
Conductor Size	**Alternating Current Chapter 9, Table 9**	**Direct Current Chapter 9, Table 8**	**AC resistance greater than DC resistance by %**
250,000	0.054 ohms per 1,000 feet	0.0515 ohms per 1,000 feet	4.85%
500,000	0.029 ohms per 1,000 feet	0.0258 ohms per 1,000 feet	12.40%
1,000,000	0.018 ohms per 1,000 feet	0.0129 ohms per 1,000 feet	39.50%
ALUMINUM – Alternating Current versus Direct Current Resistance at 75°C			
Conductor Size	**Alternating Current Chapter 9, Table 9**	**Direct Current Chapter 9, Table 8**	**AC resistance greater than DC resistance by %**
250,000	0.086 ohms per 1,000 feet	0.0847 ohms per 1,000 feet	1.5%
500,000	0.045 ohms per 1,000 feet	0.0424 ohms per 1,000 feet	6.13%
1,000,000	0.025 ohms per 1,000 feet	0.0212 ohms per 1,000 feet	17.12%

8–5 RESISTANCE ALTERNATING CURRENT [Chapter 9, Table 9 of the *NEC®*]

Alternating current conductor's resistances are listed in Chapter 9, Table 9 of the *NEC®*. The alternating current resistance of a conductor is dependent on the conductors material (copper or aluminum) and on the magnetic property of the raceway.

❑ Alternating Current Resistance

What is the alternating current resistance for a 250,000 circular mils conductor 1,000 feet long?

Copper conductor in nonmetallic raceway	= 0.052 ohms per 1,000 feet
Copper conductor in aluminum raceway	= 0.057 ohms per 1,000 feet
Copper conductor in steel raceway	= 0.054 ohms per 1,000 feet
Aluminum conductor in nonmetallic raceway	= 0.085 ohms per 1,000 feet
Aluminum conductors in aluminum raceway	= 0.090 ohms per 1,000 feet

Aluminum conductors in steel raceway = 0.086 ohms per 1,000 feet

Alternating Current Conductor Resistance Formula

The following formula can be used to determine the resistance of different lengths of conductors:

$$\textbf{Alternating Current Resistance} = \frac{\textbf{Conductor Resistance Ohms}}{\textbf{1,000 Feet}} \times \textbf{Conductor Length}$$

What is the alternating current resistance of 420 feet of No. 2/0 copper installed in a metal raceway (Figure 8–6)?

(a) 0.069 ohm (b) 0.042 ohm (c) 0.072 ohm (d) 0.021 ohm

• Answer: (b) 0.042 ohm

The resistance of No. 2/0 copper is 0.1 ohm per 1,000 feet, Chapter 9, Table 9.

The resistance of 420 feet of No. 2/0 is: (0.1 ohm/1,000 feet) × 420 feet = 0.042 ohm

❑ **Resistance Aluminum**

What is the alternating current resistance of 169 feet of 500 kcmil installed in aluminum conduit?

(a) 0.0049 ohm (b) 0.0029 ohm (c) 0.0054 ohm (d) 0.0021 ohm

• Answer: (c) 0.0054 ohm

The resistance of 500 kcmil installed in aluminum conduit is 0.032 ohm per 1,000 feet.

Resistance of 169 feet of 500 kcmil in aluminum conduit: (0.032 ohm/1,000 feet) × 169 feet = 0.0054 ohm

Converting Copper to Aluminum or Aluminum to Copper

When requested to determine the replacement conductor for copper or aluminum the following steps should be helpful:

Step 1: ➸ Determine the resistance of the existing conductor using Table 9, Chapter 9 for 1,000 feet.

Step 2: ➸ Using Table 9, Chapter 9, locate a replacement conductor that has a resistance of not more than the existing conductors.

Step 3: ➸ Verify that the replacement conductor has an ampacity [Table 310-16] sufficient for the load.

❑ **Aluminum to Copper**

A 100-ampere, 240-volt, single-phase load is wired with No. 2/0 aluminum conductors in a steel raceway. What size copper wire can we use to replace the aluminum wires and not have a greater voltage drop?

Note: The wire selected must have an ampacity of at least 100 ampere (Figure 8–7).

(a) No. 1/0 (b) No. 1 (c) No. 2 (d) No. 3

• Answer: (b) No. 1 copper

The resistance of No. 2/0 aluminum (steel raceway) is 0.16 ohms per 1,000 feet.

The resistance of No. 1 copper (steel raceway) is 0.16 ohms per 1,000 feet (ampacity of 130 ampere at 75°C).

Determining the Resistance of Parallel Conductors

The resistance total in a parallel circuit is always less than the smallest resistor. The equal resistors formula can be used to determine the resistance total of parallel conductors:

$$\textbf{Resistance Total} = \frac{\textbf{Resistance of One Conductor*}}{\textbf{Number of Parallel Conductors}}$$

*Resistance according to Chapter 9, Table 8 or Table 9 of the *NEC*®, assuming 1,000 feet unless specified otherwise.

❑ **Resistance of Parallel Conductors**

What is the direct current resistance for two 500 kcmil conductors in parallel (Figure 8–8)?

(a) 0.0129 ohms (b) 0.0258 ohms (c) 0.0518 ohms (d) 0.0347 ohms

• Answer: (a) 0.0129 ohm

$$\text{Resistance Total} = \frac{\text{Resistance of One Conductor}}{\text{Number of Parallel Conductors}} = \frac{0.0258 \text{ ohm}}{2 \text{ conductors}} = 0.0129 \text{ ohm}$$

Note: The resistance of 1,000 kcmil in Chapter 9, Table 8 is 0.0129 ohms.

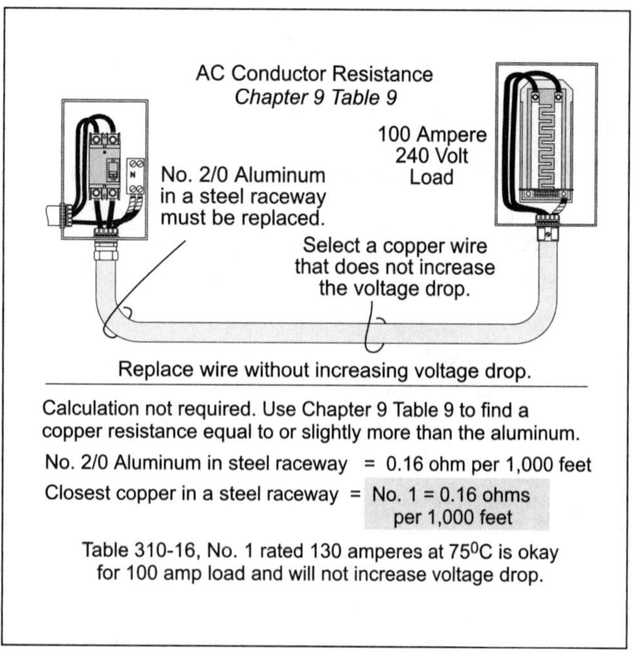

Figure 8–7
AC Conductor Resistance

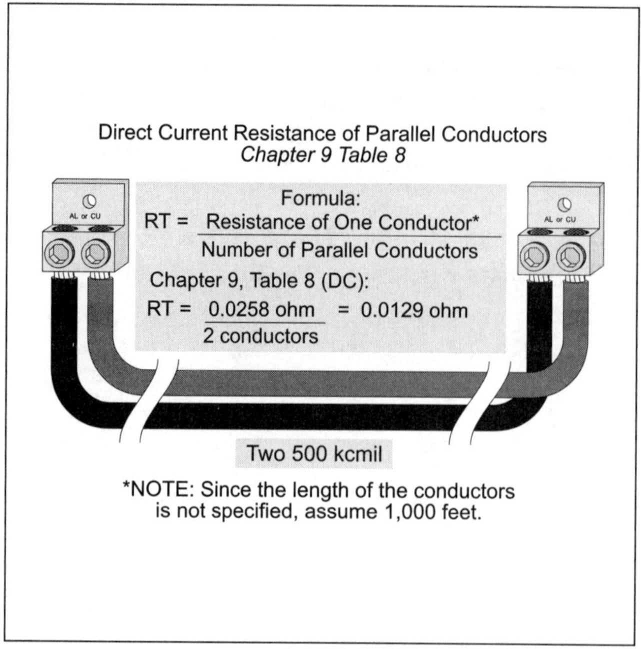

Figure 8–8
Direct Current Resistance of Parallel Conductors

PART B – VOLTAGE DROP CALCULATIONS

8–6 VOLTAGE DROP CONSIDERATIONS

The voltage drop of a circuit is in direct proportion to the conductors resistance and the magnitude (size) of the current. The longer the conductor, the greater the conductor resistance, the greater the conductor voltage drop; or, the greater the current, the greater the conductor voltage drop.

Note: Undervoltage for inductive loads can cause overheating, inefficiency, and a shorter life span for electrical equipment. This is especially true in such solid-state equipment as TVs, data processing equipment (e.g., computers), and similar equipment. When a conductor resistance causes the voltage to drop below an acceptable point, the conductor size should be increased (Figure 8–9).

8–7 *NEC®* VOLTAGE DROP RECOMMENDATIONS

Contrary to many beliefs, the *NEC®* does not contain any requirements for sizing ungrounded conductors for voltage drop. It does recommend in many areas of the *NEC®* that we consider the effects of conductor voltage drop when sizing conductors.

See some of these recommendations in the FPN to sections 210-19(a), 215-2, 230-31(c), and 310-15(a)(1). Please be aware that FPN in the *NEC®* are recommendations, *not* requirements [*Section 90-5(c)*]. The *NEC®* recommends that the maximum combined voltage drop for both the feeder and branch circuit should not exceed 5 percent, and the maximum on the feeder or branch circuit should not exceed 3 percent (Figure 8–10).

❑ *NEC®* Voltage Drop Recommendation

What are the minimum *NEC®* recommended operating volts for a 115-volt rated load that is connected to a 120/240 volt source (Figure 8–11)?

(a) 120 volts (b) 115 volts (c) 114 volts (d) 116 volts

• Answer: (c) 114 volts

The maximum conductor voltage drop recommended for both the feeder and branch circuit is 5 percent of the voltage source (120 volts). The total conductor voltage drop (feeder and branch circuit) should not exceed 120 volts × 0.05 = 6 volts. The operating voltage at the load is calculated by subtracting the conductors voltage drop from the voltage source:
120 volts – 6 volts dropped = 114 volts.

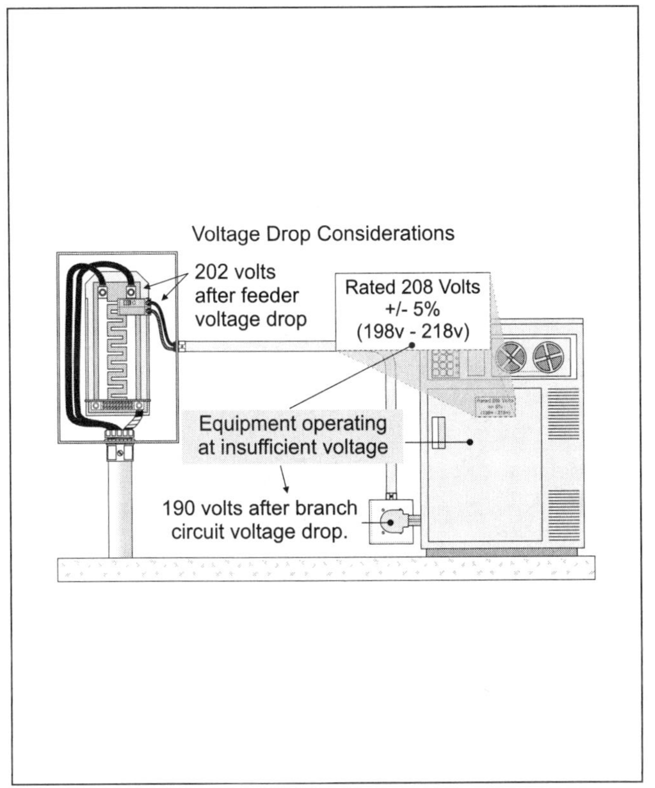

Figure 8–9
Voltage Drop Considerations

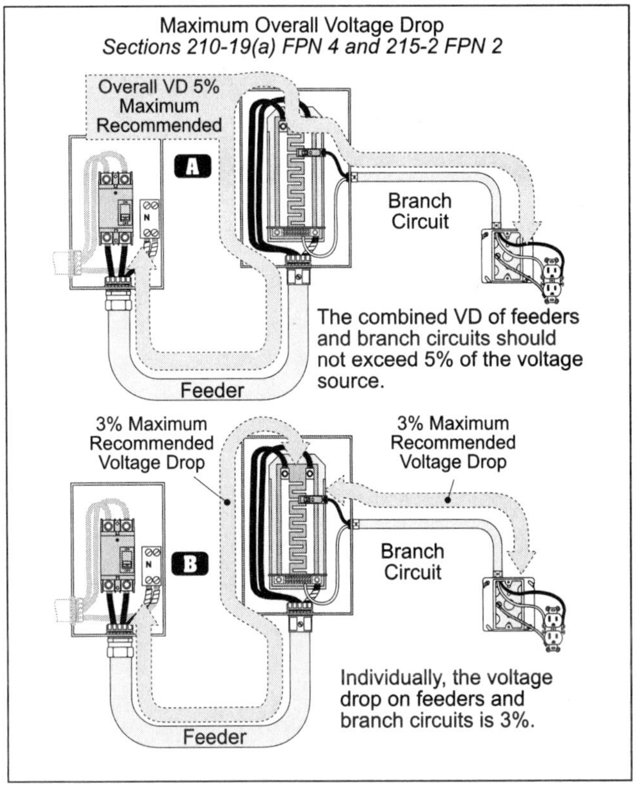

Figure 8–10
Maximum Overall Voltage Drop

8–8 DETERMINING CIRCUIT CONDUCTORS VOLTAGE DROP

When the circuit conductors have already been installed, the voltage drop of the conductors can be determined by the Ohm's Law method or by the formula method.

Ohm's Law Method – Single-Phase Only

VD = I × R

VD = Conductor voltage drop expressed in volts.

I = The load in ampere at 100 percent, not at 125 percent, for motors or continuous loads.

R* = Conductor Resistance, Chapter 9, Table 8 direct current; or Chapter 9, Table 9 for alternating current.

*For conductors No. 1/0 and smaller, the difference in resistance between direct current and alternating current circuits is so little that it can be ignored. In addition, you can ignore the small difference in resistance between stranded and solid wires.

❑ **Voltage Drop 120 Volt**

What is the voltage drop of two No. 12 THHN conductors that supply a 16-ampere, 120-volt load located 100 feet from the power supply (Figure 8–12)?

(a) 3.2 volts (b) 6.4 volts (c) 9.6 volts (d) 12.8 volts

• Answer: (b) 6.4 volts

VD = I × R

I = 16 ampere

R = 2 ohm per 1,000 feet, Chapter 9, Table 9: (2 ohm/1,000 feet) × 200 feet = 0.4 ohm

VD = 16 ampere × 0.4 ohm = 6.4 volts

❑ **Voltage Drop 240 Volt**

A single-phase, 24-ampere, 240-volt load is located 160 feet from the panelboard and is wired with No. 10 THHN. What is the voltage drop of the circuit conductors (Figure 8–13)?

(a) 4.53 volts (b) 9.22 volts (c) 3.64 volts (d) 5.54 volts

• Answer: (b) 9.22 volts

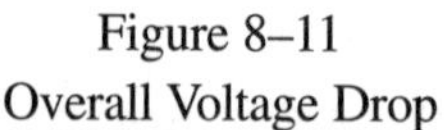

Figure 8–11
Overall Voltage Drop

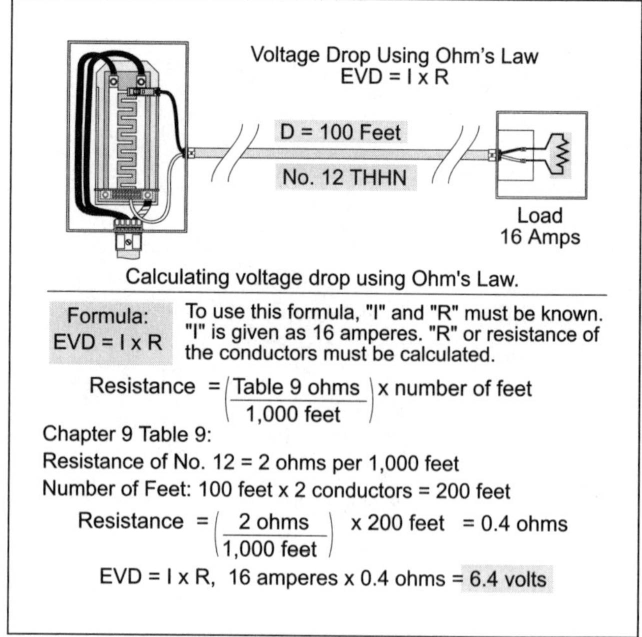

Figure 8–12
Voltage Drop Using Ohm's Law

VD = I × R

I = 24 ampere

R = 1.2 ohm per 1,000 feet, Chapter 9, Table 9: (1.2 ohm/1,000 feet) × 320 feet = 0.384 ohm

VD = 24 ampere × 0.384 ohm = 9.216 volts dropped

Voltage Drop Using the Formula Method

In addition to the Ohm's Law method, the following formula can be used to determine the conductor voltage drop:

$$\textbf{VD (single-phase)} = \frac{2 \times (K \times Q) \times I \times D}{CM} \qquad \textbf{VD (three-phase)} = \frac{\sqrt{3} \times (K \times Q) \times I \times D}{CM}$$

Note: $\sqrt{3} = 1.732$

VD = Volts Dropped: The voltage drop of the circuit expressed in volts. The *NEC®* recommends a maximum 3 percent voltage drop for either the branch circuit or feeder.

K = Direct Current Constant: This constant K represents the direct current resistance for a 1,000 circular mils conductor that is 1,000 feet long, at an operating temperature of 75°C. The constant K value is 12.9 ohm for copper, and 21.2 ohm for aluminum.

I = Ampere: The load in ampere at 100 percent, (not at 125 percent for motors or continuous loads).

Q = Alternating Current Adjustment Factor: For alternating current circuits with conductors No. 2/0 and larger, the direct current resistance constant K must be adjusted for the effects of self-induction (eddy currents). The Q-Adjustment Factor is calculated by dividing the alternating current resistance listed in Chapter 9, Table 9 by the direct current resistance listed in Chapter 9, Table 8 in the *NEC®*. For all practical exam purposes, this resistance adjustment factor can be ignored because exams rarely give alternating current voltage drop questions with conductors larger than No. 1/0.

D = Distance: The distance the load is from the power supply.

CM = Circular-Mils: The circular mils of the circuit conductor as listed in *NEC®* Chapter 9, Table 8.

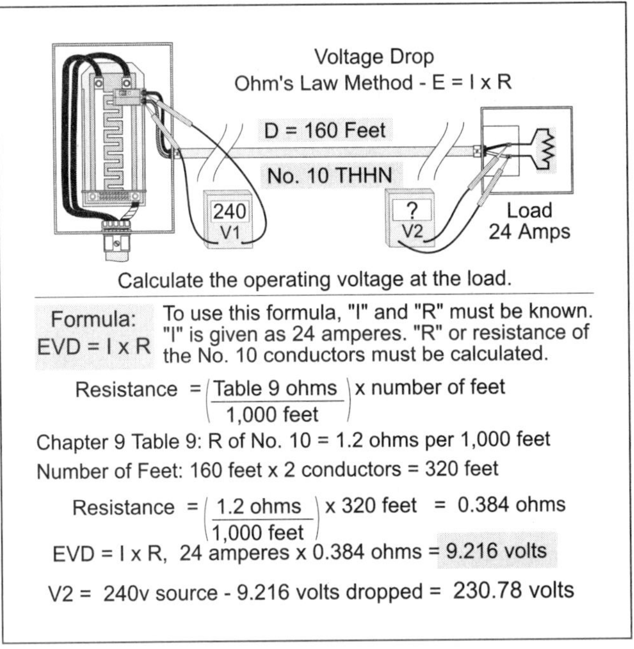

Figure 8–13
Voltage Drop Ohm's Law Method

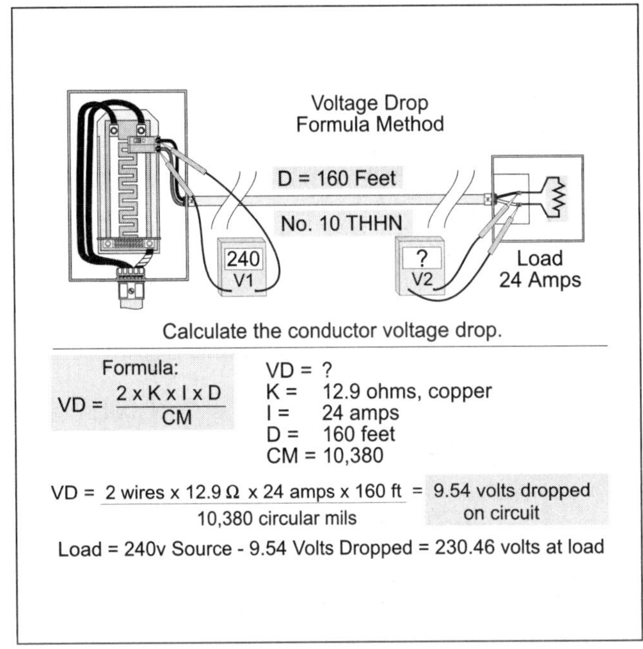

Figure 8–14
Voltage Drop Formula Method

❑ Voltage Drop – Single-Phase

A 24-ampere, 240-volt load is located 160 feet from a panelboard and is wired with No. 10 THHN. What is the approximate voltage drop of the branch circuit conductors (Figure 8–14)?

(a) 4.25 volts (b) 9.5 volts (c) 3 percent (d) 5 percent

- Answer: (b) 9.5 volts is the closest

$$VD = \frac{2 \times K \times I \times D}{CM}$$

K = 12.9 ohm, copper

I = 24 ampere

D = 160 feet

CM = 10,380 Chapter 9, Table 8 (No. 10)

$$VD = \frac{2 \text{ wires} \times 12.9 \text{ ohms} \times 24 \text{ amps} \times 160 \text{ feet}}{10{,}380 \text{ circular mils}} = 9.54 \text{ volts dropped}$$

❑ Voltage Drop – Three-Phase

A 3-phase, 36 kVA load rated 208 volts is located 80 feet from the panelboard and is wired with No. 1 THHN aluminum. What is the approximate voltage drop of the feeder circuit conductors (Figure 8–15)?

(a) 3.5 volts (b) 7 volts (c) 3 percent (d) 5 percent

- Answer: (a) 3.5 volts is the closest

$$VD = \frac{\sqrt{3} \times K \times I \times D}{CM}$$

$\sqrt{3} = 1.732$

K = 21.2 ohm, aluminum

$$I = \frac{VA}{(E \times \sqrt{3})} = \frac{36{,}000 \text{ VA}}{(208 \text{ volts} \times 1.732)} = 100 \text{ ampere}$$

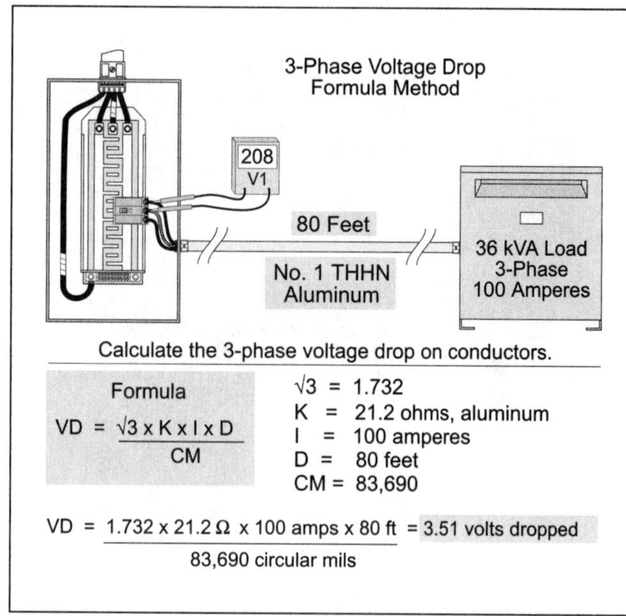

Figure 8–15
3-Phase Voltage Drop Formula Method

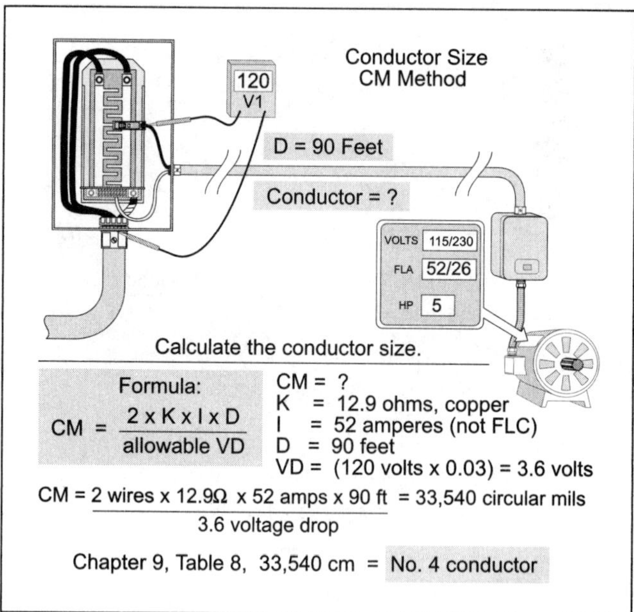

Figure 8–16
Conductor Size CM Method

D = 80 feet

CM = 83,690 Chapter 9, Table 8

$$VD = \frac{1.732 \times 21.2 \text{ ohms} \times 100 \text{ amps} \times 80 \text{ feet}}{83{,}690 \text{ circular mils}} = 3.51 \text{ volts dropped}$$

8–9 SIZING CONDUCTORS TO PREVENT EXCESSIVE VOLTAGE DROP

The size of a conductor (actually its resistance) affects voltage drop. If we want to decrease the voltage drop of a circuit, we can increase the size of the conductor (reduce its resistance). When sizing conductors to prevent excessive voltage drop, use the following formulas:

$$\textbf{CM (single-phase)} = \frac{\mathbf{2 \times K \times I \times D}}{\textbf{Volts Dropped}} \qquad \textbf{CM (three-phase)} = \frac{\mathbf{\sqrt{3} \times K \times I \times D}}{\textbf{Volts Dropped}}$$

❑ Size Conductor – Single-Phase

A 5 horsepower motor is located 90 feet from a 120/240-volt panelboard. What size conductor should be used if the motor nameplate indicates 52 ampere at 115 volts? Terminals rated for 75°C (Figure 8–16).

(a) No. 10 THHN (b) No. 8 THHN (c) No. 6 THHN (d) No. 4 THHN

• Answer: (d) No. 4 THHN

$$CM = \frac{2 \times K \times I \times D}{VD}$$

K = 12.9 ohm, copper

I = 52 ampere at 115 volts

D = 90 feet

VD = 120 volts × 0.03 = 3.6 volts

$$CM = \frac{2 \text{ wires} \times 12.9 \text{ ohms} \times 52 \text{ ampere} \times 90 \text{ feet}}{3.6 \text{ volts}} = 33{,}540 \text{ (No. 4, Chapter 9, Table 8)}$$

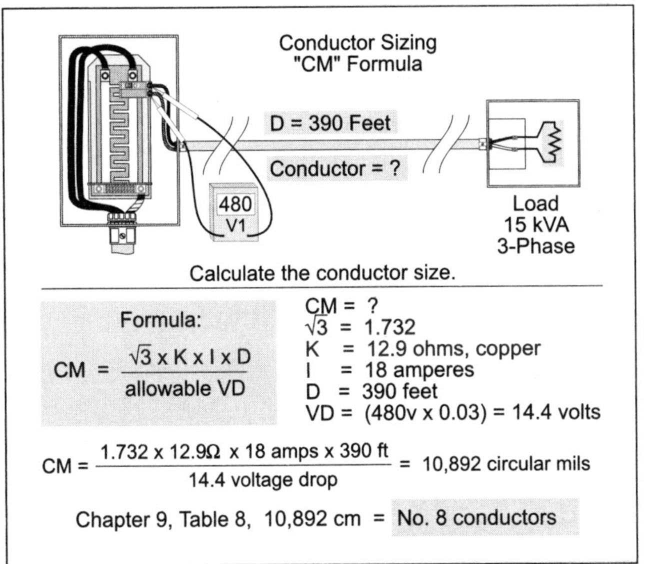

Figure 8–17
Conductor Sizing "CM" Formula

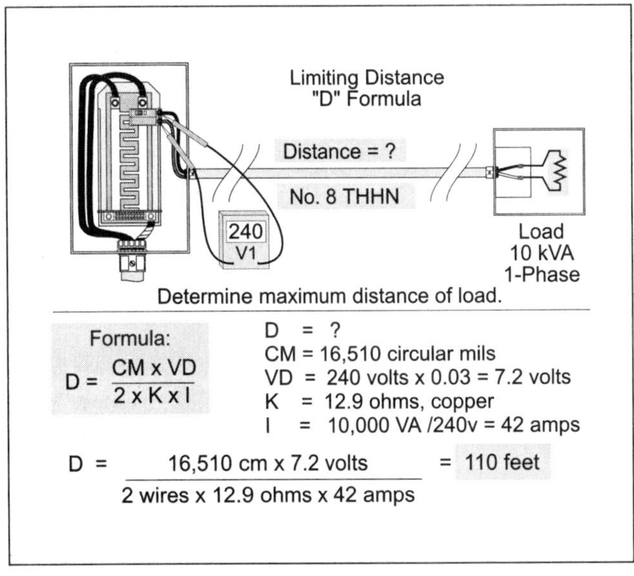

Figure 8–18
Limiting Distance "D" Formula

Note: *NEC*® Section 430-22(a), requires that the motor conductors be sized not less than 125 percent of the motor full-load currents as listed in Table 430-148. The motor full-load current is 56 ampere, and the conductor must be sized at 56 ampere × 1.25 = 70 ampere. The No. 4 THHN required for voltage drop is rated for 75 ampere at 75ºC according to Table 310-16 and Section 110-14(c).

❑ **Size Conductor – Three-Phase**

A 3-phase, 15 kVA load rated 480 volts is located 390 feet from the panelboard. What size conductor is required to prevent the voltage drop from exceeding 3 percent (Figure 8–17)?

(a) No. 10 THHN (b) No. 8 THHN (c) No. 6 THHN (d) No. 4 THHN

- Answer: (b) No. 8 THHN

$$CM = \frac{\sqrt{3} \times K \times I \times D}{VD}$$

K = 12.9 ohm, copper

$$I = 18 \text{ ampere } \frac{VA}{(E \times 1.732)} = \frac{15{,}000 \text{ VA}}{(480 \text{ volts} \times 1.732)} = 18 \text{ ampere}$$

D = 390 feet

VD = 480 volts × 0.03 = 14.4 volts

$$CM = \frac{1.732 \times 12.9 \text{ ohms} \times 18 \text{ ampere} \times 390 \text{ feet}}{14.4 \text{ volts}} = 10{,}892 \text{ (No. 8, Chapter 9, Table 8)}$$

8–10 LIMITING CONDUCTOR LENGTH TO LIMIT VOLTAGE DROP

Voltage drop can also be reduced by limiting the length of the conductors. The following formulas can be used to help determine the maximum conductor length to limit the voltage drop to *NEC*® suggestions:

$$\textbf{D (single-phase)} = \frac{CM \times VD}{2 \times K \times I \sqrt{3}} \qquad \textbf{D (three-phase)} = \frac{CM \times VD}{\sqrt{3} \times K \times I}$$

❑ **Distance – Single-Phase**

What is the maximum distance a single-phase, 10 kVA, 240-volt load can be located from the panelboard so the voltage drop does not exceed 3 percent? The load is wired with No. 8 THHN (Figure 8–18).

(a) 55 feet (b) 110 feet (c) 165 feet (d) 220 feet

- Answer: (b) 110 feet

$$D = \frac{CM \times VD}{2 \times K \times I}$$

CM = 16,510 (No. 8), Chapter 9, Table 8
VD = 240 volts × 0.03 = 7.2 volts
K = 12.9 ohm, copper
I = VA/E = 10,000 VA/240 volts = 42 amperes

$$D = \frac{16{,}510 \text{ circular mils} \times 7.2 \text{ volts}}{2 \text{ wires} \times 12.9 \text{ ohms} \times 42 \text{ ampere}} = 110 \text{ feet}$$

❑ **Distance – Three-Phase**

What is the maximum distance a 3-phase, 37.5 kVA, 480-volt transformer, wired with No. 6 THHN can be located from the panelboard so that the voltage drop does not exceed 3 percent (Figure 8–19)?

(a) 275 feet (b) 325 feet
(c) 375 feet (d) 425 feet

• Answer: (c) 375 feet

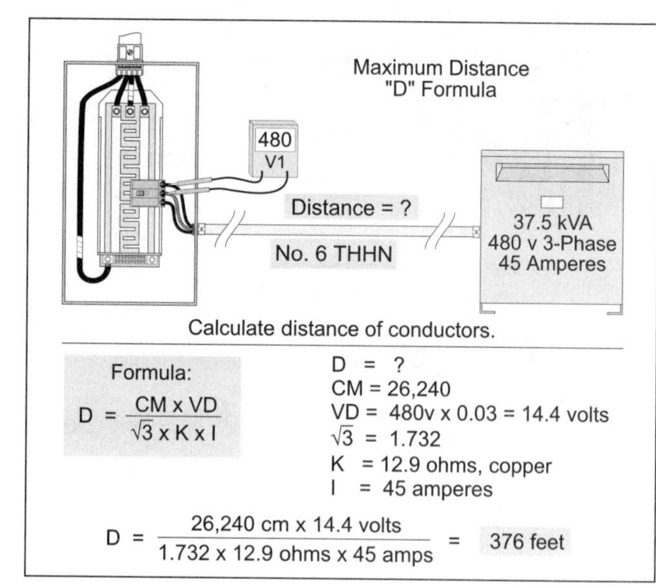

Figure 8–19
Maximum Distance "D" Formula

$$D = \frac{CM \times VD}{\sqrt{3} \times K \times I}$$

CM = 26,240 (No. 6), Chapter 9, Table 8
K = 12.9 ohm, copper
VD = 480 volts × 0.03 = 14.4 volts

$$I = \frac{VA}{(\text{Volts} \times 1.732)} = \frac{37{,}500 \text{ VA}}{(480 \text{ volts} \times 1.732)} = 45 \text{ ampere}$$

$$D = \frac{26{,}240 \text{ circular mils} \times 14.4 \text{ volts}}{1.732 \times 12.9 \text{ ohms} \times 45 \text{ ampere}} = 376 \text{ feet}$$

8–11 LIMITING CURRENT TO LIMIT VOLTAGE DROP

Sometimes the only method of limiting the circuit voltage drop is to limit the load on the conductors. The following formulas can be used to determine the maximum load.

$$\textbf{I (single-phase)} = \frac{\mathbf{CM \times VD}}{\mathbf{2 \times K \times D}} \qquad \textbf{I (three-phase)} = \frac{\mathbf{CM \times VD}}{\mathbf{\sqrt{3} \times K \times D}}$$

❑ **Maximum Load – Single-Phase**

An existing installation contains No. 1/0 THHN aluminum conductors in a nonmetallic raceway to a panelboard located 220 feet from a 230-volt power source. What is the maximum load that can be placed on the panelboard so that the *NEC®* recommendations for voltage drop are not exceeded (Figure 8–20)?

(a) 51 ampere (b) 78 ampere (c) 94 ampere (d) 115 ampere

• Answer: (b) 78 ampere

$$I = \frac{CM \times VD}{2 \times K \times D}$$

CM = 105,600 (No. 1/0), Chapter 9, Table 8
VD = 230 volts × 0.03 = 6.9 volts
K = 21.2 ohm, aluminum
D = 220 feet

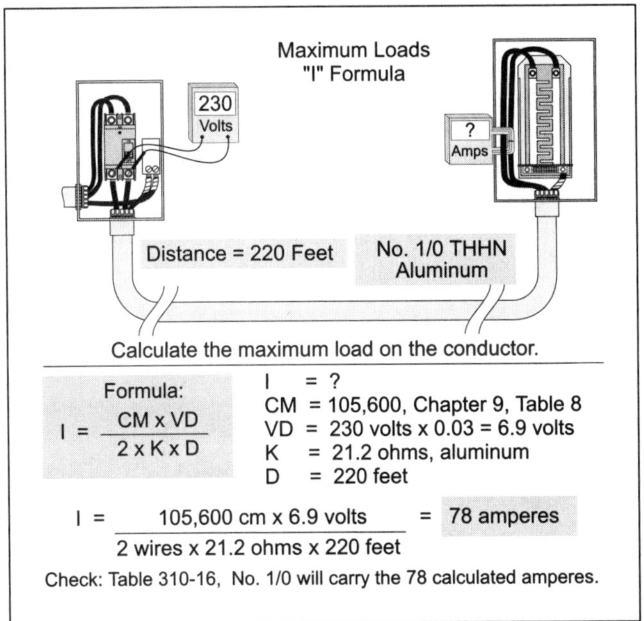

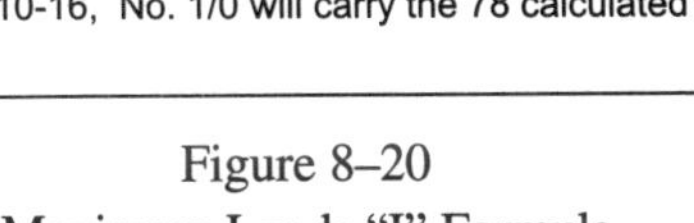

Figure 8–20
Maximum Loads "I" Formula

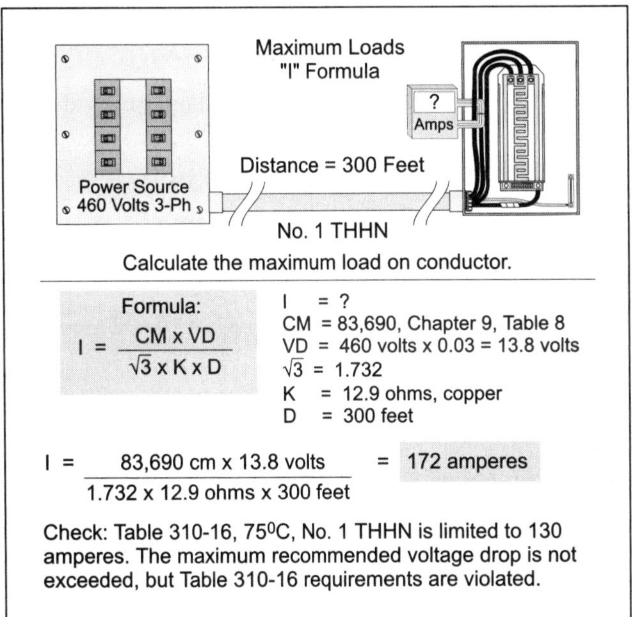

Figure 8–21
Maximum Loads "I" Formula

$$I = \frac{105{,}600 \text{ circular mils} \times 6.9 \text{ volts drop}}{2 \text{ wires} \times 21.2 \text{ ohms} \times 220 \text{ feet}} = 78 \text{ ampere}$$

Note: The maximum load permitted on 1/0 THHN Aluminum at 75°C is 120 ampere [Table 310-16].

❑ Maximum Load – Three-Phase

An existing installation contains No. 1 THHN conductors in an aluminum raceway to a panelboard located 300 feet from a 3-phase 460/230-volt power source. What is the maximum load the conductors can carry so that the *NEC®* recommendation for voltage drop is not exceeded (Figure 8–21)?

(a) 170 ampere (b) 190 ampere (c) 210 ampere (c) 240 ampere

• Answer: (a) 170 ampere

$$I = \frac{CM \times VD}{\sqrt{3} \times K \times D}$$

CM = 83,690 (No. 1), Chapter 9, Table 8

VD = 460 volts × 0.03 = 13.8 volts

K = 12.9 ohm, copper

D = 300 feet

$$I = \frac{(83{,}690 \text{ circular mils} \times 13.8 \text{ volts})}{(1.732 \times 12.9 \text{ ohms} \times 300 \text{ feet})} = 172 \text{ ampere}$$

Note: The maximum load permitted on No. 1 THHN at 75°C is 130 ampere [*Section 110-14(c)* and *Table 310-16*].

8–12 EXTENDING CIRCUITS

If you want to extend an existing circuit and you want to limit the voltage drop, follow these steps:

Step 1: ➵ Determine the voltage drop of the existing conductors.

$$\text{VD (single-phase)} = \frac{2 \times K \times I \times D}{CM}$$

$$\text{VD (three-phase)} = \frac{\sqrt{3} \times K \times I \times D}{CM}$$

Step 2: ➸ Determine the voltage drop permitted for the extension by subtracting the voltage drop of the existing conductors from the permitted voltage drop.

Step 3: ➸ Determine the extended conductor size.

$$\text{CM (single-phase)} = \frac{2 \times K \times I \times D}{VD}$$

$$\text{CM (three-phase)} = \frac{\sqrt{3} \times K \times I \times D}{VD}$$

❑ Extending Circuits

An existing junction box is located 55 feet from the panelboard and contains No. 4 THW aluminum. This circuit is to be extended 65 feet and supply a 50-ampere, 240-volt load. What size copper conductors must be used for the extension (Figure 8–22)?

(a) No. 8 THHN (b) No. 6 THHN
(c) No. 4 THHN (d) No. 5 THHN

• Answer: (b) No. 6 THHN

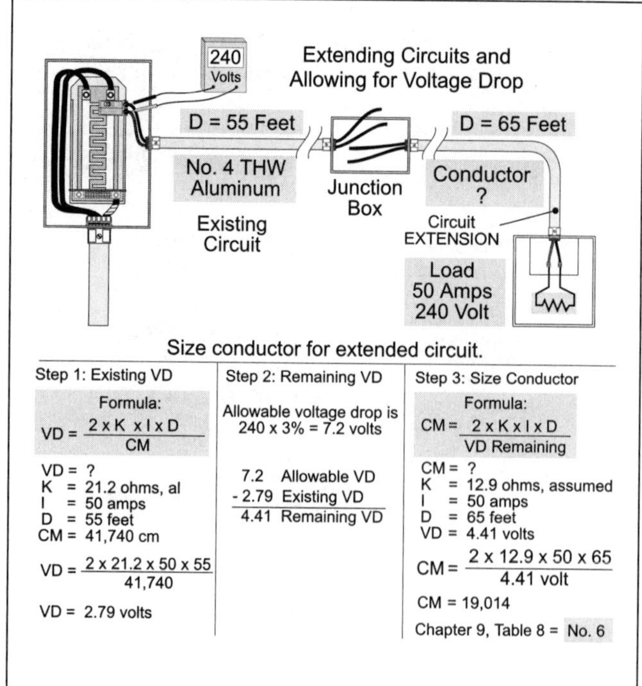

Figure 8–22
Extending Circuits

Step 1: ➸ Determine the voltage drop of the existing conductors:

$$\text{Single-Phase VD} = \frac{2 \times K \times I \times D}{CM}$$

K = 21.2 ohm, aluminum
D = 55 feet
I = 50 ampere
CM = No. 4 (41,740 circular mils, Chapter 9, Table 8)

$$VD = \frac{\text{2 wires} \times \text{21.2 ohms} \times \text{50 amps} \times \text{55 feet}}{\text{41,740 circular mils}} = \text{2.79 volts}$$

Step 2: ➸ Determine the voltage drop permitted for the extension by subtracting the voltage drop of the existing conductors from the total permitted voltage drop.

Total permitted voltage drop = 240 volts × 0.03 = 7.2 volts less 2.79 volts = 4.41 volts

Step 3: ➸ Determine the extended conductor size.

$$CM = \frac{2 \times K \times I \times D}{VD}$$

K = 12.9 ohm, copper
I = 50 ampere
D = 65 feet
VD = 7.2 volts total less 2.79 volts = 4.41 volts

$$CM = \frac{\text{2 wires} \times \text{12.9 ohms} \times \text{50 amps} \times \text{65 feet}}{\text{4.41 volts drop}} = \text{19,014 (No. 6, Chapter 9, Table 8)}$$

Unit 8 – Voltage Drop Summary Questions

8–1 Conductor Resistance

1. The _____ the number of free electrons, the better the conductivity of the conductor.
(a) greater (b) fewer

2. Conductor resistance is determined by the _____ .
(a) material type (b) cross-sectional area (c) conductor length (d) all of these

3. _____ is the best conductor, better than gold, but the high cost limits its use to special applications, such as fuse elements and some switch contacts.
(a) Silver (b) Copper (c) Aluminum (d) none of these

4. _____ conductors are often used when weight or cost are an important consideration.
(a) Silver (b) Copper (c) Aluminum (d) none of these

5. • Conductor cross-sectional area is expressed in _____ .
(a) square inches (b) mils (c) circular mils (d) none of these

6. The resistance of a conductor is directly proportional to its length.
(a) True (b) False

7. The resistance of a conductor changes with temperature. Temperature coefficient describes the effect that temperature has on the resistance of a conductor. Conductors with a _____ temperature coefficient have an increase in resistance with an increase in temperature.
(a) positive (b) negative (c) neutral (d) none of these

8–2 Conductor Resistance – Direct Current Circuits [Chapter 9, Table 8]

8. The *National Electrical Code®* lists the resistance and area in circular mils for both direct current and alternating current conductors. Direct current conductor resistances are listed in Chapter 9, Table _____ and alternating current conductor resistance are listed in Chapter 9, Table _____ .
(a) 1, 5 (b) 3, 4 (c) 9, 8 (d) 8, 9

9. What is the direct current resistance of 400 feet of No. 6?
(a) 0.2 ohm (b) 0.3 ohm (c) 0.4 ohm (d) 0.5 ohm

10. What is the direct current resistance of 150 feet of No. 1?
(a) 0.023 ohm (b) 0.031 ohm (c) 0.042 ohm (d) 0.056 ohm

11. What is the direct current resistance of 1,400 feet of No. 3 AL?
(a) 0.23 ohm (b) 0.31 ohm (c) 0.42 ohm (d) 0.56 ohm

12. What is the direct current resistance of 800 feet of No. 1/0 AL?
(a) 0.23 ohm (b) 0.16 ohm (c) 0.08 ohm (d) 0.56 ohm

13. What is the direct current resistance of 120 feet of No. 1?
(a) 0.23 ohm (b) 0.16 ohm (c) 0.02 ohm (d) 0.56 ohm

14. What is the direct current resistance of 1,249 feet of No. 3 AL?
(a) 0.23 ohm (b) 0.16 ohm (c) 0.02 ohm (d) 0.5 ohm

8–3 Conductor Resistance – Alternating Current Circuits

15. The intensity of the magnetic field is dependent on the intensity of alternating current. The greater the current flow, the greater the overall magnetic field.
(a) True (b) False

16. • The expanding and collapsing magnetic field within the conductor exerts a force on the moving electrons. This force is called counter-electromotive force (CEMF).
(a) True (b) False

17. _____ currents are small independent currents that are produced as a result of the expanding and collapsing magnetic field. They flow erratically within the conductor opposing current flow and consuming power.
(a) Lenz (b) Ohm's (c) Eddy (d) Kerchoff's

18. The expanding and collapsing magnetic field induces a counter voltage within the conductors, which repels the flowing electrons towards the conductor surface. This is known as _____ effect.
(a) inductive (b) skin (c) surface (d) watt

8–4 Alternating Current Resistance As Compared to Direct Current

19. The opposition to current flow is greater for alternating current circuits because of _____ than for direct current circuits.
(a) eddy currents (b) skin effect (c) CEMF (d) all of these

8–5 Conductor Resistance – Alternating Current [Chapter 9, Table 9 of The *NEC*®]

20. The alternating current conductor resistances listed in Chapter 9, Table 9 of the *NEC*® are different for copper and aluminum and for nonmagnetic and magnetic raceways.
(a) True (b) False

21. What is the alternating current resistance of 320 feet of No. 2/0?
(a) 0.03 ohm (b) 0.04 ohm (c) 0.05 ohm (d) 0.06 ohm

22. What is the alternating current resistance of 1,000 feet of 500 kcmil when installed in an aluminum raceway?
(a) 0.032 ohm (b) 0.027 ohm (c) 0.029 ohm (d) 0.030 ohm

23. What is the alternating current resistance of 220 feet of No. 2/0?
(a) 0.012 ohm (b) 0.022 ohm (c) 0.33 ohm (d) 0.43 ohm

24. What is the alternating current resistance of 369 feet of 500 kcmil installed in nonmetallic conduit?
(a) 0.01 ohm (b) 0.02 ohm (c) 0.03 ohm (d) 0.04 ohm

25. A 36-ampere load is located 80 feet from the panelboard and is wired with No. 1 THHN aluminum. What is the total resistance of the circuit conductors?
(a) 0.04 ohm (b) 0.25 ohm (c) 0.5 ohm (d) all of these

26. What size copper conductors can be used to replace No. 1/0 aluminum that supplies a 110-ampere load? Note: We do not want to increase the circuit voltage drop.
(a) No. 3 (b) No. 2 (c) No. 1 (d) No. 1/0

27. What is the direct current resistance in ohms for three 300 kcmil conductors in parallel, 1,000 feet long?
(a) 0.014 ohm (b) 0.026 ohm (c) 0.052 ohm (d) 0.047 ohm

28. What is the ac resistance in ohms for three 1/0 THHN aluminum conductors in parallel?
(a) 0.1 ohm (b) 0.2 ohm (c) 0.4 ohm (d) 0.067 ohm

8–6 Voltage Drop Consideration

29. Because of the great demand for electricity, utilities sometimes are required to reduce their output voltage. In addition, the utility and customer transformers, services, feeders, and branch circuit conductors oppose the flow of current. The opposition to current flow results in voltage drop. All circuits have voltage drop, simply because all conductors have resistance.
    (a) True (b) False

30. When sizing conductors for feeders and branch circuits, the *NEC®* _____ that we take voltage drop into consideration.
    (a) permits (b) suggests (c) requires (d) demands

31. _____ equipment such as motors and electromagnetic ballast can overheat at reduced voltage. This results in reduced equipment operating life and inconvenience to the customer.
    (a) Inductive (b) Electronic (c) Resistive (d) all of these

32. • _____ equipment such as computers, laser printers, copy machines, etc., can suddenly power down because of reduced voltage, resulting in data losses.
    (a) Inductive (b) Electronic (c) Resistive (d) all of these

33. Resistive loads such as incandescent lighting and electric space heating can have their power output decreased by the square of the voltage.
    (a) True (b) False

34. What is the power consumed of a 4.5 kW, 230-volt water heater operating at 200 volts?
    (a) 2,700 watts (b) 3,400 watts (c) 4,500 watts (d) 5,500 watts

35. How can conductor voltage drop be reduced?
    (a) reduce conductor resistance
    (b) increase conductor size
    (c) decrease conductor length
    (d) all of these

8–7 *NEC®* Voltage Drop Recommendations

36. If the branch circuit supply voltage is 208 volts, the maximum recommended voltage drop of the circuit should not be more than _____ .
    (a) 3.6 volts (b) 6.24 volts (c) 6.9 volts (d) 7.2 volts

37. If the feeder supply voltage is 240 volts, the maximum recommended voltage drop of the feeder should not be more than _____ .
    (a) 3.6 volts (b) 6.24 volts (c) 6.9 volts (d) 7.2 volts

8–8 Determining Circuit Conductors Voltage Drop

38. What is the voltage drop of two No. 12 THHN conductors supplying a 12-ampere continuous load? Note: The continuous load is located 100 feet from the power supply.
    (a) 3.2 volts (b) 4.75 volts (c) 6.4 volts (d) 12.8 volts

39. A single-phase, 24-ampere, 240-volt load is located 160 feet from the panelboard. The load is wired with No. 10 THHN. What is the approximate voltage drop of the branch circuit conductors?
    (a) 4.25 volts (b) 9.5 volts (c) 3 percent (d) 5 percent

40. A 3-phase, 36 kVA load, rated 208 volts is located 100 feet from the panelboard and is wired with No. 1 THHN aluminum. What is the approximate voltage drop of the circuit conductors?
    (a) 3.5 volts (b) 5 volts (c) 3 volts (d) 4.4 volts

8–9 Sizing Conductors to Prevent Excessive Voltage Drop

41. A single-phase, 5 horsepower motor is located 110 feet from a panelboard. The nameplate indicates that the voltage is 115/230 and the FLA is 52/26 ampere. What size conductor is required if the motor winding are connected in parallel and operates at 115 volts?
(a) No. 10 THHN (b) No. 8 THHN (c) No. 6 THHN (d) No. 3 THHN

42. A single-phase, 5 horsepower motor is located 110 feet from a panelboard. The nameplate indicates that the voltage is 115/230 and the FLA is 52/26 ampere. What size conductor is required if the motor winding are connected in series and operates at 230 volts?
(a) No. 10 THHN (b) No. 8 THHN (c) No. 6 THHN (d) No. 4 THHN

43. A 3-phase, 15 kW, 480-volt load is located 300 feet from the panelboard. What size conductor is required to prevent the voltage drop from exceeding 3 percent?
(a) No. 10 THHN (b) No. 8 THHN (c) No. 6 THHN (d) No. 4 THHN

8–10 Limiting Conductor Length to Limit Voltage Drop

44. What is the approximate distance a single-phase, 7.5 kVA, 240-volt load can be located from the panelboard so the voltage drop does not exceed 3 percent? The load is wired with No. 8 THHN.
(a) 55 feet (b) 110 feet (c) 145 feet (d) 220 feet

45. What is the approximate distance a 3-phase, 37.5 kVA, 460-volt transformer, wired with No. 6 THHN, can be located from the panelboard so the voltage drop does not exceed 3 percent?
(a) 250 feet (b) 300 feet (c) 345 feet (d) 400 feet

8–11 Limiting Current to Limit Voltage Drop

46. An existing installation consists of No. 1/0 THHN copper conductors in a nonmetallic raceway to a panelboard located 200 feet from a 240-volt power source. What is the maximum load that can be placed on the panelboard so that the *NEC®* recommendations for voltage drop are not exceeded?
(a) 94 ampere (b) 109 ampere (c) 71 ampere (d) 147 ampere

47. An existing installation contains No. 2 THHN aluminum conductors in an aluminum raceway to a panelboard located 300 feet from a 3-phase 460/230-volt power source. What is the maximum load the conductors can carry without exceeding the *NEC®* recommendation for voltage drop?
(a) 83 ampere (b) 85 ampere (c) 64 ampere (d) 49 ampere

8–12 Extending Circuits

48. An existing junction box is located 65 feet from the panelboard and contains No. 4 THHN aluminum conductors. What size copper conductors can be used to extend this circuit 85 feet and supply a 50-ampere, 208-volt load?
(a) No. 8 THHN (b) No. 6 THHN (c) No. 4 THHN (d) No. 5 THHN

8–13 Miscellaneous

49. What is the circuit voltage if the conductor voltage drop is 3.3 volts? Assume 3 percent voltage drop.
(a) 110 volts (b) 115 volts (c) 120 volts (d) none of these

✰ Challenge Questions

8–1 Conductor Resistance

50. The resistance of a conductor is affected by temperature change. This is called the _____ .
(a) temperature correction factor (b) temperature coefficient
(c) ambient temperature factor (d) none of these

8–3 Conductor Resistance – Alternating Current Circuits

51. The total opposition to current flow in an alternating current circuit is expressed in ohms and is called _____ .
(a) impedance (b) conductance (c) reluctance (d) resistance

8–5 Resistance Alternating Current [Chapter 9, Table 9 of the *NEC®*]

52. A 40-ampere, 240-volt, single-phase load is located 150 feet from an existing junction box. The junction box is located 50 feet from the panelboard and is wired with No. 4 THHN aluminum wire. The total resistance of the two No. 4 conductors from the panelboard to the junction box is approximately _____ ohm.
(a) 0.03 (b) 0.09 (c) 0.05 (d) 0.04

53. A load is located 100 feet from a 230-volt power supply and is wired with No. 4 THHN aluminum conductors. What size copper conductor can be used to replace the aluminum conductors and not increase the conductor voltage drop?
(a) No. 6 (b) No. 8 (c) No. 1 (d) No. 1/0

8–7 *NEC®* Voltage Drop Recommendations

54. A 40-ampere, 240-volt rated, single-phase load is wired 150 feet from a junction box, and the junction box is located 50 feet from a panelboard (for a total of 200 feet). If the voltage at the panelboard is 240 volts, what is the minimum voltage recommended by the *NEC®* at the 40-ampere load?
(a) 228.2 volts (b) 232.8 volts (c) 236.2 volts (d) 117.7 volts

8–8 Determining Circuit Conductors Voltage Drop

55. What is the voltage drop of 2 No. 4 aluminum conductors that supply a 5 horsepower single-phase, 208-volt motor that has a nameplate rating of 55 ampere? The motor is located 95 feet from the power supply.
(a) 3.25 volts (b) 5.31 volts (c) 6.24 volts (d) 7.26 volts

8–9 Sizing Conductors To Prevent Excessive Voltage Drop

56. A 40-ampere, 240-volt, single-phase load is located 150 feet from an existing junction box. The junction box is located 50 feet from the panelboard. When the 40-ampere load is on, the voltage at the junction box would be calculated to be 236 volts. The *NEC®* recommends voltage drop for this branch circuit not exceed 3 percent of the 240 voltage source (7.2 volts). What size conductor could be installed from the junction box to the load and still meet *NEC®* recommendations?
(a) No. 3 (b) No. 1 (c) No. 1/0 (d) No. 6

8–10 Limiting Conductor Length To Limit Voltage Drop

57. How far can a 50-ampere, 3-phase, 230-volt load be located from the panel, if fed with No. 3 THHN, and still meet *NEC®* recommendations for voltage drop?
(a) 275 feet (b) 300 feet (c) 325 feet (d) 350 feet

8–11 Limiting Current To Limit Voltage Drop

58. Two No. 8 THHN supply a 120-volt load that is located 225 feet from the panelboard. What is the maximum load, in ampere, that can be applied to these conductors without exceeding the *NEC®* recommendation on conductor voltage drop?
(a) 0 ampere (b) 5 ampere (c) 10 ampere (d) 15 ampere

8–13 Miscellaneous Voltage Drop Questions

59. An 8 ohm resistor is connected to a 120-volt power supply. Using a voltmeter, we measure 112 volts across the resistor. What is the current of the 8 ohm resistor in ampere?
(a) 14 ampere (b) 13 ampere (c) 15 ampere (d) 19 ampere

60. A 480-volt, single-phase feeder carries 400 ampere and has a 7.2 voltage drop. What is the total resistance of the conductors in this circuit?
(a) .1880 ohm (b) .1108 ohm (c) .0190 ohm (d) .0180 ohm

61. An 8 ohm resistor operates at 112 volts and is connected to a 115-volt power supply. The percent of voltage drop for the circuit is _____ percent.
(a) 2.6 (b) 3.0 (c) 5.0 (d) 7.0

Unit 9

Single-Family Dwelling Unit Load Calculations

OBJECTIVES

After reading this unit, the student should be able to briefly explain the following concepts:

- Air-conditioning versus heat
- Appliance (small) circuits
- Appliance demand load
- Clothes dryer demand load
- Cooking equipment calculations
- Dwelling-unit calculations –Optional method
- Dwelling-unit calculations –Standard method
- Laundry circuit
- Lighting and receptacles

After reading this unit, the student should be able to briefly explain the following terms:

- General lighting
- General use receptacles
- Optional method
- Rounding
- Standard method
- Unbalanced demand load
- Voltages

PART A – GENERAL REQUIREMENTS

9–1 GENERAL REQUIREMENTS

Article 220 provides the requirements for residential branch circuits, feeders, and service calculations. Other applicable Articles are Branch Circuits – 210, Feeders – 215, Services – 230, Overcurrent Protection – 240, Wiring Methods – 300, Conductors – 310, Appliances – 422, Electric Space Heating Equipment – 424, Motors – 430, and Air-Conditioning – 440.

9–2 VOLTAGES [*SECTION 220-2(a)*]

Unless other voltages are specified, branch circuit, feeder, and service loads shall be computed at nominal system voltages of 120, 120/240, 208Y/120, or 240 (Figure 9–1).

9–3 FRACTION OF AN AMPERE [*SECTION 220-2(b)*]

The rules for rounding an ampere require that when the calculation results in a fraction of an ampere of 0.50 and more, we “round up” to the next ampere. If the calculation is 0.49 of an ampere or less, we “round down” to the next lower ampere.

❑ **Rounding**

What size THHN conductor is required to supply a 9.6 kW, 230-volt, single-phase fixed space heater that has a 3 ampere blower motor [*Section 424-3(b)*]? Terminal rating of 75°C (Figure 9–2).

(a) No. 8 (b) No. 6 (c) No. 4 (d) none of these

• Answer: (b) No. 6

According to Section 424-3(b), the conductors and overcurrent protection device to electric space heating equipment shall be sized no less than 125 percent of the total load (heat plus motors).

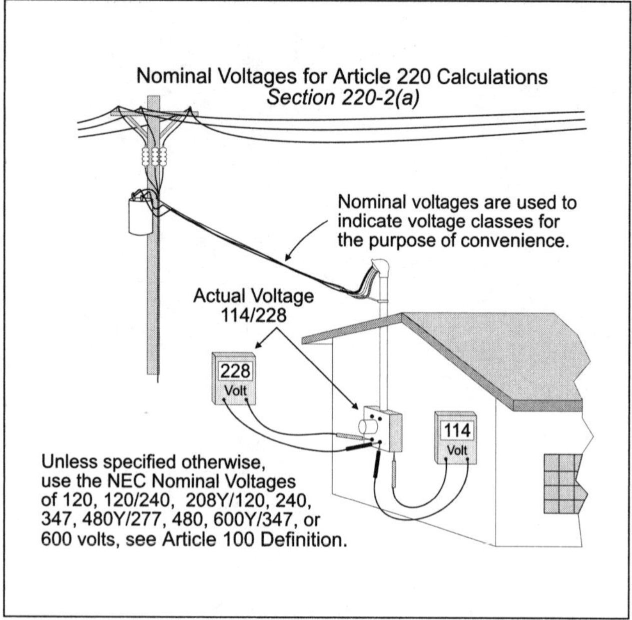

Figure 9–1
Nominal Voltages for Article 220 Calculations

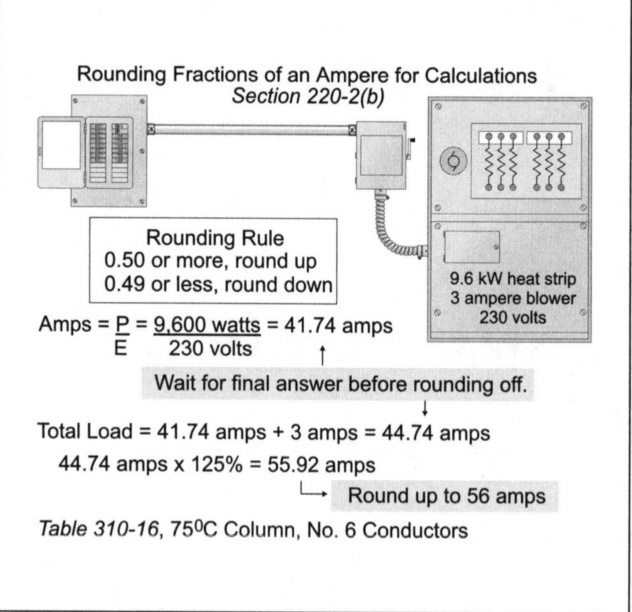

Figure 9–2
Rounding Fractions of an Ampere for Calculations

Space Heater Ampere = Watts/Volts = 9,600 watts/230 volts = 41.74 ampere

Total Load (heat + motor) = 41.74 ampere + 3 ampere = 44.74 ampere

Conductor Sized 44.74 ampere × 1.25 = 55.92 ampere, round up to 56 ampere

No. 6 THHN is rated 65 ampere at 75°C [Table 310-16].

9–4 APPLIANCE (SMALL) CIRCUITS [*SECTION 220-11(c)(1)*]

A minimum of two 20-ampere *small appliance branch circuits* is required for receptacle outlets in the kitchen, dining room, breakfast room, pantry, or similar dining areas. In general, no other receptacle or lighting outlet can be connected to these 20-ampere small appliance branch circuits [*Section 210-52(b)(2)* Exceptions] (Figure 9–3).

Feeder and Service

When sizing the feeder or service, each dwelling-unit shall have two 20-ampere small appliance branch circuits with a feeder load of 1,500 volt-ampere for each circuit [*Section 220-16(a)*].

Other Related *Code* Sections

Areas that the small appliance circuits supply [*Section 210-52(b)(1)*].

Receptacle outlets required for kitchen counter tops [*Section 210-52(c)*].

15- or 20-ampere rated receptacles can be used on 20-ampere circuit [*Section 210-21(b)(3)*].

9–5 COOKING EQUIPMENT – BRANCH CIRCUIT [Table 220-19, Note 4]

To determine the branch circuit demand load for cooking equipment, we must comply with the requirements of Table 220-19, Note 4. The branch circuit demand load for one range shall be according to the demand loads listed in Table 220-19, but a minimum 40-ampere circuit is required for 8.75 kW and larger ranges [*Section 210-19(c)*].

❑ **Less than 12 kW**

What is the branch circuit demand load (in ampere) for one 9 kW range (Figure 9–4)?

(a) 21 ampere (b) 27 ampere (c) 38 ampere (d) 33 ampere

• Answer: (d) 33 ampere

The demand load for one range in Table 220-19 Column A is 8 kW.

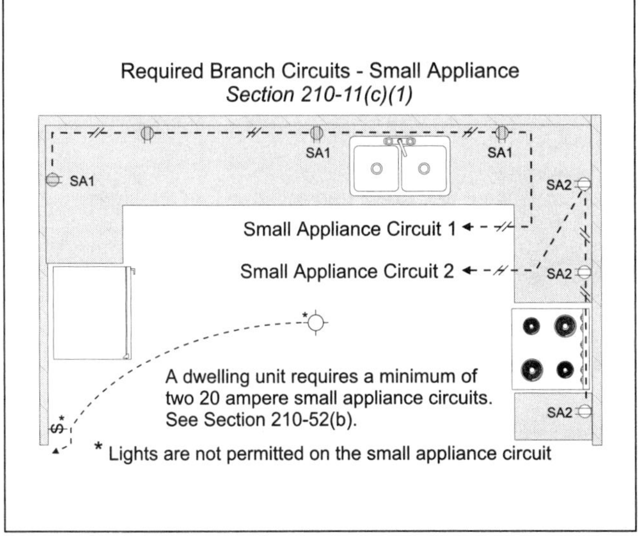

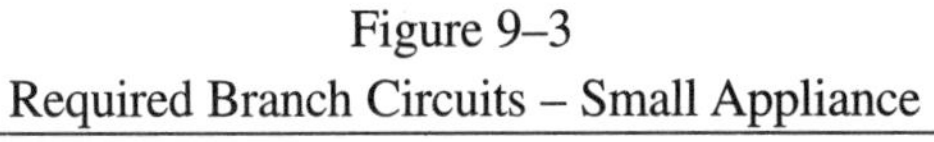
Figure 9–3
Required Branch Circuits – Small Appliance

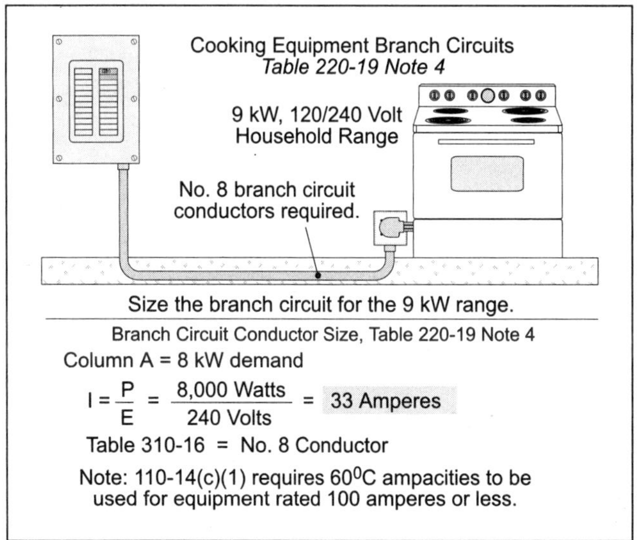

Figure 9–4
Cooking Equipment Branch Circuits

This can be converted to ampere by dividing the power by the voltage, $I = \frac{P}{E} = \frac{8 \text{ kW} \times 1{,}000}{240 \text{ volts*}} = 33$ ampere

*Assume 120/240 volt, single-phase for all calculations unless the question gives a specific voltage and system.

❑ More Than 12 kW

What is the branch circuit load (in ampere) for one 14 kW range (Figure 9–5)?

(a) 33 ampere (b) 37 ampere (c) 50 ampere (d) 58 ampere

• Answer: (b) 37 ampere

Step 1:➛ Since the range exceeds 12 kW, follow Note 1 of Table 220-19. The first step is to determine the demand load as listed in Column A of Table 220-19 for one unit = 8 kW.

Step 2:➛ We must increase the Column A value (8 kW) by 5 percent for each kW that the average range (in this case 14 kW) exceeds 12 kW. This results is an increase of the Column A value (8 kW) by 10 percent,

8 kW x 1.1 = 8.8 kW or 8,800 Watts.

Step 3:➛ Convert the demand load in kW to ampere. $I = \frac{P}{E} = \frac{8{,}800 \text{ watts}}{240 \text{ volts}} = 36.7$ ampere

❑ One Counter-Mounted Cooking Unit

What is the branch circuit load (in ampere) for one 6 kW counter-mounted cooking unit (Figure 9–6)?

(a) 15 ampere (b) 20 ampere (c) 25 ampere (d) 30 ampere

• Answer: (c) 25 ampere

$$I = \frac{P}{E} = \frac{6{,}000 \text{ watts}}{240 \text{ volts}} = 25 \text{ ampere}$$

One Counter-Mounted Cooking Unit and up to Two Ovens [Section 220-19 Note 4]

To calculate the load for one counter-mounted cooking unit and up to two wall-mounted ovens, complete the following steps:

Step 1:➛ Total Load. Add the nameplate ratings of the cooking appliances and treat this total as one range.

Step 2:➛ Table 220-19 Demand. Determine the demand load for one unit from Table 220-19, Column A.

Step 3:➛ Net Computed Load. If the total nameplate rating exceeds 12 kW, increase Column A (8 kW) 5 percent for each kW, or major fraction (0.5 kW), that the combined rating exceeds 12 kW.

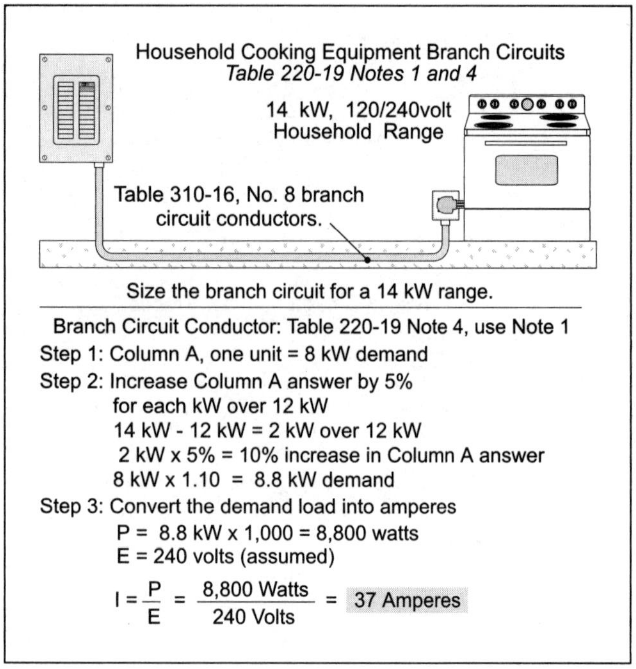

Household Cooking Equipment Branch Circuits
Table 220-19 Notes 1 and 4

14 kW, 120/240volt Household Range

Table 310-16, No. 8 branch circuit conductors.

Size the branch circuit for a 14 kW range.

Branch Circuit Conductor: Table 220-19 Note 4, use Note 1

Step 1: Column A, one unit = 8 kW demand

Step 2: Increase Column A answer by 5% for each kW over 12 kW

14 kW - 12 kW = 2 kW over 12 kW

2 kW x 5% = 10% increase in Column A answer

8 kW x 1.10 = 8.8 kW demand

Step 3: Convert the demand load into amperes

P = 8.8 kW x 1,000 = 8,800 watts

E = 240 volts (assumed)

$$I = \frac{P}{E} = \frac{8{,}800 \text{ Watts}}{240 \text{ Volts}} = 37 \text{ Amperes}$$

Figure 9–5
Household Cooking Equipment Branch Circuits

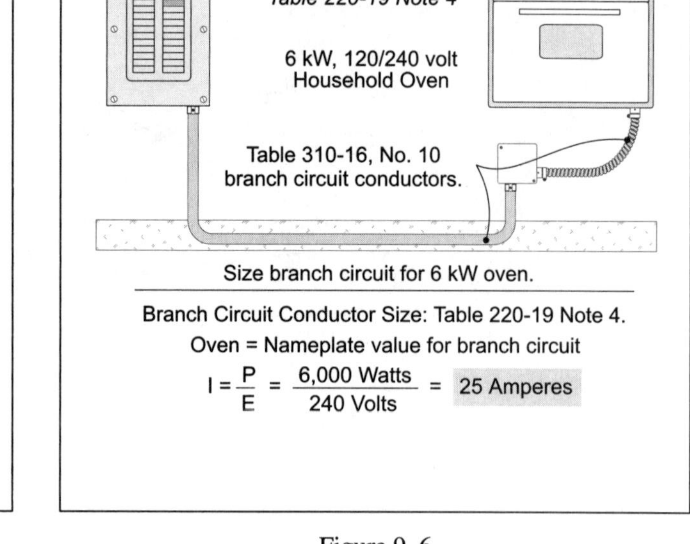

Figure 9–6
Oven Branch Circuit

❑ Cooktop and One Oven

What is the branch circuit load (in ampere) for one 6 kW counter-mounted cooking unit and one 3 kW wall-mounted oven (Figure 9–7)?

(a) 25 ampere (b) 38 ampere (c) 33 ampere (d) 42 ampere

- Answer: (c) 33 ampere

Step 1: Total connected load: 6 kW + 3 kW = 9 kW.

Step 2: Demand load for one range, Column A of Table 220-19: 8 kW.

Step 3: $I = \frac{P}{E} = \frac{8{,}000 \text{ watts}}{240 \text{ volts}} = 33.3 \text{ ampere}$

❑ Cooktop and Two Ovens

What is the branch circuit load (in ampere) for one 6 kW counter-mounted cooking unit and two 4 kW wall-mounted ovens (Figure 9–8)?

(a) 22 ampere (b) 27 ampere (c) 33 ampere (d) 37 ampere

- Answer: (d) 37 ampere

Step 1: The total connected load: 6 kW + 4 kW + 4 kW = 14 kW.

Step 2: Demand load for one range, Column A or Table 220-19: 8 kW.

Step 3: Since the total (14 kW) exceeds 12 kW, we must increase Column A (8 kW) 5 percent for each kW that the total exceeds 12 kW.

14 kW exceeds 12 kW by 2 kW, which results in a 10 percent increase of the Column A value.

8 kW × 1.1 = 8.8 kW

Step 4: $I = \frac{P}{E} = \frac{8{,}800 \text{ watts}}{240 \text{ volts}} = 36.7 \text{ ampere}$

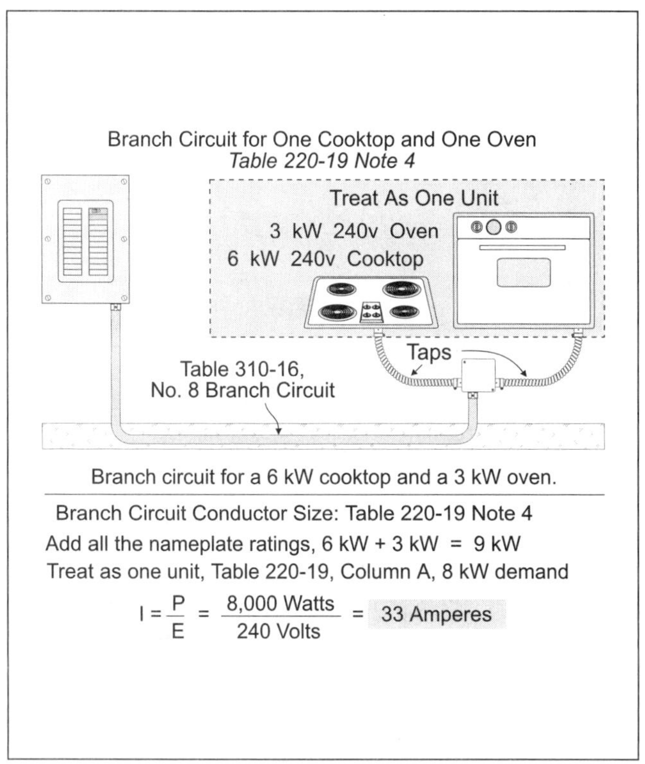

Figure 9–7
Branch Circuit for One Cooktop and One Oven

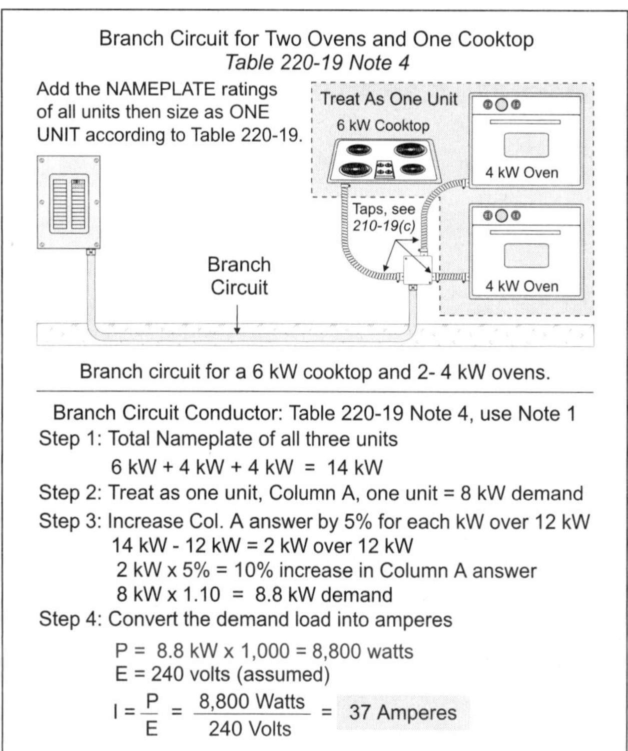

Figure 9–8
Branch Circuit for Two Ovens and One Cooktop

9–6 LAUNDRY RECEPTACLE(S) CIRCUIT [*SECTION 220-11(c)(2)*]

One 20-ampere branch circuit for the laundry receptacle outlet (or outlets) is required, and this laundry circuit cannot serve any other outlet such as the laundry room lights [*Section 210-52(f)*]. The *NEC®* does not require a separate circuit for the washing machine, but does require a separate circuit for the laundry room receptacle or receptacles (Figure 9–9). If the washing machine receptacle(s) are located in the garage or basement, GFCI protection might be required, see Section 210-8(a)(2) and (5) and their exceptions.

Feeder and Service

Each dwelling-unit shall have a feeder load consisting of 1,500 volt-ampere for the 20-ampere laundry circuit [*Section 220-16(b)*].

Other Related *Code* Rules

A laundry outlet must be within 6 feet of a washing machine [*Section 210-50(c)*].

Laundry area receptacle outlet required [*Section 210-52(f)*].

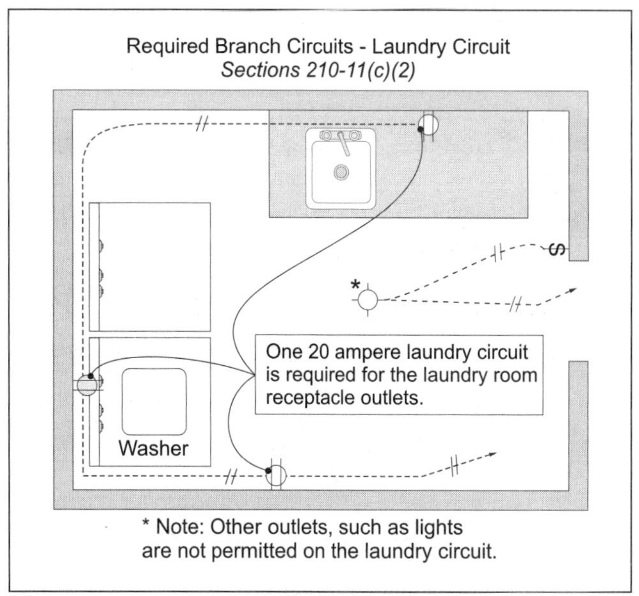

Figure 9–9
Required Branch Circuits – Laundry Circuit

9–7 LIGHTING AND RECEPTACLES

General Lighting Volt-Ampere Load [*Section 220-3(a)*]

The *NEC®* requires a minimum 3 volt-ampere each square foot [Table 220-3(a)] for the *general lighting* and general-use receptacles. The dimensions for determining the area shall be computed from the outside dimensions of the building and shall not include open porches, garages, or spaces not adaptable for future use (Figure 9–10).

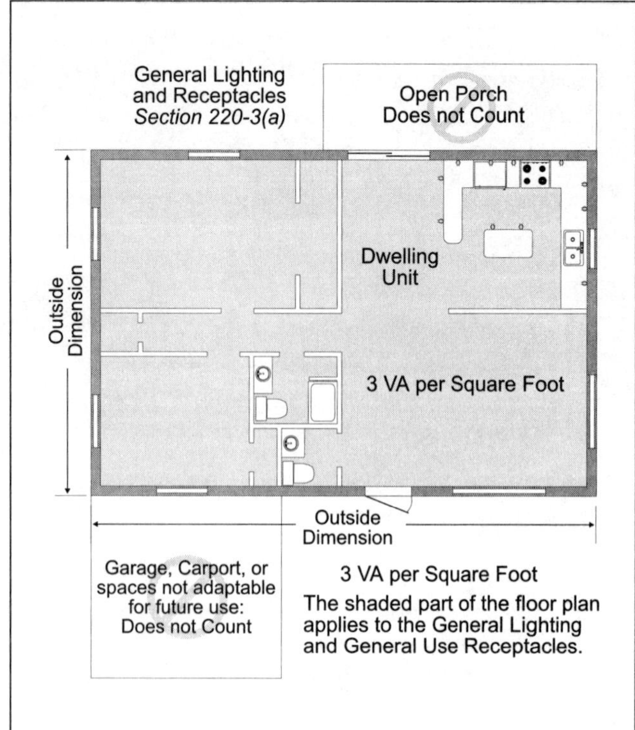

Figure 9–10
General Lighting and Receptacles

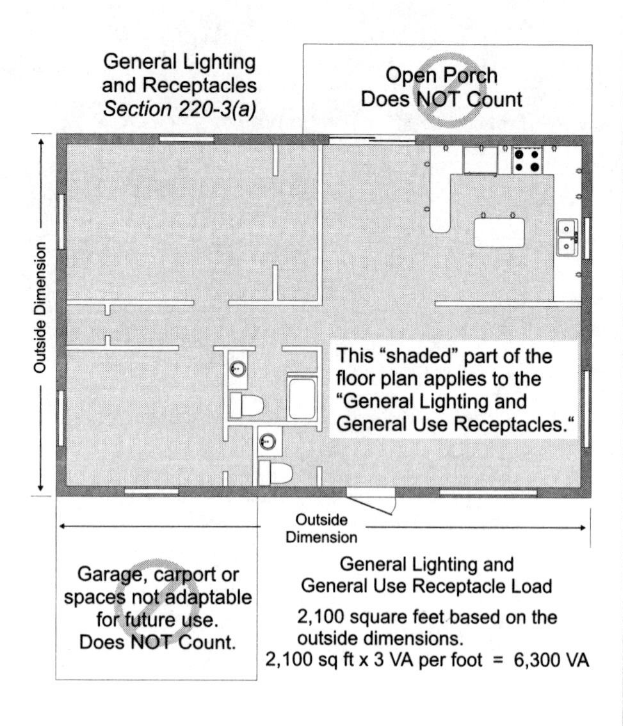

Figure 9–11
General Lighting and Receptacles

Note: The 3 volt-ampere each square foot rule for general lighting includes all 15- and 20-ampere general-use receptacles; but it does not include the appliance or laundry circuit receptacles, see *Section 220-3(10)* for details.

❑ **General Lighting Load [Table 220-3(a)]**

What is the general lighting and receptacle load for a 2,100 square-foot dwelling-unit that has 34 convenience receptacles and 12 lighting fixtures rated 100 watts each (Figure 9–11)?

(a) 2,100 VA (b) 4,200 VA (c) 6,300 VA (d) 8,400 VA

• Answer: (c) 6,300 VA

2,100 square feet × 3 VA = 6,300 VA

Note. No additional load is required for general-use receptacles and lighting outlets, see *Section 220-3(a)(10).*

Number of Circuits Required [Chapter 9, Example No. D1(a)]

The number of branch circuits required for general lighting and receptacles shall be determined from the general lighting load and the rating of the circuits [*Section 210-11(a)*].

To determine the number of branch circuits for general lighting and receptacles, follow these steps:

Step 1:➛ Determine the general lighting VA load: Living area square footage × 3 VA.

Step 2:➛ Determine the general lighting ampere load: Ampere = $^{VA}/_{E}$.

Step 3:➛ Determine the number of branch circuits:

$$\frac{\text{General Light Ampere (Step 2)}}{\text{Circuit Ampere (100\%)}}$$

❑ **Number of 15-Ampere Circuits**

How many 15-ampere circuits are required for a 2,100 square-foot dwelling-unit (Figure 9–12)?

(a) 2 circuits (b) 3 circuits (c) 4 circuits (d) 5 circuits

• Answer: (c) 4 circuits

Step 1: ➛ General lighting VA:
2,100 square feet × 3 VA = 6,300 VA

Step 2: ➛ General lighting ampere: $I = \frac{VA}{E}$

$$I = \frac{6{,}300\ VA}{120\ volts^*} = 53\ ampere$$

*Use 120 volts single-phase unless specified otherwise.

Step 3: ➛ Determine the number of circuits:

$$\frac{\text{General Light Ampere}}{\text{Circuit Ampere}}$$

$$= \frac{53\ ampere}{15\ ampere} = 3.53 \text{ or } 4 \text{ circuits}$$

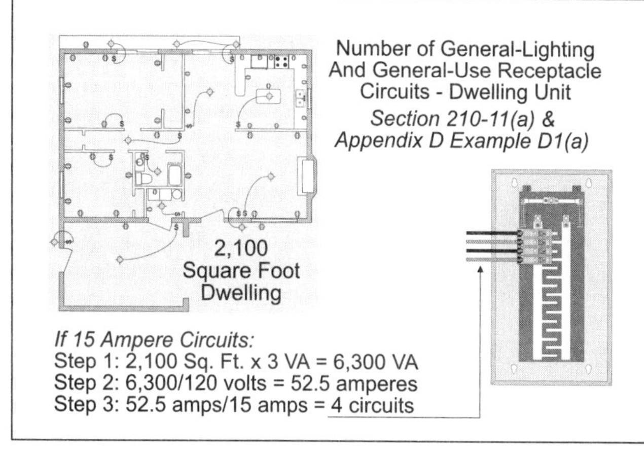

Figure 9–12
Number of General Lighting and General Use Receptacle Circuits

❑ **Number of 20-Ampere Circuits**

How many 20-ampere circuits are required for a 2,100 square-foot home?

(a) 2 circuits (b) 3 circuits (c) 4 circuits (d) 5 circuits

• Answer: (b) 3

Step 1: ➛ General lighting VA: 2,100 square foot × VA = 6,300 VA

Step 2: ➛ General lighting ampere: $I = \frac{VA}{E} = \frac{6{,}300\ VA}{120\ volts} = 53\ ampere$

Step 3: ➛ Determine the number of circuits: $\frac{\text{General Lighting Ampere}}{\text{Circuit Ampere}} = \frac{53\ ampere}{20\ ampere} = 2.65 \text{ or } 3 \text{ circuits}$

PART B – STANDARD METHOD – FEEDER/SERVICE LOAD CALCULATIONS

9–8 DWELLING UNIT FEEDER/SERVICE LOAD CALCULATIONS (Part B of Article 220)

When determining a dwelling-unit feeder or service using the standard method, use the following steps:

Step 1: ➛ **General Lighting and Receptacles, Small Appliance and Laundry Circuits [Table 220-11]**

The *NEC®* recognizes that the general lighting and receptacles, small appliance, and laundry circuits will not all be on, or loaded, at the same time and permits a demand factor to be applied [*Section 220-16*]. To determine the feeder demand load for these loads, follow these steps:

(a) Total Connected Load: Determine the total connected load for (1) general lighting and receptacles (3 VA per square foot), (2) two small appliance circuits each at 1,500 VA, and (3) one laundry circuit at 1,500 VA.

(b) Demand Factor: Apply Table 220-11 demand factors to the total connected load (Step 1).

First 3,000 VA at 100 percent demand

Remainder VA at 35 percent demand

Step 2: ➛ **Air-Conditioning versus Heat**

Because the air-conditioning and heating loads are not on at the same time, it is permissible to omit the smaller of the two loads [*Section 220-21*]. The air-conditioning demand load is sized at 125 percent of the largest air-conditioning VA, plus the sum of the other air-conditioning VA's [*Section 440-34*]. Fixed electric space-heating demand loads are calculated at 100 percent of the total heating load [*Section 220-15*].

Step 3: **Appliances [*Section 220-17*]**
A demand factor of 75 percent is permitted to be applied when four or more appliances fastened in place, such as dishwasher, waste disposal, trash compactor, water heater, etc. are on the same feeder. This does not apply to motors [*Section 220-14*], space-heating equipment [*Section 220-15*], clothes dryers [*Section 220-18*], cooking appliances [*Section 220-19*], or air-conditioning equipment [*Section 440-34*].

Step 4: **Clothes Dryer [*Section 220-18*]**
The feeder or service demand load for electric clothes dryers located in a dwelling-unit shall not be less than: (1) 5,000 watts or (2) the nameplate rating if greater than 5,000 watts. A feeder or service dryer load is not required if the dwelling-unit does not contain an electric dryer.

Step 5: **Cooking Equipment [*Section 220-19*]**
Household cooking appliances rated over $1^3/_4$ kW can have the feeder and service loads calculated according to the demand factors of Section 220-19, Table and Notes 1, 2, and 3.

Step 6: **Feeder and Service Conductor Size**
400 Ampere and Less: The feeder or service conductors are sized according to Table 310-15(b)(6) for 3-wire, single-phase, 120/240-volt systems up to 400 ampere. The grounded (neutral) conductor is sized based on the maximum unbalanced load [*Section 220-22*] according to conductors as listed in Table 310-16.
Over 400 Ampere: The ungrounded and grounded (neutral) conductors are sized according to Table 310-16.

9–9 DWELLING UNIT FEEDER/SERVICE CALCULATIONS EXAMPLES

Step 1: General Lighting, Small Appliance, and Laundry Demand [Section 220-11]

General Lighting No. 1

What is the general lighting, small appliance, and laundry demand load for a 2,700 square-foot dwelling-unit (Figure 9–13)?

(a) 8,100 VA (b) 12,600 VA (c) 2,700 VA (d) 6,360 VA

• Answer: (d) 6,360 VA

General Lighting/Receptacles	(2,700 square feet × 3 VA)	8,100 VA		
Small Appliance Circuits	(1,500 VA × 2)	3,000 VA		
Laundry Circuit	(1,500 VA × 1)	1,500 VA		
Total Connected Load		12,600 VA		
First 3,000 VA at 100%		–3,000 VA	× 1.00 =	3,000 VA
Remainder at 35%		9,600 VA	× 0.35 =	+ 3,360 VA
				6,360 VA

General Lighting No. 2

What is the general lighting, small appliance, and laundry demand load for a 6,540 square-foot dwelling-unit?

(a) 8,100 VA (b) 12,600 VA (c) 2,700 VA (d) 10,392 VA

• Answer: (d) 10,392 VA

General Lighting/Receptacles	(6,540 square feet × 3 VA)	19,620 VA		
Small Appliance Circuits	(1,500 VA × 2)	3,000 VA		
Laundry Circuit	(1,500 VA × 1)	1,500 VA		
Total Connected Load		24,120 VA		
First 3,000 VA at 100%		–3,000 VA	× 1.00 =	3,000 VA
Remainder at 35%		21,120 VA	× 0.35 =	+ 7,392 VA
				10,392 VA

Step 2: Air-Conditioning versus Heat [Section 220-15]

When determining the cooling versus heating load, compare the air-conditioning load at 125 percent against the heating load at 100 percent. Section 220-15 only refers to the heating load and Section 440-32 contains the requirement of the air-conditioning load. You are permitted to omit the smaller of the two loads according to Section 220-21.

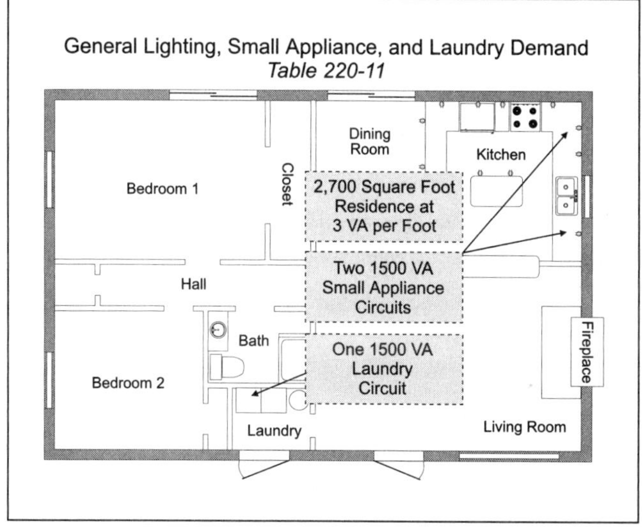

Figure 9–13
General Lighting, Small Appliance, and Laundry Demand

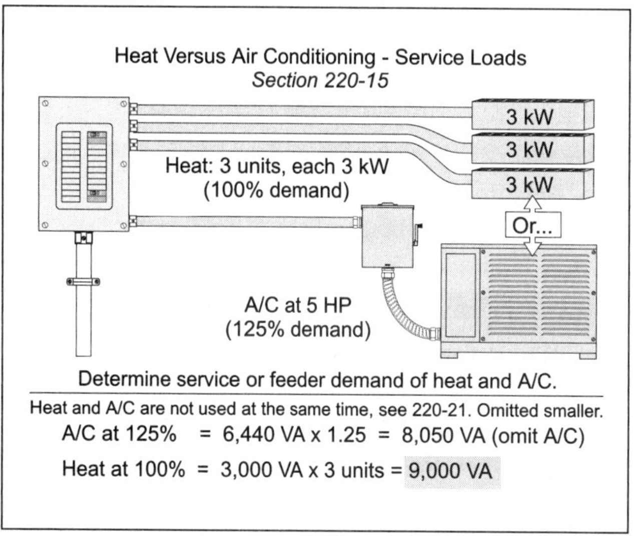

Figure 9–14
Heat versus Air Conditioning – Service Loads

➛ Air-Conditioning versus Heat No. 1

What is the service demand load for a air-conditioning (5 horsepower 230 volt) versus 3 baseboard heaters (each 3 kW) (Figure 9–14)?

(a) 6,400 VA (b) 3,000 watts (c) 8,050 VA (d) 9,000 watts

- Answer: (d) 9,000 watts

A/C [*Section 440-34*] 230 volts × 28 ampere*
A/C = 6,440 × 1.25 = 8,050 VA, omit [*Section 220-21*]
Heat [*Section 220-15*] 3,000 watts × 3 units
Heat = 9,000 watts
*Table 430-148

➛ Air-Conditioning versus Heat No. 2

What is the service or feeder demand load for A/C (4 horsepower 230 volt) and electric space heater (10 kW)?

(a) 10,000 watts (b) 3,910 watts (c) 6,440 watts (d) 10,350 watts

- Answer: (a) 10,000 kW heat

A/C: [*Section 440-34*] Since a 4 horsepower 230 volt motor is not listed in Table 430-148, we must determine the approximate VA rating by averaging the values for a 3 horsepower and 5 horsepower motor.

3 horsepower = 230 volts × 17 ampere* = 3,910 VA
5 horsepower = 230 volts × 28 ampere* = 6,440 VA
Total VA for 3 horsepower + 5 horsepower (8 horsepower) = 10,350 VA
VA for 4 horsepower = 10,350 VA (8 horsepower)/2 = 5,175 VA × 1.25 = 6,469 VA
Omit because it is less than 10 kW heat [*Section 220-21*] Note: This type of question is rarely given on an exam.

Step 3: Appliance Demand Load [Section 220-17]

➛ Appliance No. 1

What is the demand load for a waste disposal (940 VA), a dishwasher (1,250 VA), and a water heater (4,500 VA)?

(a) 5,018 VA (b) 6,690 VA (c) 8,363 VA (d) 6,272 VA

- Answer: (b) 6,690 VA

Waste disposal	940 VA
Dishwasher	1,250 VA
Water heater	4,500 VA
	6,690 VA

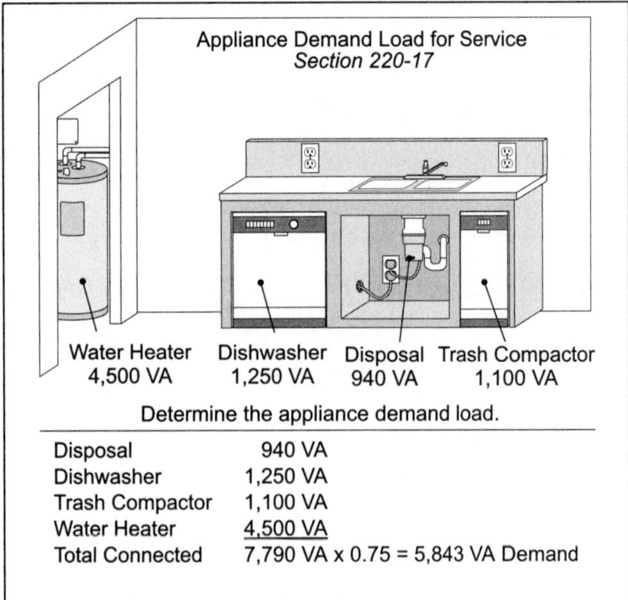

Figure 9–15
Appliance Demand Load for Service

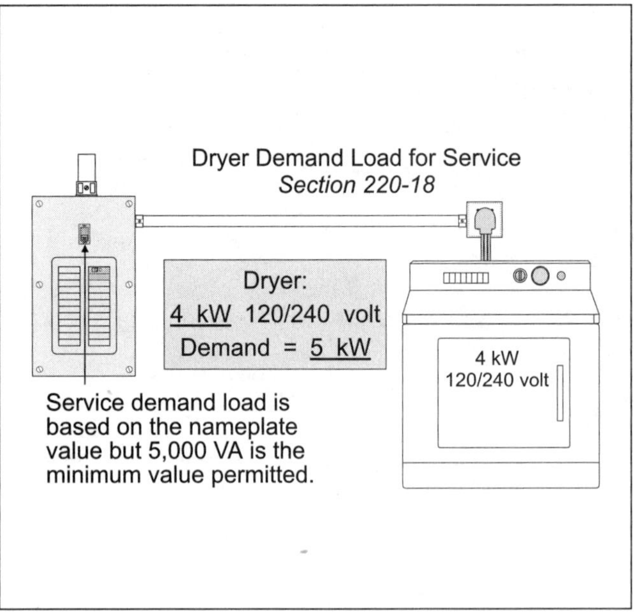

Figure 9–16
Dryer Demand Load for Service

➣ Appliance No. 2

What is the demand load for a waste disposal (940 VA), a dishwasher (1,250 VA), a trash compactor (1,100 VA), and a water heater (4,500 VA) (Figure 9–15)?

(a) 7,790 VA (b) 5,843 VA (c) 7,303 VA (d) 9,738 VA

- Answer: (b) 5,843 VA

Waste disposal	940 VA
Dishwasher	1,250 VA
Trash Compactor	1,100 VA
Water Heater	4,500 VA
	7,790 VA × 0.75 = 5,843 VA

Step 4: Dryer Demand Load [Section 220-18]

➣ Dryer No. 1

What is the service and feeder demand load for a 4 kW dryer (Figure 9–16)?

(a) 4,000 watts (b) 3,000 watts (c) 5,000 watts (d) 5,500 watts

- Answer: (c) 5,000 VA

The dryer load must not be less than 5,000 VA.

➣ Dryer No. 2

What is the service and feeder demand load for a 5.5 kW dryer?

(a) 4,000 watts (b) 3,000 watts (c) 5,000 watts (d) 5,500 watts

- Answer: (d) 5,500 VA

The dryer load must not be less than the nameplate rating if greater than 5,000 VA.

Step 5: Cooking Equipment Demand Load [Section 220-19]

➣ Note 3 – Not over $3^1/_2$ kW – Column B

What is the service and feeder demand load for two 3 kW cooking appliances in a dwelling-unit?

(a) 3 kW (b) 4.8 kW (c) 4.5 kW (d) 3.9 kW

- Answer: (c) 4.5 kW

3 kW × 2 units = 6 kW × 0.75 = 4.5 kW

➛ Note 3 – Not over $8^3/_4$ – Column C

What is the service and feeder demand load for one 6 kW cooking appliance in a dwelling unit?

(a) 6 kW (b) 4.8 kW
(c) 4.5 kW (d) 3.9 kW

• Answer: (b) 4.8 kW

6 kW × 0.8 = 4.8 kW

➛ Note 3 – Column B and C

What is the service and feeder demand load for two 3 kW ovens and one 6 kW cooktop in a dwelling unit (Figure 9–17)?

(a) 6 kW (b) 4.8 kW
(c) 4.5 kW (d) 9.3 kW

• Answer: (d) 9.3 kW

Column B demand: (3 kW × 2) × 0.75	=	4.5 kW
Column C demand: 6 kW × 0.8	=	4.8 kW
Total demand		9.3 kW

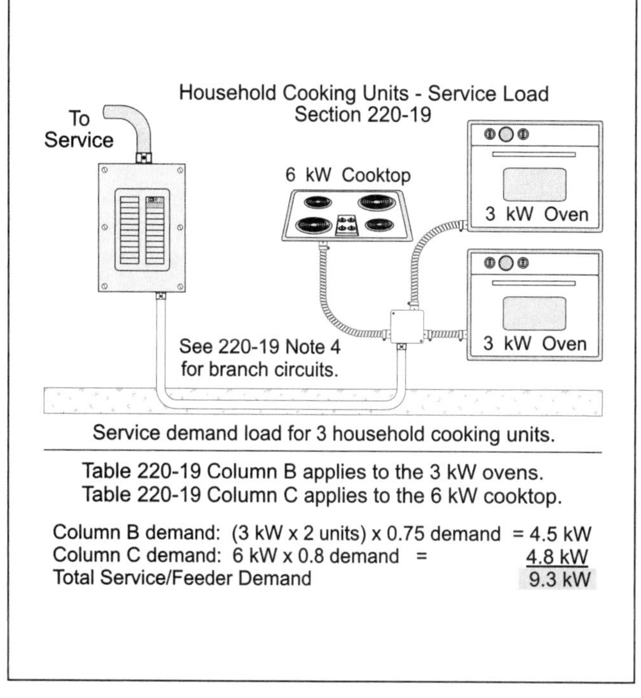

Figure 9–17
Household Cooking Units – Service Load

➛ Column A – Not over 12 kW

What is the service and feeder demand load for an 11.5 kW range in a dwelling unit?

(a) 11.5 kW (b) 8 kW
(c) 9.2 kW (d) 6 kW

• Answer: (b) 8 kW

Table 220-19 Column A: 8 kW

➛ Over 12 kW – Note 1

What is the service and feeder demand load for a 13.6 kW range in a dwelling unit?

(a) 8.8 kW (b) 8 kW (c) 9.2 kW (d) 6 kW

• Answer: (a) 8.8 kW

Step 1:➛ 13.6 kW exceeds 12 kW by one kW and one major fraction of a kW. Column A value (8 kW) must be increased 5 percent for each kW or major fraction of a kW (0.5 kW or larger) over 12 kW.

Step 2:➛ 8 kW × 1.1 = 8.8 kW

Step 6: Service Conductor Size [Table 310-15(b)(6)]

➛ Service Conductor No. 1

What size THHN feeder or service conductor (120/240 volt, single-phase) is required for a 225-ampere service demand load?

(a) No. 1/0 (b) No. 2/0 (c) No. 3/0 (d) No. 4/0

• Answer: (c) No. 3/0

➛ Service Conductor No. 2

What size THHN feeder or service conductor (208Y/120 volt, single-phase) is required for a 225-ampere service demand load?

(a) No. 1/0 (b) No. 2/0 (c) No. 3/0 (d) No. 4/0

• Answer: (d) No. 4/0

Table 310-15(b)(6), does not apply to 208Y/120-volt systems because the grounded (neutral) conductor carries current, even if the loads are balanced.

9–10 DWELLING UNIT OPTIONAL FEEDER/SERVICE CALCULATIONS [*Section 220-30*]

Instead of sizing the dwelling-unit feeder and/or service conductors according to the standard method (Part B of Article 220), an optional method [*Section 220-30*] can be used. The following Steps can be used to determine the demand load:

Step 1: **Section 220-30(b) General Loads**

The calculated load shall not be less 100 percent for the first 10 kW, plus 40 percent of the remainder of the following loads:

(1) Small Appliance and Laundry Branch Circuits: 1,500 volt-ampere for each 20-ampere small appliance and laundry branch circuit.

(2) General Lighting and Receptacles: 3 volt-ampere per square-foot.

(3) Appliances: The nameplate VA rating of all appliances and motors that are fastened in place (permanently connected), or located on a specific circuit.

Note: Be sure to use the range and dryer nameplate rating.

Step 2: **Section 220-30(c) Heating and Air-Conditioning Load**

Include the largest of the following:

(1) 100 percent of the nameplate rating of the air-conditioning equipment.

(2) 100 percent of the heat pump compressors and supplemental heating unless the controller prevents both the compressor and supplemental heating from operating at the same time.

(3) 100 percent of the nameplate ratings of electric thermal storage and other heating systems where the usual load is expected to be continuous at the full nameplate value. Systems qualifying under this selection shall not be figured under any other selection in this table.

(4) 65 percent of the nameplate rating(s) of the central electric space heating, including integral supplemental heating in heat pumps where the controller prevents both the compressor and supplemental heating from operating at the same time.

(5) 65 percent of the nameplate rating(s) of electric space heating, if there are less than four separately controlled units.

(6) 40 percent of the nameplate rating(s) of electric space heating of four or more separately controlled units.

Step 3: **Size Service/Feeder Conductors [Section 310-15(b)(6)].**

400 Ampere and Less: The ungrounded conductors are permitted to be sized to Table 310-15(b)(6) for 120/240-volt single-phase systems up to 400 ampere. The grounded (neutral) conductor must be sized to carry the unbalanced load, per Table 310-16.

Over 400 Ampere: The ungrounded and grounded neutral conductors are sized according to Table 310-16.

9–11 DWELLING UNIT OPTIONAL CALCULATION EXAMPLES

❑ **Optional Load Calculation No. 1**

What size service conductor is required for a 1,500 square foot dwelling-unit containing the following loads?

Dishwasher (1,200 VA) 115 volt
Disposal (900 VA) 115 volt
Cooktop (6,000 VA) 115/230 volt
A/C (5 horsepower) 230 volt
Water heater (4,500 VA) 230 volt
Dryer (4,000 VA) 230 volt
Two ovens (each 3,000 VA) 115/230 volt
Electric heating (10 kW), 230 volt

(a) No. 6 (b) No. 4 (c) No. 3 (d) No. 2

• Answer: (c) No. 3

Step 1: General Loads [*Section 220-30(b)*]

Small Appliance Circuits (1,500 VA × 2 circuits)	3,000 VA	
General Lighting (1,500 square feet × 3 VA)	4,500 VA	
Laundry Circuit	1,500 VA	
Appliances (nameplate)		
Dishwasher	1,200 VA	
Water Heater	4,500 VA	
Disposal	900 VA	
Dryer	4,000 VA	
Cooktop	6,000 VA	
Ovens 3,000 VA × 2 units	6,000 VA	
Total Connected Load	31,600 VA	
First 10,000 at 100%	10,000 VA × 1.00 =	10,000 VA
Remainder at 40%	21,600 VA × 0.40 =	8,640 VA
Demand load		18,640 VA

Step 2: Air-Conditioning versus Heat [*Section 220-30(c)*]

Air-conditioning at 100 percent versus 65 percent electric space-heating

Air Conditioner: 230 volts × 28 ampere = 6,440 VA (omit)

Electric heat: 10,000 VA × 0.65 = 6,500 VA

Step 3: Service/Feeder Conductors [*Section 310-15(c)(6)*]

General Loads	18,640 VA
Heat	6,500 VA
Total Demand Load	25,140 VA

I = VA/Volts, I = 25,140 VA/230 volts = 109 ampere

Note: The feeder/service conductor is sized to 110 ampere, No. 3 AWG.

❑ Optional Load Calculation No. 2

What size service conductor is required for a 2,330 square-foot residence with a 300 square-foot open porch and a 150 square-foot carport that contains the following:

Note: System voltage 120/240.

Dishwasher (1.5 kVA)	Waste disposal (1 kVA)	Trash compactor (1.5 kVA)
Water heater (6 kW)	Range (14 kW)	Dryer (4.5 kVA)
A/C (5 horsepower 230 volt)	Electric heat (four, each 2.5 kW)	

(a) No. 5 (b) No. 4 (c) No. 3 (d) No. 2

- Answer: (d) No. 2

Step 1: Total Connected Load

			Totals
Step 1:	General Loads [*Section 220-30(b)*]		
	Small Appliance Circuits (1,500 VA × 2 circuits)	3,000 VA	
	General Lighting (2,330 square feet × 3 VA)	6,990 VA	
	Laundry circuit	1,500 VA	
	Appliances (nameplate)		
	Dishwasher	1,500 VA	
	Disposal	1,000 VA	
	Trash Compactor	1,500 VA	
	Water Heater	6,000 VA	
	Range	14,000 VA	
	Dryer (nameplate)	4,500 VA	
	Total Connected Load	39,900 VA	
	First 10,000 at 100%	10,000 VA × 1.00 =	10,000 VA
	Remainder at 40%	29,000 VA × 0.40 =	11,996 VA
	Demand load		21,996 VA

Step 2: Air-conditioning versus Heat [*Section 220-30(c)*]

Air-conditioning at 100 percent versus 40 percent electric space-heating

Air-conditioner: 230 volts × 28 ampere [Table 430-148] = 6,440 VA

Electric heat: 10,000 VA × 0.40 = 4,000 VA, omit [*Section 220-21*]

Step 3: Service/Feeder Conductors [*Section 310-15(c)(6)*]

General Loads	21,996 VA
A/C	6,440 VA
Total Demand Load	28,436 VA

I = VA/Volts,

I = 28,436 VA/240 volts

I = 118 ampere, No. 2

9–12 NEUTRAL CALCULATIONS – GENERAL [*SECTION 220-22*]

The feeder and service neutral load is the maximum unbalanced demand load between the grounded (neutral) conductor and any one ungrounded conductor as determined by Article 220 Part B (standard calculations). This means that since 240-volt loads are not connected to the neutral conductor, they are not considered for sizing the neutral conductor.

❑ **Neutral Not over 200 Ampere**

What size 120/240-volt single-phase ungrounded and grounded (neutral) conductor is required for a 375 ampere, two-family dwelling-unit demand load, of which 175 ampere consist of 240-volt loads (Figure 9–18)?

(a) Two – 400 kcmil and 1 – 350
(b) Two – 350 kcmil and 1 – 350
(c) Two – 500 kcmil and 1 – 500
(d) Two – 400 kcmil and 1 – No. 3/0

• Answer: (d) Two – 400 kcmil and 1 – No. 3/0

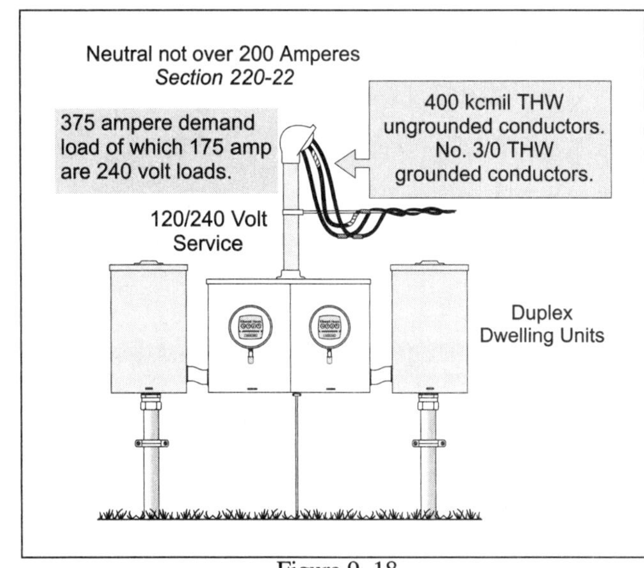

Figure 9–18
Neutral Not over 200 Amperes

The ungrounded conductor is sized at 375 ampere (400 kcmil THHN) according to Table 310-15(b)(6). The grounded (neutral) conductor must be sized to carry the maximum unbalanced load of 200 ampere (120-volt loads only).

Cooking Appliance Feeder/Service Neutral Load [Section 220-22]

The feeder or service cooking appliance neutral load for household cooking appliances such as electric ranges, wall-mounted ovens, or counter-mounted cooking units shall be calculated at 70 percent of the demand load as determined by Section 220-19.

❑ **Range Neutral**

What is the neutral load (in ampere) for one 14 kW range?

(a) 33 ampere (b) 26 ampere (c) 45 ampere (d) 18 ampere

• Answer: (b) 26 ampere

Step 1:➛ Since the range exceeds 12 kW, we must comply with Note 1 of Table 220-19. The first step is to determine the demand load as listed in Column A of Table 220-19 for one unit = 8 kW.

Step 2:➛ We must increase the Column A value (8 kW) by 5 percent for each kW that the average range (in this case 14 kW) exceeds 12 kW. This results is an increase of the Column A value (8 kW) by 10 percent.

8 kW × 1.1 = 8.8 kW × 1,000 = 8,800 Watts

Step 3:➛ Convert the demand load to ampere.

$$I = \frac{P}{E} = \frac{8{,}800 \text{ VA}}{240 \text{ volts}} = 36.7 \text{ ampere}$$

Step 4:➛ Neutral load at 70 percent: 36.7 ampere × 0.7 = 26 ampere

Dryer Feeder/Service Neutral Load [Section 220-22]

The feeder and service dryer neutral demand load for electric clothes dryers shall be calculated at 70 percent of the demand load as determined by Section 220-18.

❑ **Dryer over 5 kW**

What is the feeder/service neutral load for one 5.5 kW dryer?

(a) 3.85 kW (b) 5.5 kW (c) 4.6 kW (d) none of these

• Answer: (a) 3.85 kW

5.5 kW × 0.7 = 3.85 kW

Unit 9 – Single-Family Dwelling Unit Load Calculations Summary Questions

Part A – General Requirements

9–2 Voltages

1. Unless other voltages are specified, branch circuit, feeder, and service loads shall be computed at nominal system voltages of _____ .
(a) 120 (b) 120/240 (c) 208Y/120 (d) any of these

9–3 Fraction of an Ampere

2. When a calculation results in _____ of an ampere or larger, such fractions may be rounded up to the next ampere.
(a) 0.15 (b) 0.5 (c) 0.49 (d) 0.051

9–4 Appliance (Small) Circuits [*Section 220–11(c)(1)*]

3. A minimum of _____ 20-ampere small appliance branch circuits are required for receptacle outlets in the kitchen, dining room, breakfast room, pantry, or similar areas of dining.
(a) one (b) two (c) three (d) none of these

4. When sizing the feeder or service, each dwelling-unit shall have a minimum feeder load of _____ VA for the two small appliance branch circuits.
(a) 1,500 (b) 2,000 (c) 3,000 (d) 4,500

9–5 Cooking Equipment – Branch Circuit [Table 220-19, Note 4]

Table 220-19 Column A

5. What is the branch circuit demand load (in ampere) for one 11 kW range?
(a) 21 ampere (b) 27 ampere (c) 38 ampere (d) 33 ampere

6. What is the branch circuit load (in ampere) for one 13.5 kW range?
(a) 33 ampere (b) 37 ampere (c) 50 ampere (d) 58 ampere

Wall Mounted Oven and/or Counter-Mounted Cooking Unit

7. • What is the branch circuit load (in ampere) for one 6 kW wall-mounted oven, 240 volts, single-phase?
(a) 15 ampere (b) 25 ampere (c) 30 ampere (d) 40 ampere

8. • What is the branch circuit load (in ampere) for one 4.8 kW counter-mounted cooking unit, 240 volts, single-phase?
(a) 15 ampere (b) 20 ampere (c) 25 ampere (d) 30 ampere

9. What is the branch circuit load (in ampere) for one 6 kW counter-mounted cooking unit and one 3 kW wall-mounted oven, 240 volts, single-phase?
(a) 25 ampere (b) 38 ampere (c) 33 ampere (d) 42 ampere

10. What is the branch circuit load (in ampere) for one 6 kW counter-mounted cooking unit and two 4 kW wall-mounted ovens, 240 volts, single-phase?
(a) 22 ampere (b) 27 ampere (c) 33 ampere (d) 37 ampere

9–6 Laundry Receptacle(s) Circuit [*Section 220-11(c)(2)*]

11. The *NEC*® does not require a separate circuit for the washing machine, but requires a separate circuit for the laundry room receptacle or receptacles, of which one can be for the washing machine.
(a) True (b) False

12. Each dwelling-unit shall have a feeder load consisting of _____ VA for the 20-ampere laundry circuit.
(a) 1,500 (b) 2,000 (c) 3,000 (d) 4,500

9–7 Lighting and Receptacles

13. The *NEC*® requires a minimum 3 volt-ampere each square foot for the required general lighting and receptacles . The dimensions for determining the area shall be computed from the outside of the building and shall not include _____ .
(a) open porches (b) garages
(c) spaces not adaptable for future use (d) all of these

14. The 3 volt-ampere each square foot for general lighting includes all 15 and 20 ampere general use receptacles, but not the appliance or laundry circuit receptacles.
(a) True (b) False

15. What is the general lighting load and receptacle load for a 2,100 square-foot home that has 14 extra convenience receptacles, and nine extra recessed fixtures rated 100 watts each?
(a) 2,100 VA (b) 4,200 VA (c) 6,300 VA (d) 8,400 VA

Number of General Lighting Circuits

16. The number of branch circuits required for general lighting and receptacles shall be determined from the general lighting load and the rating of the circuits.
(a) True (b) False

17. How many 15-ampere general lighting circuits are required for a 2,340 square-foot home?
(a) 2 (b) 3 (c) 4 (d) 5

18. How many 20-ampere general lighting circuits are required for a 2,100 square-foot home?
(a) 2 (b) 3 (c) 4 (d) 5

19. How many 20-ampere circuits are required for the general lighting and receptacles for a 1,500 square-foot dwelling-unit?
(a) 2 (b) 3 (c) 4 (d) 5

Part B – Standard Method – Feeder/Service Load Calculation

9–8 Dwelling Unit Feeder/Service Load Calculations (Part B of Article 240)

20. The *NEC*® recognizes that the general lighting and receptacles and small appliance and laundry circuits will not all be on or loaded at the same time and permits a _____ to be applied to the total of these loads.
(a) demand factor (b) adjustment factor (c) correction factor (d) correction

21. Because the air-conditioning and heating loads are not on at the same time (simultaneously), it is permissible to omit the smaller of the two loads when determining the air-conditioning versus heat demand load.
(a) True (b) False

22. A load demand factor of 75 percent is permitted for _____ or more appliances fastened in place such as dishwasher, waste disposal, trash compactor, water heater, etc.
(a) one (b) two (c) three (d) four

23. The feeder or service demand load for electric clothes dryers located in dwelling-units shall not be less than _____ .
(a) 5,000 watts (b) nameplate rating (c) a or b (d) the greater of a or b

24. • A feeder or service dryer load is required, even if the dwelling-unit does not contain an electric dryer.
(a) True (b) False

25. Household cooking appliances rated $1^3/_4$ kW can have the feeder and service loads calculated according to the demand factors of Section 220-19, Table and Notes.
(a) True (b) False

26. Dwelling-unit feeder and service ungrounded conductors are sized according to Note 3 of Table 310-16 for _____ .
(a) 3-wire, single-phase, 120/240-volt systems up to 400 ampere
(b) 3-wire, single-phase, 120/208-volt systems up to 400 ampere
(c) 4-wire, single-phase, 120/240-volt systems up to 400 ampere
(d) none of these

27. What is the general lighting load for a 2,700 square-foot dwelling-unit?
(a) 8,100 VA (b) 12,600 VA (c) 2,700 VA (d) 6,840 VA

28. What is the total connected load for general lighting and receptacles and small appliance and laundry circuits for a 6,540 square-foot dwelling-unit?
(a) 8,100 VA (b) 12,600 VA (c) 2,700 VA (d) 24,120 VA

29. What is the feeder or service demand load for an air-conditioner (5 horsepower 230 volt) and three baseboard heaters (3 kW)?
(a) 6,400 VA (b) 3,000 VA (c) 8,050 VA (d) 9,000 VA

30. What is the feeder or service demand load for an air-conditioner (4 horsepower 230 volt) and electric space heater (10 kW)?
(a) 10,000 VA (b) 3,910 VA (c) 6,440 VA (d) 10,350 VA

31. What is the feeder or service demand load for a waste disposal (940 VA), dishwasher (1,250 VA), and a water heater (4,500 VA)?
(a) 5,018 VA (b) 6,690 VA (c) 8,363 VA (d) 6,272 VA

32. • What is the feeder or service demand load for a waste disposal (940 VA), dishwasher (1,250 VA), water heater (4,500 VA), and a trash compactor (1,100 VA)?
(a) 7,790 VA (b) 5,843 VA (c) 7,303 VA (d) 9,738 VA

33. What is the feeder or service demand load for a 4 kW dryer?
(a) 4,000 VA (b) 3,000 VA (c) 5,000 VA (d) 5,500 VA

34. What is the feeder or service demand load for a 5.5 kW dryer?
(a) 4,000 VA (b) 3,000 VA (c) 5,000 VA (d) 5,500 VA

35. • What is the feeder or service demand load for two 3 kW ranges?
(a) 3 kW (b) 4.8 kW (c) 4.5 kW (d) 3.9 kW

36. What is the feeder or service demand load for one 6 kW cooking appliance?
(a) 6 kW (b) 4.8 kW (c) 4.5 kW (d) 3.9 kW

37. What is the feeder or service demand load for one 6 kW and two 3 kW cooking appliances?
(a) 6 kW (b) 4.8 kW (c) 4.5 kW (d) 9.3 kW

38. What is the feeder or service demand load for an 11.5 kW range?
(a) 11.5 kW (b) 8 kW (c) 9.2 kW (d) 6 kW

39. What is the feeder or service demand load for a 13.6 kW range?
(a) 8.8 kW (b) 8 kW (c) 9.2 kW (d) 6 kW

40. What size 120/240-volt single-phase service or feeder THHN conductors are required for a dwelling-unit that has a 190-ampere service demand load?
(a) No. 1/0 (b) No. 2/0 (c) No. 3/0 (d) No. 4/0

41. • What is the net computed demand load for the general lighting and small appliance and laundry circuits for a 1,500 square foot-dwelling-unit?
(a) 4,500 VA (b) 9,000 VA (c) 5,100 VA (d) none of these

42. What is the feeder or service demand load for an air-conditioner (3 horsepower 230 volt) with a blower (1/4 horsepower) and electric space heating (4 kW)?
(a) 3,910 VA (b) 5,555 VA (c) 4,000 VA (d) none of these

43. What is the feeder or service demand load for a dwelling-unit that has one water heater (4 kVA) and one dishwasher (1.5 kVA)?
(a) 4 kW (b) 5.5 kW (c) 4.13 kW (d) none of these

44. What is the net computed load for a 4.5 kW dryer?
(a) 4.5 kW (b) 5 kW (c) 5.5 kW (d) none of these

45. What is the total net computed demand load for a 14 kW range?
(a) 8 kW (b) 8.4 kW (c) 8.8 kW (d) 14 kW

46. If the total demand load is 30 kW for a 120/240-volt dwelling-unit, what is the service and feeder conductor size?
(a) No. 4 (b) No. 3/0 (c) No. 2 (d) No. 1/0

47. If the service raceway contains No. 2 service conductors, what size bonding jumper is required for the service raceway?
(a) No. 8 (b) No. 6 (c) No. 4 (d) No. 3

48. If the service contains No. 2 conductors, what is the minimum size grounding electrode conductor required?
(a) No. 8 (b) No. 6 (c) No. 4 (d) No. 3

Part C – Optional Method – Load Calculation Questions

9–10 Dwelling Unit Optional Feeder/Service Calculations [*Section 220-30*]

49. When sizing the feeder or service conductor according to the optional method we must:
(a) Determine the total connected load of the general lighting and receptacles, small appliance and laundry branch circuits, and the nameplate VA rating of all appliances and motors.
(b) Determine the demand load by applying the following demand factor to the total connected load: First 10 kVA at 100 percent, remainder at 40 percent.
(c) Determine the air-conditioning versus heat demand load.
(d) all of these.

Use this information for Question 50.
A 1,500 square-foot dwelling-unit containing the following loads:

Dishwasher (1.5 kVA)
Waste disposal (1 kVA)
Cooktop (6 kVA)
A/C (3 horsepower 230 volts)
Water heater (5 kVA)
Dryer (5.5 kVA)
Two ovens (each 3 kVA)
Electric heat (10 kW)

50. • Using the optional method, what size aluminum conductor is required for the 120/240-volt single-phase service?
(a) No. 1 (b) No. 1/0 (c) No. 2/0 (d) No. 3/0

Use this information for Question 51.
A 2,330 square-foot residence with a 300 square-foot porch and 150 square-foot carport contains the following:

Dishwasher (1.5 kVA)
Trash compactor (1.5 kVA)
Range (14 kW)
A/C (5 horsepower 230 volts)
Waste disposal (1 kVA)
Water heater (6 kW)
Dryer (4.5 kW)
Baseboard heat (four each 2.5 kW)

51. Using the optional method, what size conductors are required for the 120/230-volt single-phase service?
(a) No. 6 (b) No. 4 (c) No. 3 (d) No. 2

9–12 Neutral Calculations – General

52. The feeder and service neutral load is the maximum unbalanced demand load between the grounded (neutral) conductor and any one ungrounded conductor as determined by Article 220 Part B. Since 240-volt loads cannot be connected to the neutral conductor, 240-volt loads are not considered for sizing the feeder neutral conductor.
(a) True (b) False

53. • What size single-phase, 3-wire feeder is required for a 475-ampere demand load of which 275 ampere consists of 240-volt loads? Size the conductors based on 75°C terminals, in accordance with Section 110-14(c).
(a) 2 – 750 and 1 – 350 (b) 2 – 500 and 1 – 350
(c) 2 – 750 and 1 – 500 (d) 2 – 750 and 1 – No. 3/0

54. The feeder or service neutral load for household cooking appliances such as electric ranges, wall-mounted ovens, or counter-mounted cooking units shall be calculated at _____ of the demand load as determined by Section 220-19.
(a) 50% (b) 60% (c) 70% (d) 80%

55. What is the dwelling-unit service neutral load for one 9 kW ranges?
(a) 5.6 kW (b) 6.5 kW (c) 3.5 kW (d) 12 kW

56. The service neutral demand load for household electric clothes dryers shall be calculated at _____ of the demand load as determined by Section 220-18.
(a) 50% (b) 60% (c) 70% (d) 80%

57. What is the feeder neutral load for one 6 kW household dryers?
(a) 4.2 kW (b) 4.7 kW (c) 5.4 kW (d) 6.6 kW

✯ Challenge Questions

9–5 Cooking Equipment – Branch Circuit [Table 220-19 Note 4]

58. The branch circuit demand load for one 18 kW range is _____ .
(a) 12 kW (b) 8 kW (c) 10.4 kW (d) 18 kW

59. A dwelling-unit kitchen has the following appliances: one 9 kW cooktop and one wall-mounted oven rated 5.3 kW. The branch circuit demand load for these appliances is _____ .
(a) 14.3 kW (b) 12 kW (c) 8.8 kW (d) 8 kW

9–8 Dwellling Unit Feeder/Service Load Calculation (Part B of Article 220)

General Lighting [Table 220-11]

60. An apartment building contains 20 units, each 840 square feet. What is the general lighting feeder demand load for each dwelling-unit? Note: Laundry facilities are provided on the premises for all tenants and no laundry circuit is required in each unit, see 210-52(f) Exception No. 1.
(a) 3,520 VA (b) 3,882 VA (c) 4,220 VA (d) 6,300 VA

61. An 1,800 square-foot residence has a 300 square-foot open porch and a 450 square-foot carport. What is the demand load for the general lighting, receptacles, small appliance circuits, and laundry circuits?
(a) 5,400 VA (b) 7,900 VA (c) 5,415 VA (d) 6,600 VA

62. How many 15-ampere, 120-volt branch circuits are required for general use receptacle and lighting for a 1,800 square-foot dwelling-unit?
(a) 1 circuit (b) 2 circuits (c) 3 circuits (d) 4 circuits

63. How many 20-ampere, 120-volt branch circuits are required for general use receptacle and lighting for a 2,800 square-foot dwelling-unit?
(a) 2 circuits (b) 6 circuits (c) 7 circuits (d) 4 circuits

Appliance Demand Factors [*Section 220-17*]

64. A dwelling-unit contains one of each of the following: washing machine (1.2 kW), water heater (4 kW), dishwasher (1.2 kW), trash compactor (1.5 kW). The appliance demand load added to the service is _____ .
(a) 5.9 kW (b) 6.7 kW (c) 7.7 kW (d) 8.8 kW

65. • A dwelling-unit contains one of each of the following: water heater (4 kW), dishwasher ($^1/_2$ horsepower), dryer, pool pump ($^3/_4$ horsepower), cooktop (6 kW), oven (6 kW), A/C (4 horsepower 230 volts), heat (6 kW). What is the appliance demand load for the dwelling-unit?
(a) 6.7 kW (b) 4 kW (c) 9 kW (d) 11 kW

Dryer Calculation [*Section 220-18*]

66. A dwelling-unit contains a 5.5 kW electric clothes dryer. What is the feeder demand load for the dryer?
(a) 3.38 kW (b) 4.5 kW (c) 5 kW (d) 5.5 kW

Cooking Equipment Demand Load [*Section 220-19*]

67. The feeder load for twelve 1$^3/_4$ kW cooktops is _____ .
(a) 21 kW (b) 10.5 kW (c) 13.5 kW (d) 12.5 kW

68. The minimum neutral for a branch circuit to an 8 kW household range is _____ . See Article 210.
(a) No. 10 (b) No. 12 (c) No. 8 (d) No. 6

69. What is the maximum dwelling-unit feeder or service demand load for fifteen 8 kW cooking units?
(a) 38.4 kW (b) 33 kW (c) 88 kW (d) 27 kW

Note 2 of Table 220-19

70. What is the feeder demand load for five 10 kW, five 14 kW, and five 16 kW household ranges?
(a) 70 kW (b) 23 kW (c) 14 kW (d) 33 kW

Note 3 to Table 220-19

71. A dwelling-unit has one 6 kW cooktop and one 6 kW oven. What is the minimum feeder demand load for the cooking appliances?
(a) 13 kW (b) 8.8 kW (c) 7.8 kW (d) 8.2 kW

72. The maximum feeder demand load for an 8 kW range is _____ .
(a) 8.5 kW (b) 8 kW (c) 6.3 kW (d) 12 kW

73. The minimum feeder demand load for five 5 kW ranges, two 4 kW ovens, and four 7 kW cooking units is _____ .
(a) 61 kW (b) 22 kW (c) 19.5 kW (d) 18 kW

74. The minimum feeder demand load for two 3 kW wall-mounted ovens and one 6 kW cooktop is _____ .
(a) 9.3 kW (b) 11 kW (c) 8.4 kW (d) 9.6 kW

9–9 Dwelling Unit Feeder/Service Calculations Examples

75. • Both units of a duplex apartment require a 100-ampere main; the resulting 200-ampere service would require _____ THHN.
(a) No. 1/0 (b) No. 2/0 (c) No. 3/0 (d) No. 4

76. After all demand factors have been taken into consideration, the demand load for a dwelling-unit is 21,560 VA. The minimum service size for this residence would be _____ if the optional method of service calculations were used for a single-phase 120/240-volt system. Be sure to read Section 220-30 carefully.
(a) 90 ampere (b) 100 ampere (c) 110 ampere (d) 125 ampere

77. After all demand factors have been taken into consideration, the net computed load for a service is 24,221 VA, 120/240-volt single-phase system. The minimum service size is _____ .
(a) 150 ampere (b) 175 ampere (c) 125 ampere (d) 110 ampere

9–10 Optional Method – Load Calculation Questions [*Section 220-30*]

78. Using the optional calculation method, the demand load for a 6 kW central space heating and a 4 kW air-conditioning unit would be _____ .
(a) 9,000 watts (b) 3,900 watts (c) 4,000 watts (d) 5,000 watts

79. The total connected load of a dwelling-unit is 25 kVA, not including heat or air-conditioning. If the heat is separately controlled in five rooms (10 kW), and the air-conditioning is 6 kW, then the total service demand load is _____ if the optional method is used.
(a) 22 kVA (b) 29 kVA (c) 35 kVA (d) 36 kVA

80. Using the optional method, the service size would be _____ for the following loads: 1,200 square feet first floor, plus 600 square feet on the second floor, and a 200 square-foot open porch, water heater (4 kW), dishwasher (1/2 horsepower), clothes dryer (4 kW), A/C (4 horsepower), heat (6 kW), pool pump (3/4 horsepower), oven (6 kW), and cooktop (6 kW). The service source voltage is 230/115-volt single-phase.
(a) 200 ampere (b) 175 ampere (c) 125 ampere (d) 110 ampere

81. An 1,800 square-foot residence contains the following: a 300 square-foot porch, a 150 square-foot carport, a 4 kW water heater, 10 kW heat separated in five rooms, one 1.5 kW dishwasher, one 6 kW range, one 4.5 kW dryer, two 3 kW ovens, and one 6 kW air conditioner. The 120/240-volt single-phase service size for the loads would be _____ . The optional method of service calculations were used.
(a) 175 ampere (b) 110 ampere (c) 125 ampere (d) 150 ampere

82. A dwelling-unit has 1,200 square feet on the first floor, 600 square feet upstairs (unfinished but adaptable for future use), and a 200 square-foot open porch with the following loads: pool pump (3/4 horsepower), range (13.9 kW), dishwasher (1.2 kW), water heater (4 kW), dryer (4 kW), A/C (5 horsepower), and heat (6 kW). Using the optional calculation, what size service and feeder conductor is required for a 120/240-volt single-phase service?
(a) No. 4 (b) No. 3 (c) No. 2 (d) No. 1

CHAPTER 3
Advanced *NEC*® Calculations

Scope of Chapter 3

Unit 10 Multifamily Dwelling Unit Load Calculations

Unit 11 Commercial Load Calculations

Unit 12 Delta/Delta and Delta/Wye Transformer Calculations

Unit 10

Multifamily Dwelling Unit Load Calculations

OBJECTIVES

After reading this unit, the student should be able to explain the following concepts:

Air-conditioning versus heat
Appliance (small) branch circuits
Appliance demand load
Clothes dryer demand load
Cooking equipment calculations
Dwelling-unit calculations – optional method
Dwelling-unit calculations – standard method
Laundry circuit
Lighting and receptacles

After reading this unit, the student should be able to explain the following terms:

General lighting
General use receptacles
Optional method
Rounding
Standard method
Unbalanced demand load
Voltages

10–1 MULTIFAMILY DWELLING-UNIT CALCULATIONS – STANDARD METHOD

When determining the ungrounded and grounded (neutral) conductor for multifamily dwelling units (Figure 10–1), apply the following standard method steps:

Step 1: ➼ **General Lighting and Receptacles, Small Appliance and Laundry Circuits [*Section 220-11*]**

The *NEC*® recognizes that the general lighting and receptacles, small appliance, and laundry circuits will not all be on, or loaded, at the same time. The *NEC*® permits the following demand factors to be applied to these loads [*Section 220-16*]:

(a) Total Connected Load: Determine the total connected general lighting and receptacles (3 VA per square-foot), small appliance (3,000 VA), and the laundry (1,500 VA) circuit load of all dwelling-units. The laundry load (1,500 VA) can be omitted if laundry facilities which are available to all building occupants, are provided on the premises [*Section 210-52(f)* Exception 1].

(b) Demand Factor: Apply Table 220-11 demand factors to the total connected load (Step 1a).
First 3,000 VA at 100% demand
Next 117,000 VA at 35% demand
Remainder at 25% demand

Step 2: ➼ **Air-Conditioning versus Heat [*Section 220-15, 220-21,* and *440-34*]**

Because the air-conditioning and heating loads are not on at the same time (simultaneously), it is permissible to omit the smaller of the two loads.

(a) Air-conditioning: The air-conditioning demand load shall be calculated at 125 percent of the largest air-conditioning motor VA, plus the sum of the other air-conditioning motor VAs [*Section 440-34*].

(b) Heat: Electric space-heating loads shall be computed at 100 percent of the total connected load [*Section 220-15*].

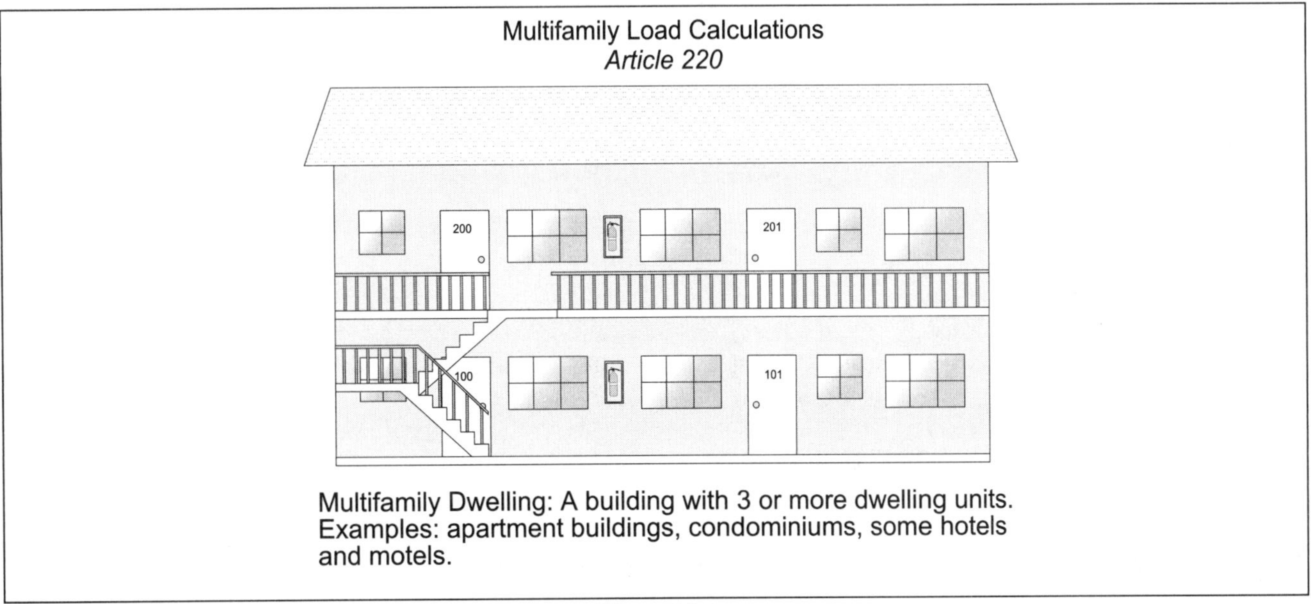

Figure 10-1
Multifamily Load Calculations

Step 3: ➛ **Appliances [*Section 220-17*]**

A demand factor of 75 percent is permitted for four or more appliances fastened in place such as a dishwasher, waste disposal, trash compactor, water heater, etc. This does not apply to space-heating equipment [*Section 220-15*], clothes dryers [*Section 220-18*], cooking appliances [*Section 220-19*], or air-conditioning equipment [*Section 440-34*].

Step 4: ➛ **Clothes Dryers [*Section 220-18*]**

The feeder or service demand load for electric clothes dryers located in dwelling units shall not be less than: 5,000 watts, or the nameplate rating (whichever is greater) adjusted according to the demand factors listed in Table 220-18.
Note: A dryer load is not required if the dwelling unit does not contain an electric dryer. Laundry room dryers shall not have their loads calculated according to this method. This is covered in Unit 11.

Step 5: ➛ **Cooking Equipment [*Section 220-19*]**

Household cooking appliances rated over $1^3/_4$ kW can have their feeder and service loads calculated according to the demand factors of Section 220-19, Table and Notes.

Step 6: ➛ **Feeder and Service Conductor Size**

400 ampere and less: The ungrounded conductors are sized according to Table 310-15(b)(6) for 120/240-volt single-phase systems.
Over 400 ampere: The ungrounded conductors are sized according to Table 310-16 to the calculated unbalanced demand load.

10–2 MULTIFAMILY DWELLING-UNIT CALCULATION EXAMPLES – STANDARD METHOD

Step 1. General Lighting, Small Appliance, and Laundry Demand [Section 220-11]

➛ **General Lighting Load No. 1**

What is the demand load for an apartment building that contains 20 units? Each apartment is 840 square feet.
Note: Laundry facilities are provided on the premises for all tenants [*Section 210-52(f)*].

(a) 5,200 VA (b) 40,590 VA (c) 110,400 VA (d) none of these

• Answer: (b) 40,590 VA

General lighting	(840 square feet × 3 VA)	2,520 VA	
Small appliance circuits		3,000 VA	
Laundry circuit		0 VA	
Total connected load for one unit		5,520 VA	
Demand factor [*Section 220-11*] (5,520 VA × 20 units)		110,400 VA	
First 3,000 VA at 100%		− 3,000 VA × 1.00 =	3,000 VA
Next 117,000 VA at 35%		107,400 VA × 0.35 =	37,590 VA
Total demand load			40,590 VA

➛ General Lighting Load No. 2

What is the general lighting net computed demand load for a 20-unit apartment building? Each unit is 990 square feet.

(a) 74,700 VA (b) 149,400 VA (c) 51,300 VA (d) 105,600 VA

• Answer: (c) 51,300 VA

General lighting	(990 square feet × 3 VA)	2,970 VA	
Small appliance circuits		3,000 VA	
Laundry circuit		1,500 VA	
Total connected load for one unit		7,470 VA	
Demand factor [*Section 220-11*] (7,470 VA × 20 units)		149,400 VA	
First 3,000 VA at 100%		− 3,000 VA × 1.00 =	3,000 VA
		146,400 VA	
Next 117,000 VA at 35%		−117,000 VA × 0.35 =	40,950 VA
Remainder VA at 25%		29,400 VA × 0.25 =	+ 7,350 VA
Total demand load			51,300 VA

Step 2. Air-Conditioning versus Heat [Section 220-15 and 440-34]

➛ Air- Conditioning versus Heat No. 1

What is the net computed air-conditioning versus heat load for a 40-unit multifamily building that has an air conditioner (3 horsepower 230 volt) and two baseboard heaters (3 kW) in each unit?

(a) 160 kW (b) 240 kW (c) 60 kW (d) 50 kW

• Answer: (b) 240 kW

Air-conditioning [*Section 440-34*] 230 volts × 17 ampere* = 3,910 VA

(3,910 VA × 1.25) + (3,910 VA × 39 units) = 157,378 VA, Omit smaller than heat [*Section 220-21*]

Heat [*Section 220-15*] 3,000 watts × 2 units = 6,000 watts × 40 units = 240,000 watts/1,000 = 240 kW

➛ Air-Conditioning versus Heat No. 2

What is the air-conditioning versus heat demand load for a 25-unit multifamily building that has a air-conditioning (3 horsepower 230 volt) and electric heat (5 kW)?

(a) 160 kVA (b) 125 kVA (c) 6 kVA (d) 5 kVA

• Answer: (b) 125 kVA

Air-conditioning [*Section 440-34*] 230 volts × 17 ampere* = 3,910 VA

(3,910 VA × 1.25) + (3,910 × 24 units) = 98,728 VA, Omit

Heat [*Section 220-15*] 5,000 watts × 25 units = 125,000 watts/1,000 = 125 kW

* Table 430-148

Step 3. Appliance Demand Load [Section 220-17]

➛ Appliance Load No. 1

What is the appliance demand load for a 20-unit multifamily building that contains a 940 VA waste disposal, a 1,250 VA dishwasher, and a 4,500 VA water heater?

(a) 100 kVA (b) 134 kVA (c) 7 kVA (d) 5 kVA

• Answer: (a) 100 kVA

Waste disposal	940 VA
Dishwasher	1,250 VA
Water heater	4,500 VA
Total demand load	6,690 VA × 20 units × 0.75 = 100,350 VA

➵ Appliance Load No. 2

What is the appliance demand load for a 35-unit multifamily building that contains a 900 VA waste disposal, a 1,200 VA dishwasher, and a 5,000 VA water heater?

(a) 71 kVA (b) 142 kVA (c) 107 kVA (d) 186 kVA

• Answer: (d) 186 kVA

Waste disposal	900 VA
Dishwasher	1,200 VA
Water heater	5,000 VA
Total demand load	7,100 VA × 35 units × 0.75 = 186,375 VA/1,000 = 186 kVA

Step 4. Dryer Demand Load [Section 220-18]

➵ Dryer Load No. 1

A multifamily dwelling (12-unit building) contains a 4.5 kVA electric clothes dryer in each unit. What is the feeder and service dryer demand load for the building?

(a) 5 kVA (b) 27 kVA (c) 60 kVA (d) none of these

• Answer: (b) 27 kVA

5 kVA × 12 units = 60 kVA × 0.45 = 27 kVA

➵ Dryer Load No. 2

What is the demand load for twenty 5.25 kW dryers installed in dwelling units of a multifamily building?

(a) 5 kVA (b) 27 kVA (c) 60 kVA (d) 37 kVA

• Answer: (d) 37 kVA

5.25 kVA × 20 units = 105 kVA × 0.35 = 36.75 kVA

Step 5. Cooking Equipment Demand Load [Section 220-19]

❑ Column A

What is the feeder and service demand load for five 9 kW ranges?

(a) 9 kW (b) 45 kW (c) 20 kW (d) none of these

• Answer: (c) 20 kW

❑ Over 12 kW – Note 1

What is the feeder and service demand load for three ranges rated 16 kW each?

(a) 15 kW (b) 14 kW (c) 17 kW (d) 21 kW

• Answer: (c) 17 kW (closest answer)

Step 1: ➵ "Column A" demand load: 14 kW (3 units)

Step 2: ➵ The average range (16 kW) exceeds 12 kW by 4 kW. Increase "Column A" demand load (14 kW) by 20 percent: 14 kW × 1.2 = 16.8 kW.

❑ Unequal Ratings over 12 kW – Note 2

What is the feeder and service demand load for three ranges rated 9 kW and three ranges rated 14 kW?

(a) 36 kW (b) 42 kW (c) 78 kW (d) 22 kW

• Answer: (d) 22 kW

Step 1: ➼ Determine the total connected load.

9 kW (minimum 12 kW)	3 ranges × 12 kW =	36 kW
14 kW	3 ranges × 14 kW =	42 kW
Total connected load		78 kW

Step 2: ➼ Determine the average range rating, 78 kW/6 units = 13 kW average rating.

Step 3: ➼ Demand load Table 220-19 Column A: 6 ranges = 21 kW.

Step 4: ➼ The average range (13 kW) exceeds 12 kW by 1 kW. Increase Column A demand load (21 kW) by 5 percent: 21 kW × 1.05 = 22.05 kW.

❑ **Note 3 – Less than $3^1/_2$ kW – Column B**

What is the feeder and service demand load for ten 3 kW ovens?

(a) 10 kW (b) 30 kW (c) 15 kW (d) 20 kW

• Answer: (c) 15 kW closest answer

3 kW × 10 units = 30 kW × 0.49 = 14.70 kW

❑ **Note 3 – Less than $8^3/_4$ kW – Column C**

What is the feeder and service demand load for eight 6 kW cooktops?

(a) 10 kW (b) 17 kW (c) 14.7 kW (d) 48 kW

• Answer: (b) 17 kW

6 kW × 8 units = 48 kW × 0.36 = 17.28 kW

Step 6. Service Conductor Size [Table 310-15(b)(6)]

➼ **Service Conductor Size No. 1**

What size aluminum service conductors are required for a 120/240-volt, single-phase multifamily building that has a total demand load of 93 kVA (Figure 10–2)?

(a) 300 kcmil (b) 350 kcmil (c) 500 kcmil (d) 600 kcmil

• Answer: (d) 600 kcmil aluminum

I = VA/E

I = 93,000 VA/240 volts = 388 ampere,

600 kcmil aluminum

➼ **Service Conductor Size No. 2**

What size service conductors are required for a multifamily building that has a total demand load of 270 kVA for a 120/208-volt 3-phase system?

Note: Service conductors are run in parallel.

(a) 2 – 300 kcmil per phase (b) 2 – 350 kcmil per phase

(c) 2 – 500 kcmil per phase (d) 2 – 600 kcmil per phase

• Answer: (c) 2 – 500 kcmil per phase

I = VA/(E × 1.732)

I = 270,000 VA/(208 volts × 1.732) = 750 ampere

Ampere per parallel set:

750 ampere/2 raceways = 375 ampere per phase.

500 kcmil conductor has an ampacity of 380 ampere [Table 310-16 at 75°C].

Two sets of 500 kcmil conductors (380 ampere × 2) can be protected by an 800-ampere protection device [*Section 240-3(b)*].

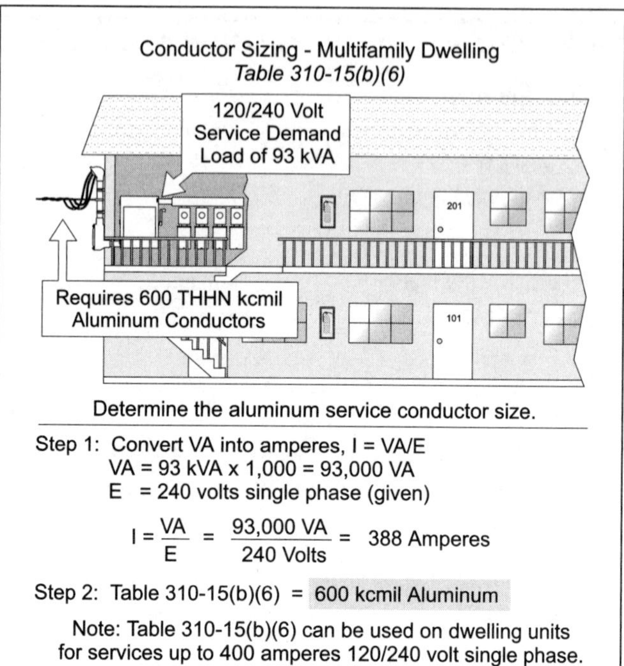

Figure 10-2
Conductor Sizing – Multifamily Dwelling

10–3 MULTIFAMILY DWELLING-UNIT CALCULATION SAMPLE – STANDARD METHOD

What is the demand load for a 25 unit apartment building? Each apartment is 1,000 square feet and contains the following: air-conditioning (28 ampere, 240 volts), heat (7.5 kW), dishwasher (1.2 kVA), waste disposal (1.5 kVA), water heater (4.5 kW), dryer (4.5 kW), and range (15.5 kW). System voltage 120/240 volt, single-phase.

Step 1. General Lighting, Small Appliance, and Laundry Demand [Section 220-11]

Each unit contains:

General lighting	(1,000 square feet × 3 VA)	3,000 VA	
Small appliance circuits		3,000 VA	
Laundry circuit		1,500 VA	
		7,500 VA	
Total demand load (7,500 VA × 25 units)		187,500 VA	
First 3,000 VA at 100% =		– 3,000 VA × 1.00 =	3,000 VA
		184,500 VA	
Next 117,000 VA at 35%		– 117,000 VA × 0.35 =	40,950 VA
Remainder at 25%		67,500 VA × 0.25 =	16,875 VA
			60,875 VA

Step 2. Air-Conditioning versus Heat [Section 220-15 and 440-34]

Air-conditioning [*Section 440-34*], 230 volts × 28 ampere = 6,440 VA

6,440 VA × 1.25 = 8,050 VA + (6,440 × 24 units) = 162,610 VA, omit [*Section 220-21*]

Heat [*Section 220-15*]

7,500 VA × 25 units = 187,500 VA

Step 3. Appliance Demand Load [Section 220-17]

Waste disposal	1,200 VA
Dishwasher	1,500 VA
Water heater	4,500 VA
Total connected load	7,200 VA × 25 units × 0.75 = 135,000 VA

Step 4. Dryer Demand Load [Section 220-18]

Total connected load = 5 kVA* × 25 units = 125 kW

Demand factor for 25 units = 32.5%

Total demand dryer load = 125 kW × 0.325 = 40.625 kW

* The minimum dryer load for standard load calculations is 5 kW

Step 5. Cooking Equipment Demand Load [Section 220-19]

Step 1: ➛ Column A demand load for 25 units = 40 kW.

Step 2: ➛ The average range (15.5 kW) exceeds 12 kW by 3.5 kW. Increase Column A demand load (40 kW) by 20%.
Range demand load: 40 kW × 1.2 = 48 kW

Step 6. Service Conductor Size [Table 310-15(b)(6)]

Step 1: ➛ Total general lighting, small appliance, laundry demand load	60,875 VA
Step 2: ➛ Total heat demand load [*Section 220-15*] (7,500 VA × 25 units)	187,500 VA
Step 3: ➛ Total appliance demand load	135,000 VA
Step 4: ➛ Total demand dryer load	40,625 VA
Step 5: ➛ Range demand load	48,000 VA
Total demand load	472,000 VA

Service conductor ampere = VA/E = 472,000 VA/240 volts = 1,967 ampere

10–4 MULTIFAMILY DWELLING-UNIT CALCULATIONS [*SECTION 220-32*] – OPTIONAL METHOD

Instead of sizing the ungrounded conductors according to the standard method (Part B of Article 220), the optional method can be used for feeders and service conductors in multifamily dwelling-units. Follow these steps for determining the demand load:

Step 1: ➻ **Total Connected Load**

Add up the following loads:

(a) General lighting: 3 volt-ampere per square foot.

(b) Small appliance and laundry branch circuit. 1,500 volt-ampere for each 20-ampere small appliance and laundry branch circuit.

(c) Appliances: The *nameplate* VA rating of all appliances and motors fastened in place (permanently connected), but not the air-conditioning or heating load. Be sure to use the range and dryer nameplate ratings.

(d) Air-conditioning versus heat: Determine which load is larger, air-conditioning or heat.
(1) Air-conditioning or heat-pump compressors at 100 percent.
(2) Heat at 100 percent.

Step 2: ➻ **Demand Load**

The net computed demand load is determined by applying the demand factor from Table 220-32 to the total connected load (Step 1). The net computed demand load (kVA) can be converted to ampere by:

$$I\text{ (single-phase)} = \frac{VA}{E} \qquad I\text{ (three-phase)} = \frac{VA}{(E \times 1.732)}$$

Step 3: ➻ **Feeder and Service Conductor Size**

400 ampere and less: The ungrounded conductors are sized according to Table 310-15(b)(6) for 120/240-volt, single-phase systems up to 400 ampere.

Over 400 ampere: The ungrounded conductors are sized according to Table 310-16 based on the calculated demand load.

When do you use the standard method verses the optional method? For the purpose of exam preparation, always use the standard load calculation unless the question specifies the optional method. In the field, you will probably want to use the optional method because it results in a smaller service.

10–5 MULTIFAMILY DWELLING-UNIT EXAMPLE QUESTIONS [*SECTION 220-32*] – OPTIONAL METHOD

A multifamily building has 12 units on each phase, each is 1,500 square feet, and contain the following:

Dryer (4.5 kVA)
Washing machine (1.2 kVA)
Range (14.4 kW)
Dishwasher (1.5 kVA)
Water heater (4 kW)
Heat (5 kW)
A/C (3 horsepower with $^1/_8$ horsepower compressor fan)

➻ **General Lighting Demand Question**

Using the optional calculations, what is the net computed load for the building general lighting and general use receptacles, small appliance, and laundry circuits? The system voltage is 115/230 volts, single-phase.

(a) 45 kVA (b) 90 kVA (c) 108 kVA (d) 60 kVA

• Answer: (a) 45 kVA [Table 220-32]

(a) General lighting (1,500 square feet × 3 VA)	4,500 VA
(b) Small appliance circuits	3,000 VA
(c) Laundry circuit	1,500 VA
	9,000 VA × 12 units × 0.41 = 44,280 VA

Note: The washing machine is calculated as part of the 1,500 VA laundry circuit.

➻ Air-Conditioning versus Heat Question

Using the optional calculations, what is the demand load for the building air-conditioning versus heat?

(a) 52 kVA (b) 60 kVA (c) 30 kVA (d) 25 kVA

- Answer: (d) 25 kVA [*Section 220-32*]

A/C (3 horsepower) 230 volts × 17 ampere 3,910 VA

Compressor (1/8 horsepower) 230 volts × 1.6 ampere* 368 VA

*One-half the value of a 1/4 horsepower motor. 4,278 VA × 12 units × 0.41 = 21,048 VA, omit

Heat [*Section 220-15*]: 5,000 watts × 12 units = 60,000 VA × 0.41 = 24,600 watts.

➻ Appliance Demand Question

Using the optional calculations, what is the 12-unit building net demand load for the water heaters and dishwashers?

(a) 41 kVA (b) 53 kVA (c) 33 kVA (d) 27 kVA

- Answer: (d) 27 kVA [*Section 220-32*]

Water heater 4,000 VA

Dishwasher 1,500 VA

5,500 VA × 12 units = 66,000 VA × 0.41 = 27,060 VA

Note: The washing machine is calculated as part of the laundry circuit.

➻ Dryer Demand Load Question

Using the optional calculations, what is the 12-unit building net computed load for the 4.5 kVA dryer?

(a) 60 kVA (b) 25 kVA (c) 55 kVA (d) 22 kVA

- Answer: (d) 22 kVA

4.5 kW* × 12 units × 0.41 = 22.14 kVA

* Nameplate rating

➻ Range Demand Load Question

Using optional calculations, what is the 12-unit building demand load for the 14.4 kW ranges?

(a) 71 kW (b) 42 kW

(c) 78 kW (d) 67 kW

- Answer: (a) 71 kW

14.4 kW × 12 units × 0.41 = 70.85 kW

➻ Service Conductor Size Question

If the total demand load is 189 kVA, what is the service conductor size? Service 120/240 volt, single-phase (Figure 10–3).

(a) 600 ampere (b) 800 ampere

(c) 1,000 ampere (d) 1,200 ampere

- Answer: (b) 800 ampere

I = VA/E

I = 189,000 VA/208 volts

I = 788 amperes

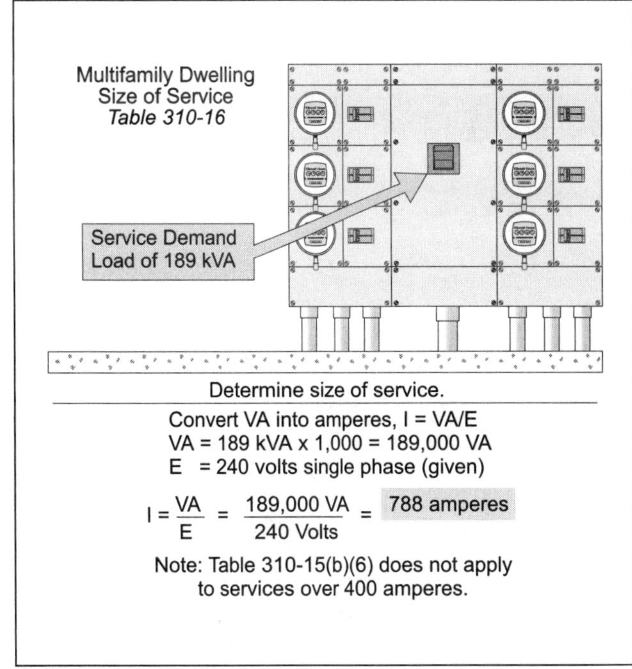

Figure 10-3
Multifamily Dwelling – Size of Service

Unit 10 – Multifamily Dwelling Unit Load Calculations Summary Questions

10–1 Multifamily Dwelling-Unit Load Calculations – Standard Method

1. • Each unit of a 20-unit apartment building is 840 square feet. What is the general lighting feeder demand load for the building? Note: Laundry facilities are provided on the premises for all tenants.
(a) 5,200 VA (b) 40,590 VA (c) 110,400 VA (d) none of these

2. Each unit of a 20-unit apartment building is 900 square feet. What is the general lighting net computed demand load for the building?
(a) 74,700 VA (b) 149,400 VA (c) 51,300 VA (d) 105,600 VA

3. A 40-unit multifamily building has a air-conditioning (3 horsepower 230 volt) and two baseboard heaters (3 kW) in each unit. What is the air-conditioning versus heat net computed demand load?
(a) 160 kW (b) 240 kW (c) 60 kW (d) 50 kW

4. A 25-unit multifamily building has a air-conditioning (3 horsepower 230 volt) and electric heat (5 kW). What is the air-conditioning versus heat net computed demand load?
(a) 160 kW (b) 125 kW (c) 6 kW (d) 5 kW

5. Each 16-unit multifamily building contains a waste disposal (940 VA), a dishwasher (1,250 VA), and a water heater (4,500 VA). What is the service demand load for these appliances?
(a) 100 kVA (b) 134 kVA (c) 80 kVA (d) 5 kVA

6. A 28-unit apartment building contains a waste disposal (900 VA), dishwasher (1,200 VA), and a water heater (5,000 VA) in each unit. What is the feeder and service demand load for the appliances?
(a) 149 kVA (b) 142 kVA (c) 107 kVA (d) 186 kVA

7. A multifamily dwelling (40 unit) contains a 4.5 kW electric clothes dryer in each unit. What is the feeder and service demand load for all the dryers?
(a) 50 kW (b) 27 kW (c) 60 kW (d) none of these

8. What is the demand load for ten 5.25 kW dryers installed in dwelling units of a multifamily building?
(a) 26 kW (b) 37 kW (c) 60 kW (d) 37 kW

9. What is the demand load for twelve 3.25 kW ovens?
(a) 10 kW (b) 18 kW (c) 15 kW (d) 20 kW

10. What is the demand load for eight 7 kW cooktops?
(a) 20 kW (b) 17 kW (c) 14.7 kW (d) 48 kW

11. • What is the demand load for five 12.4 kW ranges?
(a) 9 kW (b) 45 kW (c) 20 kW (d) none of these

12. What is the demand load for three ranges rated 15.5 kW?
(a) 15 kW (b) 14 kW (c) 17 kW (d) 21 kW

13. What is the feeder and service demand load for three ranges rated 11 kW and three ranges rated 14 kW?
(a) 36 kW (b) 42 kW (c) 78 kW (d) 22 kW

14. • What size aluminum service conductors are required for a 115/230-volt, single-phase multifamily building that has a total demand load of 90 kW?
(a) 500 kcmil (b) 600 kcmil (c) 700 kcmil (d) 800 kcmil

15. What size service conductors are required for a multifamily building that has a total demand load of 260 kW for a 120/208-volt, wye 3-phase system?
(a) 2 – 300 kcmil (b) 2 – 350 kcmil (c) 2 – 500 kcmil (d) 2 – 600 kcmil

For the Next Five Questions

A multifamily building has 12 units. Each is 1,500 square feet and contains the following:
System voltage 120/240 volts, single-phase.

Dryer (4.5 kW) Washing machine (1.2 kVA)
Range (14.45 kW) Dishwasher (1.5 kVA)
Water heater (4 kW) Heat (5 kW)
A/C (3 horsepower with $^1/_8$ horsepower compressor fan)

16. • What is the building's net computed load for the general lighting, small appliance, and laundry in VA?
(a) 40 kVA (b) 108 kVA (c) 105 kVA (d) 90 kVA

17. What is the building's demand load in kVA for an air-conditioner (3 horsepower with a $^1/_8$ horsepower compressor fan) versus electric heat (5 kW)?
(a) 52 kW (b) 60 kW (c) 30 kW (d) 105 kW

18. What is the building's demand load for the appliances in VA?
(a) 40 kVA (b) 55 kVA (c) 43 kVA (d) 50 kVA

19. What is the building's demand load for the 4.5 kW dryer?
(a) 60 kW (b) 27 kW (c) 55 kW (d) 101 kW

20. What is the building's demand load for the 14.45 kW range?
(a) 27 kW (b) 35 kW (c) 168 kW (d) 30 kW

21. If the total demand load of a multifamily dwelling-unit is 206 kVA, what is the service and feeder conductor size? The system voltage is 120/240, single-phase.
(a) 600 ampere (b) 800 ampere (c) 1,000 ampere (d) 1,200 ampere

22. • If a service contains three sets of parallel 400 kcmil conductors, what size bonding jumper is required for each service raceway?
(a) No. 1 (b) No. 1/0 (c) No. 2/0 (d) No. 3/0

23. • If 400 kcmil service conductors are in parallel in three raceways, what is the minimum size grounding electrode conductor required?
(a) No. 1/0 (b) No. 2/0 (c) No. 3/0 (d) No. 4/0

10–3 Multifamily Dwelling Calculation [*Section 220-32*] – Optional Method

24. • When determining the service (using optional calculations) for a multifamily dwelling, the total connected load shall have the demand factors of Table 220-32 applied. When determining the total connected load, the largest of the _____ shall be used.
(a) 100 percent of the air-conditioning
(b) 125 percent of the air-conditioning
(c) 100 percent of the heat
(d) a or c

25. A multifamily building has 60 units. Each unit is 1,500 square feet. Using the optional dwelling-unit calculations, what is the net computed load for the building general lighting and general use receptacles, small appliance and laundry circuits?
(a) 145 kVA (b) 190 kVA (c) 108 kVA (d) 130 kVA

26. A 60-unit multifamily dwelling has an air-conditioner (3 horsepower with a 1/8 horsepower blower) and electric heat (5 kW) in each unit. Using the optional method, what is the air-conditioning versus heat demand load?
(a) 50 kVA (b) 61 kVA (c) 30 kVA (d) 72 kVA

27. Using the optional method for dwelling-unit calculations, what is a 60-unit multifamily building net demand load if each unit has a water heater (4 kW and a dishwasher (1.5 kVA)?
(a) 80 kVA (b) 50 kVA (c) 30 kVA (d) 60 kVA

28. Each unit of a 60-unit apartment building has a 4 kW dryer. Using the optional method, the demand load that would be added to the service is _____ kW.
(a) 75 (b) 240 (c) 72 (d) 58

29. Using the optional method for dwelling-unit calculations, what is the 60-unit multifamily building net computed load for a 4.5 kW dryer?
(a) 65 kW (b) 25 kW (c) 55 kW (d) 75 kW

30. Using the optional method for dwelling-unit calculations, what is a 60-unit multifamily building demand load if each apartment has a 14 kW range?
(a) 150 kW (b) 50 kW (c) 100 kW (d) 200 kW

31. If the total demand load is 270 kVA, what is the service and feeder conductor size? The service is 208/120-volt, wye 3-phase.
(a) 600 ampere (b) 800 ampere (c) 1,000 ampere (d) 1,200 ampere

32. Each unit of a 20-unit multifamily dwelling has 900 square feet of living space and contains one of each of the following: air-conditioning (5 horsepower), heat (5 kW), water heater (5 kW), range (14 kW). The service for this apartment building is approximately _____ if the optional method calculation is used.
(a) 200 kVA (b) 250 kVA (c) 280 kVA (d) 320 kVA

33. • If the service conductors are in parallel in two raceways (500 kcmil), what size bonding jumper is required for each service raceway?
(a) No. 2 (b) No. 1 (c) No. 1/0 (d) No. 2/0

34. • If the service conductors are in parallel in two raceways (500 kcmil), what is the minimum size grounding electrode conductor required?
(a) No. 1/0 (b) No. 2/0 (c) No. 3/0 (d) No. 4/0

✯ Challenge Questions

General Lighting and Receptacle Calculations [Table 220- 11]

35. Each dwelling-unit of a 20-unit multifamily building has 900 square feet of living space. What is the general lighting load for the multifamily dwelling-unit apartment building?
(a) 45 kVA (b) 60 kVA (c) 37 kVA (d) 54 kVA

36. An multifamily apartment building contains 20 units and each unit has 840 square feet of living space. What is the general lighting feeder demand load for the multifamily building, if laundry facilities are provided on the premises for all tenants and no laundry circuit is installed in each unit?
(a) 35 kVA (b) 41 kVA (c) 45 kVA (d) 63 kVA

Appliance Demand Factors [*Section 220-17*]

37. An multifamily apartment building contains 20 units. Each unit contains a waste disposal (900 VA), a dishwasher (1,200 VA), and a water heater (5,000 VA). What is the feeder demand load for the appliances in this building?
(a) 106,500 VA (b) 117,100 VA (c) 137,000 VA (d) 60,000 VA

Dryer Calculation [*Section 220-18*]

38. The nameplate rating for each household dryer in a 10-unit apartment building is 4 kW. This would add _____ to the service size.
(a) 20 kW (b) 25 kW (c) 40 kW (d) 50 kW

Ranges – Note 1 of Table 220-19

39. • The demand load for thirty 15.8 kW household ranges is _____ kW.
(a) 31 (b) 47 (c) 54 (d) 33

Ranges – Note 2 of Table 220-19

40. What is the feeder demand load for five 10 kW, five 14 kW, and five 16 kW household ranges?
(a) 210 kW (b) 30 kW (c) 14 kW (d) 33 kW

41. What kW would be added to service loads for ten 12 kW, eight 14 kW, and two 9 kW household ranges?
(a) 33 kW (b) 35 kW (c) 36.75 kW (d) 29.35 kW

Ranges – Note 3 to Table 220-19

42. What is the minimum demand load for five 5 kW cooktops, two 4 kW ovens, and four 7 kW ranges?
(a) 15.5 kW (b) 8.8 kW (c) 19.5 kW (d) 18.2 kW

43. • What is the maximum dwelling-unit feeder or service demand load for fifteen 8 kW cooking units?
(a) 38.4 kW (b) 30 kW (c) 120 kW (d) none of these

Ranges – Note 5 of Table 220-19

44. A school has twenty 10 kW ranges installed in the home economics class. The minimum load this would add to the service is _____ .
(a) 35 kW (b) 44.8 kW (c) 56 kW (d) 160 kW

10–12 Neutral Calculation [*Section 220-22*]

45. The service neutral demand load for household electric clothes dryers shall be calculated at _____ of the demand load as determined by Section 220-18.
(a) 50% (b) 60% (c) 70% (d) 80%

46. The feeder neutral demand for fifteen 8 kW cooking units would be _____ .
(a) 38.4 kW (b) 30 kW (c) 120 kW (d) 21 kW

47. What is the feeder neutral load in ampere for fifteen 6 kW household dryers?
(a) 105 ampere (b) 125 ampere (c) 150 ampere (d) none of these

48. A 12-unit multifamily dwelling contains a 4 kVA electric clothes dryer in each unit. What is the feeder or service neutral demand load (in ampere)?
(a) 115 ampere (b) 225 ampere (c) 80 ampere (d) 55 ampere

49. • The feeder and service neutral demand load can be reduced 70 percent for that portion of the unbalanced load over 200 ampere. This applies to _____ systems.
(a) 3-wire, single-phase, 120/240 volt (b) 3-wire, single-phase, 120/208 volt
(c) 4-wire, 3-phase, 120/208 volt (d) a and c

50. • What is the dwelling-unit service neutral load (in ampere) for ten 9 kW ranges?
(a) 40 ampere (b) 55 ampere (c) 75 ampere (d) 105 ampere

Unit 11

Commercial Load Calculations

OBJECTIVES

After reading this unit, the student should be able to briefly explain the following concepts:

Part A – General
Conductor overcurrent protection [*Section 240–3*]
Conductor ampacity
Fraction of an ampere
General requirements

Part B – Loads
Air-conditioning
Dryers
Electric heat
Kitchen equipment
Laundry circuit
Lighting demand factors
Lighting without demand factors
Lighting miscellaneous
Multioutlet assembly [*Section 220–3(c) Exception No. 1*]
Receptacles [*Section 220–3(b)(7)*]
Banks and offices general lighting and receptacles
Signs [*Section 600–6(c)*]

Part C – Load Calculations
Marina [*Section 555–5*]
Mobile home park [*Section 550–22*]
Recreational vehicle park [*Section 551–73*]
Restaurant – Optional method [*Section 220–36*]
School – optional method [*Section 220–34*]

After reading this unit, the student should be able to briefly explain the following terms:

General lighting
General lighting demand factors
Nonlinear loads
VA rating
Ampacity
Continuous load
Next size up protection device
Overcurrent protection
Rounding
Standard ampere ratings for overcurrent protection devices
Voltages

PART A – GENERAL

11–1 GENERAL REQUIREMENTS

Article 220 provides the requirement for branch circuits, feeders, and services. In addition to this Article, other Articles are applicable such as Branch Circuits – 210, Feeders – 215, Services – 230, Overcurrent Protection – 240, Wiring Methods – 300, Conductors – 310, Appliances – 422, Electric Space-Heating Equipment – 424, Motors – 430, and Air-Conditioning – 440.

11–2 CONDUCTOR AMPACITY [ARTICLE 100]

The ampacity of a conductor is the rating, in ampere, that a conductor can carry continuously without exceeding its insulation temperature rating [*NEC*® Definition – Article 100]. The allowable ampacities listed in Table 310-16 are affected by ambient temperature, conductor insulation, and conductor bundling [*Section 310-10*] (Figure 11–1).

Continuous Loads

Conductors are sized at 125 percent of the continuous load before any derating factor and the overcurrent protection devices are sized at 125 percent of the continuous load [*Section 210-19(a), 215-2,* and *230-42*].

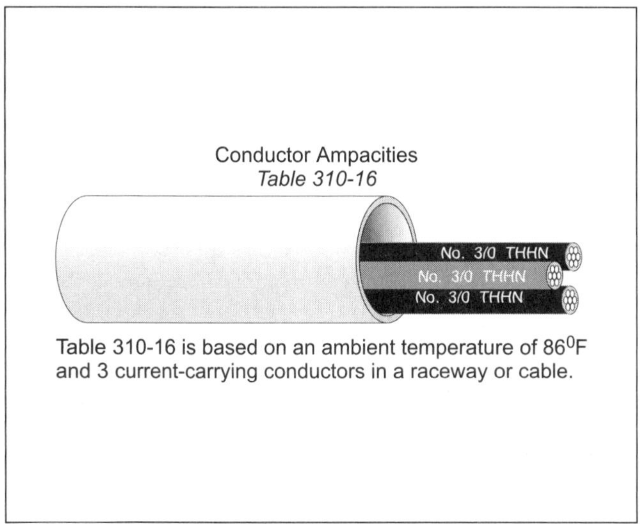

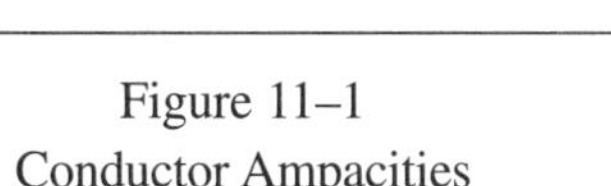

Figure 11–1
Conductor Ampacities

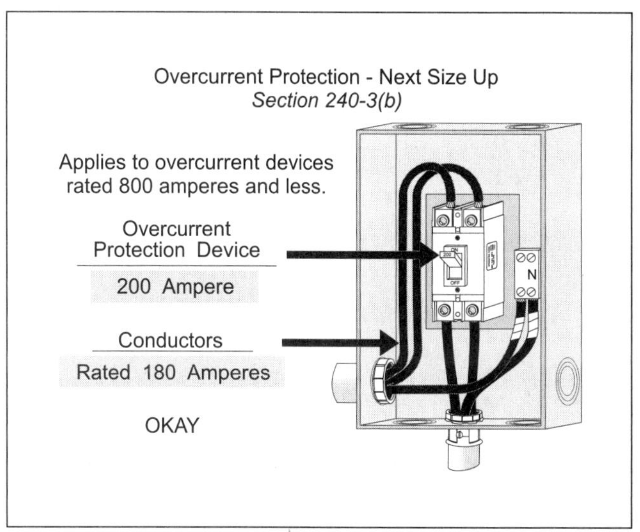

Figure 11–2
Overcurrent Protection – Next Size Up

11–3 CONDUCTOR OVERCURRENT PROTECTION [*SECTION 240-3*]

The purpose of overcurrent protection devices is to protect conductors and equipment against excessive or dangerous temperatures [*Section 240-1 FPN*]. There are many rules in the *National Electrical Code*® for conductor protection, and there are many different installation applications where the general rule of protecting the conductor at its ampacity does not apply. Examples would be motor circuits, air-conditioning, tap conductors, etc. See Section 240-3 for specific rules on conductor overcurrent protection.

Next Size Up Okay [*Section 240-3(b)*] If the ampacity of a conductor does not correspond with the standard ampere rating of a fuse or circuit breaker, as listed in Section 240-6(a), the next size up protection device is permitted. This practice only applies if the conductors do not supply multioutlet receptacles and if the next size up overcurrent protection device does not exceed 800 ampere (Figure 11–2).

Standard Size Overcurrent Devices [*Section 240-6(a)*] The following is a list of some of the standard ampere ratings for fuses and inverse time circuit breakers: 15, 20, 25, 30, 35, 40, 45, 50, 60, 70, 80, 90, 100, 110, 125, 150, 175, 200, 225, 250, 300, 350, 400, 500, 600, 800, 1,000, 1,200, and 1,600 ampere.

11–4 VOLTAGES [*SECTION* 220-2(a)]

Unless other voltages are specified, branch-circuit, feeder, and service loads shall be computed at a nominal system voltage of 120, 120/240, 208Y/120, 240, 480Y/277, 480 volts, or 600Y/347 (Figure 11–3).

11–5 ROUNDING AN AMPERE [*SECTION 220-2(b)*]

The rules for rounding an ampere require that when the calculation results in a fraction of an ampere of 0.50 and more, we "round up" to the next ampere. If the calculation is 0.49 of an ampere or less, we "round down" to the next lower ampere.

PART B – LOADS

11–6 AIR-CONDITIONING

Air-Conditioning Branch Circuit Branch circuit conductors and protection for air-conditioning equipment is marked on the equipment nameplate [*Section 110-3(b)*]. The nameplate values are determined by the use of Sections 440-32 for conductor sizing and 440-22(a) for short-circuit protection. Section 440-32 specifies that branch circuit conductors must be sized no less than 125 percent of the air-conditioner rating, and Section 440-22(a) specifies that the short-circuit protection device must be sized between 175 percent and up to 225 percent of the air-conditioning rating.

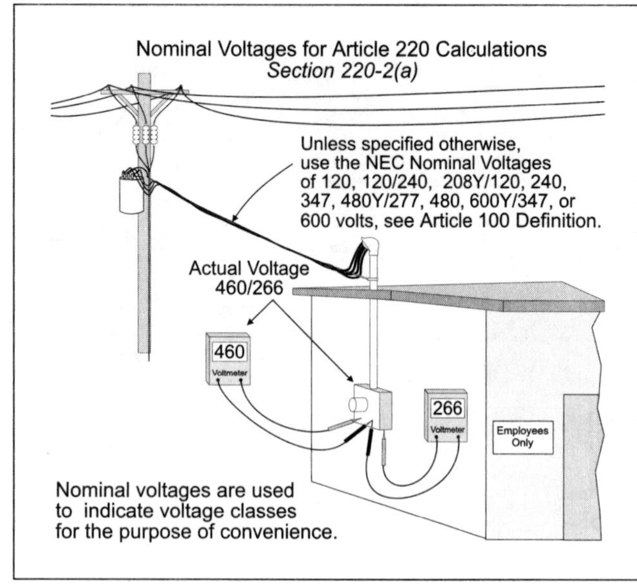

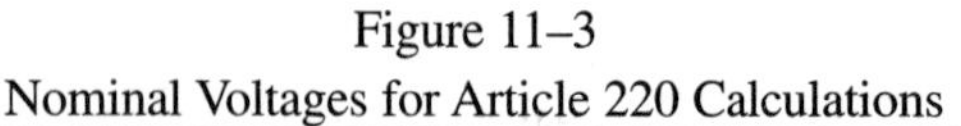
Figure 11–3
Nominal Voltages for Article 220 Calculations

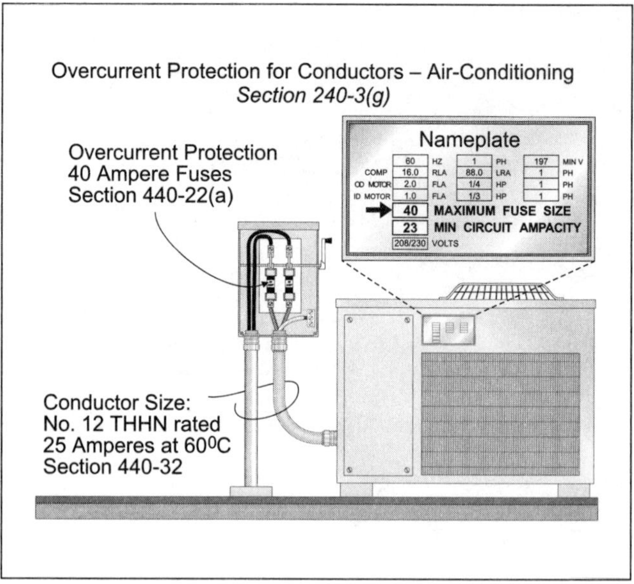

Figure 11–4
Overcurrent Protection for Conductors – Air-Conditioning

❏ Air-Conditioning Branch Circuit [*Section 240-3(g)*]

An air conditioner has a motor-compressor rated-load ampere of 18 ampere. The nameplate indicates the minimum circuit ampacity of 23 ampere and the maximum fuse size of 40 ampere. What size conductor and protection is required (Figure 11–4)?

(a) No. 12 with a 40-ampere breaker (b) No. 12 with a 40-ampere fuse
(c) No. 14 with a 30-ampere breaker (d) No. 10 with a 40-ampere breaker

• Answer: (b) No. 12 with a 40-ampere fuse

Conductor [*Section 440-32*] The conductor must be sized at no less than 125 percent of the motor-compressor rated-load current.

18 ampere × 1.25 = 22.5 ampere, No. 12 THHN rated at 25 ampere at 60ºC

Short-Circuit Protection [*Section 440-22(a)*] The short-circuit protection must be a fuse [*Section 110-3(b)*] because the nameplate specified “fuse.” However, the protection device shall not be greater than 225 percent of the motor-compressor current.

18 ampere × 2.25 = 40.5 ampere, next size down, 40 ampere [*Section 240-6(a)*]

Note: I know many feel the conductor should be larger or the short-circuit protection smaller, but this is not the *NEC®* requirement.

Air-Conditioning Feeder/Service Conductors [*Section 440-34*]. Feeder circuit conductors that supply air-conditioning equipment must be sized no less than 125 percent of the largest air conditioner VA rating plus 100 percent of the other air conditioners VA ratings.

VA Rating – The VA rating of an air conditioner is determined by multiplying the voltage rating of the unit by its ampere rating. For the purpose of this book, we will determine the unit ampere rating by using motor full-load current ratings as listed in Article 430, Table 430-148, and Table 430-150.

❏ Air-Conditioning Feeder/Service Conductor

What is the demand load required for the air-conditioning equipment of a 12-unit office building where each unit contains one air conditioner rated 5 horsepower 230 volts?

(a) 77 kVA (b) 79 kVA (c) 45 kVA (d) none of these

• Answer: (b) 79 kVA

A/C 5 horsepower 230 volts × 28 ampere = 6,440 VA

(6,440 VA × 1.25) + (6,440 VA × 11 units) = 78,890 VA/1,000 = 79 kVA

11–7 DRYERS

The branch circuit conductors and overcurrent protection device for commercial dryers are sized to the appliance nameplate rating. The feeder demand load for dryers is calculated at 100 percent of the appliance rating. Section 220-18 demand factors do not apply to commercial dryers.

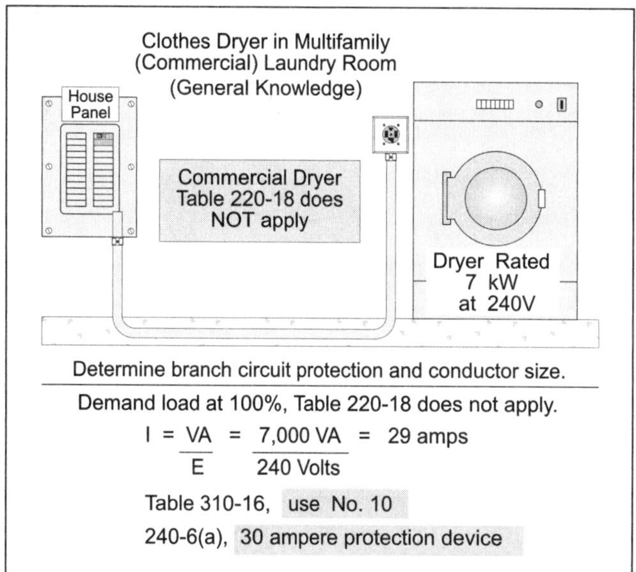

Figure 11–5
Clothes Dryer in Multifamily Laundry Room

❑ Dryer Branch Circuit

What size branch circuit conductor and overcurrent protection is required for a 7 kW dryer rated 240 volts when the dryer is located in the laundry room of a multifamily dwelling (Figure 11–5)?

(a) No. 12 with a 20-ampere breaker
(b) No. 10 with a 20-ampere breaker
(c) No. 12 with a 30-ampere breaker
(d) No. 10 with a 30-ampere breaker

- Answer: (d) No. 10 with a 30-ampere breaker

I = P/E = 7,000 watts/240 volts = 29 ampere

The ampacity of the conductor and overcurrent device must not be less than 29 ampere [*Section 240-3*].

Table 310-16, No. 10 conductor at 60ºC is rated 30 ampere protected with a 30-ampere protection device [*Section 240-6(a)*].

❑ Dryer Feeder

What is the service demand load for ten 7 kW dryers located in a laundry room?

(a) 70 kW (b) 52.5 kW (c) 35 kW (d) none of these

- Answer: (a) 70 kW

The *NEC®* does not permit a demand factor for commercial dryers, therefore the dryer demand load must be calculated at 100 percent: 7 kW × 10 units = 70 kW.

Note: If the dryers are on continuously, the conductor and protection device must be sized at 125 percent of the load [*Section 210-19(a), 215-3, 230-42,* and *384-16(d)*].

11–8 ELECTRIC HEAT

Electric Heat Branch Circuit [*Section 424-3(b)*]

Branch circuit conductors and the overcurrent protection device for electric heating shall be sized to not be less than 125 percent of the total heating load, including blower motors.

❑ Electric Heat

What size conductors and protection are required for a 3-phase, 240-volt, 15 kW heat strip with a 5.4-ampere blower motor (Figure 11–6)?

(a) No. 10 with 30-ampere protection
(b) No. 6 with 50-ampere protection
(c) No. 8 with 40-ampere protection
(d) No. 6 with 60-ampere protection

- Answer: (d) No. 6 with 60-ampere protection [*Section 424-3(b)*]

Conductors and protection must not be less than 125 percent of the total load.

I = VA/(E × $\sqrt{3}$) = 15,000 VA/(240 volts × 1.732) = 36 ampere

36 ampere + 5.4 ampere = 41.4 ampere × 1.25 = 51.75 ampere

Conductor sized to the 60ºC column ampacities of Table 310-16 [*Section 110-14(c)(1)*], No. 6 rated 55 ampere.

Overcurrent protection device sized no less than 52 ampere, 60 ampere protection device [*Section 240-6(a)*].

Electric Heat Feeder/Service Demand Load [*Section 220-15*]

The feeder/service demand load for electric heating equipment is calculated at 100 percent of the total heating load.

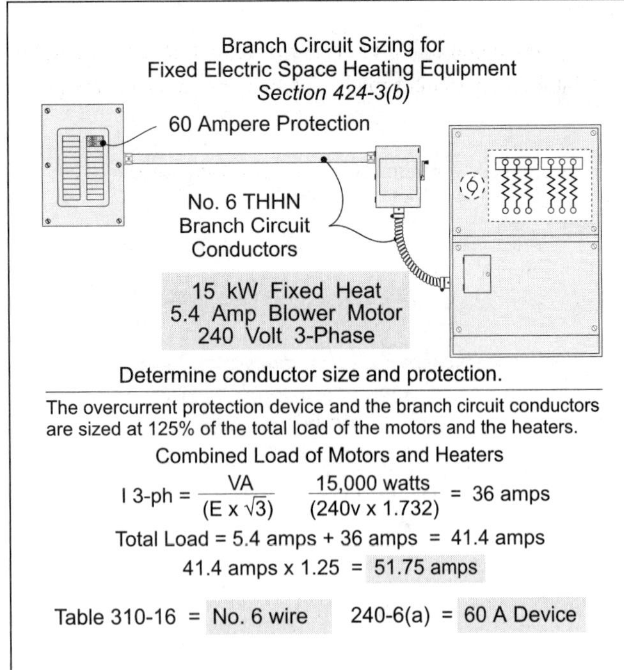

Figure 11–6
Branch Circuit Sizing for Fixed Electric Space-Heating Equipment

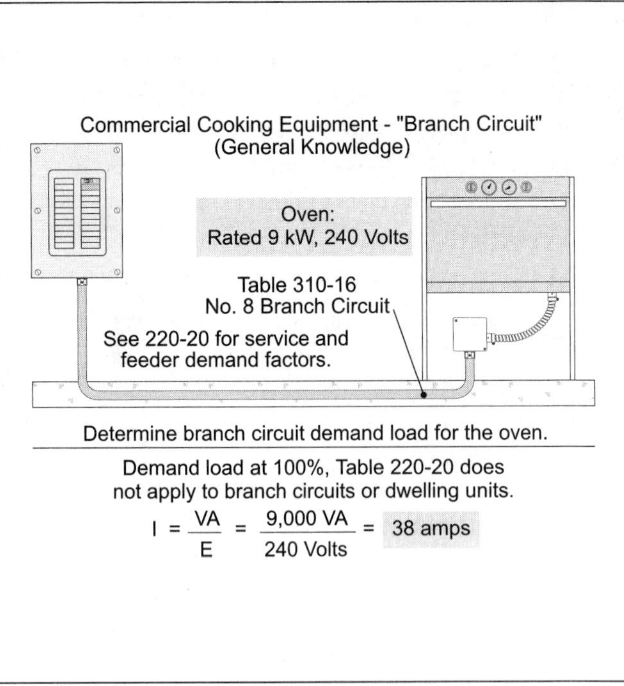

Figure 11–7
Commercial Cooking Equipment – "Branch Circuit" (General Knowledge)

❑ **Electric Heat Feeder/Service Demand Load**

What is the feeder/service demand load for a building that has seven 3-phase, 208-volt, 10 kW heat strips with a 5.4-ampere blower motor (1,945 VA) for each unit?

(a) 84 kVA (b) 53 kVA (c) 129 kVA (d) 154 kVA

- Answer: (a) 84 kVA

10,000 VA + 1,945 VA = 11,945 VA × 7 units = 83,615 VA/1,000 = 83.615 kVA

11–9 KITCHEN EQUIPMENT

Kitchen Equipment Branch Circuit

Branch circuit conductors and overcurrent protection for commercial kitchen equipment are sized according to the appliance nameplate rating.

❑ **Kitchen Equipment Branch Circuit No. 1**

What is the branch circuit demand load (in ampere) for one 9 kW oven rated 240 volts (Figure 11–7)?

(a) 38 ampere (b) 27 ampere (c) 32 ampere (d) 33 ampere

- Answer: (a) 38 ampere

I = P/E

I = 9,000 watts/240 volts

I = 38 ampere

❑ **Kitchen Equipment Branch Circuit No. 2**

What is the branch circuit load for one 14.47 kW range rated 208 volts, 3-phase?

(a) 60 ampere (b) 40 ampere (c) 50 ampere (d) 30 ampere

- Answer: (b) 40 ampere

I = P/(E × 1.732)

I = 14,470 watts/(208 volts × 1.732)

I = 40 ampere

Kitchen Equipment Feeder/Service Demand Load [*Section 220-20*]

The service demand load for thermostatic control or intermittent use for commercial kitchen equipment is determined by applying the demand factors from Table 220-20, to the total connected kitchen equipment load. *The feeder or service demand load cannot be less than the two largest appliances.*

Note: The demand factors of Table 220-20 do not apply to space heating, ventilating, or air-conditioning equipment.

❑ **Kitchen Equipment Feeder/Service No. 1**

What is the demand load for the following kitchen equipment loads (Figure 11–8)?

Water heater	5 kW	Booster heater	7.5 kW
Mixer	3 kW	Oven	5 kW
Dishwasher	1.5 kW	Waste disposal	1 kW

(a) 15 kW (b) 23 kW (c) 12.5 kW (d) none of these

• Answer: (a) 15 kW

Water heater	5 kW
Booster heater	7.5 kW
Mixer	3 kW
Oven	5 kW
Dishwasher	1.5 kW
Waste disposal	1 kW
Total connected	23 kW × 0.65 = 14.95 kW

❑ **Kitchen Equipment Feeder/Service No. 2**

What is the demand load for the following kitchen equipment loads?

Water heater	10 kW*	Booster heater	15 kW*
Mixer	4 kW	Oven	6 kW
Dishwasher	1.5 kW	Waste disposal	1 kW

(a) 24.4 kW (b) 38.2 kW (c) 25.0 kW (d) 18.9 kW

• Answer: (c) 25 kW*

Water heater	10 kW
Booster heater	15 kW
Mixer	4 kW
Oven	6 kW
Dishwasher	1.5 kW
Waste disposal	1 kW
Total connected	37.5 kW × 0.65 = 24.4 kW

* The demand load cannot be less than the two largest appliances 10 kW + 15 kW = 25 kW.

11–10 LAUNDRY EQUIPMENT

Laundry equipment circuits are sized to the appliance nameplate rating. For exam purposes, it is generally accepted that a laundry circuit is not considered a continuous load and assume all commercial laundry circuits to be rated 1,500 VA unless noted otherwise in the question.

❑ **Laundry Equipment**

What is the demand load for 10 washing machines located in a laundry room (Figure 11–9)?

(a) 1,500 VA (b) 15,000 VA
(c) 1,125 VA (d) none of these

• Answer: (b) 15,000 VA

1,500 VA × 10 units = 15,000 VA

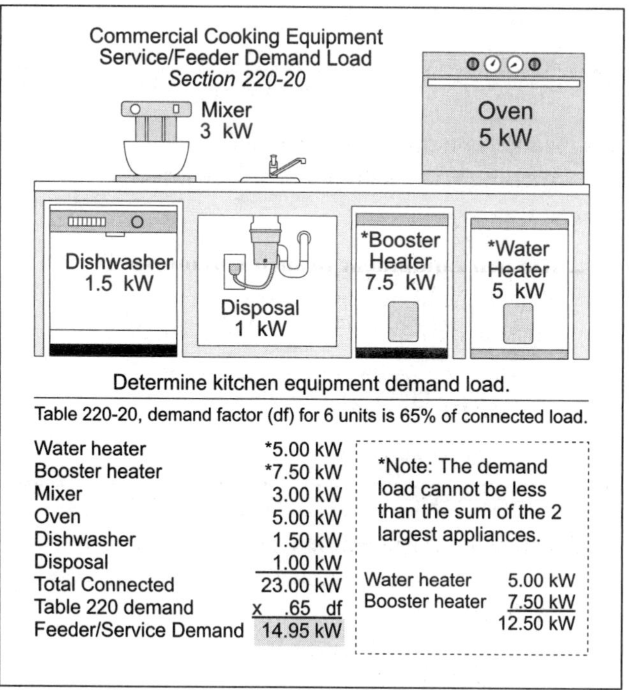

Figure 11–8
Commercial Cooking Equipment Service/Feeder Demand Load

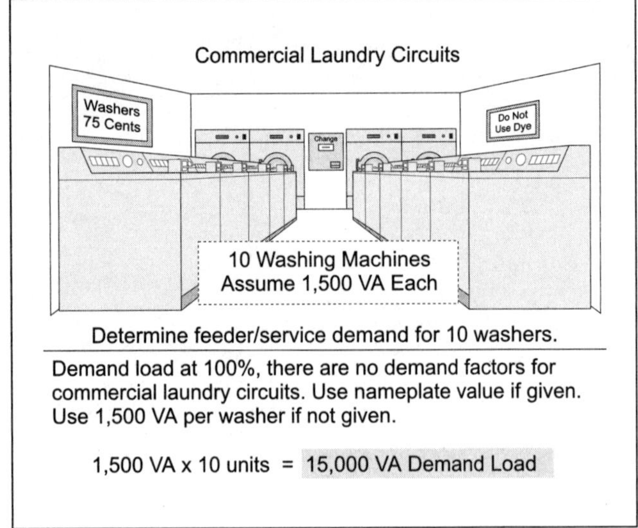

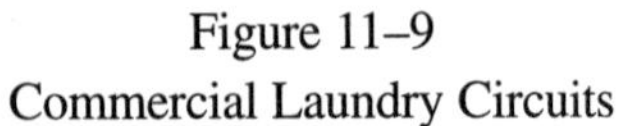

Figure 11–9
Commercial Laundry Circuits

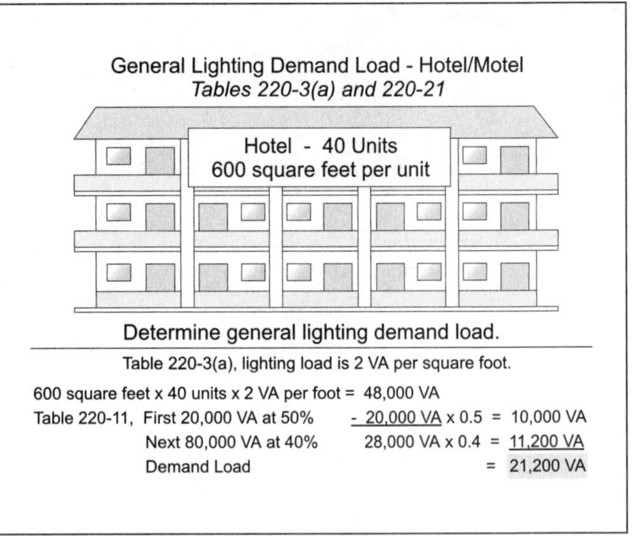

Figure 11–10
General Lighting Demand Load – Hotel/Motel

11–11 LIGHTING – DEMAND FACTORS [Table 220-3(a) and 220-11]

The *NEC®* requires a minimum load per square foot for general lighting depending on the type of occupancy [Table 220-3(a)]. For the guest rooms of hotels and motels, hospitals, and storage warehouses, the general lighting demand factors of Table 220-11 can be applied to the general lighting load.

Hotel or Motel Guest Rooms – General Lighting [Table 220-3(a) and Table 220-11]

The general lighting demand load of 2 VA each square foot [Table 220-3(a)] for the guest rooms of hotels and motels is permitted to be reduced according to the demand factors listed in Table 220-11.

General Lighting Demand Factors

First 20,000 VA at 50 percent demand factor
Next 80,000 VA at 40 percent demand factor
Remainder VA at 30 percent demand factor

❑ Hotel General Lighting Demand

What is the general lighting demand load for a 40-room hotel? Each unit contains 600 square feet of living area (Figure 11–10).

(a) 48 kVA (b) 24 kVA (c) 20 kVA (d) 21 kVA

• Answer: (d) 21 kVA

40 units × 600 square feet (24,000 square feet × 2 VA)	48,000 VA	
First 20,000 VA at 50 %	– 20,000 VA × 0.5 =	10,000 VA
Next 80,000 VA at 40 %	28,000 VA × 0.4 =	11,200 VA
		21,200 VA

11–12 LIGHTING WITHOUT DEMAND FACTORS [Table 220-3(a), 215-2 and 230-42]

The general lighting load for commercial occupancies other than guest rooms of motels and hotels, hospitals, and storage warehouses is assumed continuous and shall be calculated at 125 percent of the general lighting load as listed in Table 220-3(a) [Table 220-11].

❑ Store Lighting

What is the general lighting load load for a 21,000 square-foot store (Figure 11–11)?

(a) 40 kVA (b) 63 kVA (c) 79 kVA (d) 81 kVA

• Answer: (c) 79 kVA

21,000 square feet × 3 VA × 1.25 = 78,750 VA

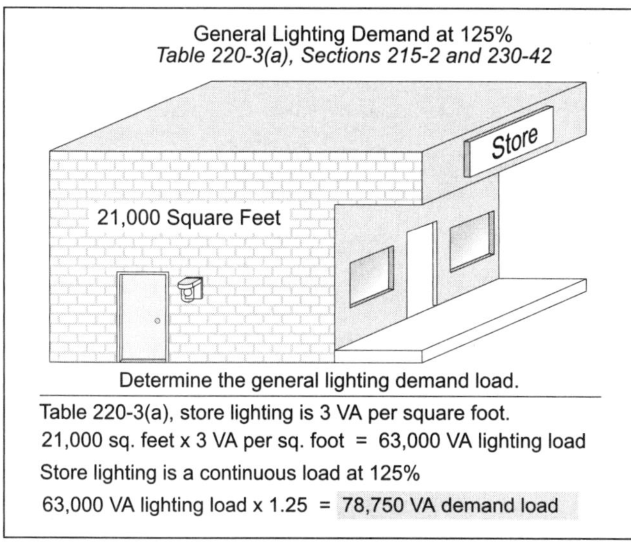

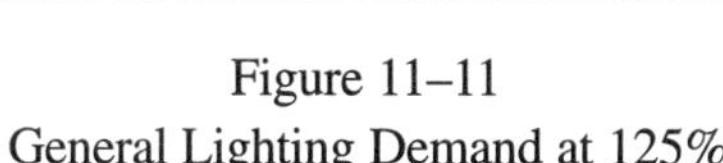

Figure 11–11
General Lighting Demand at 125%

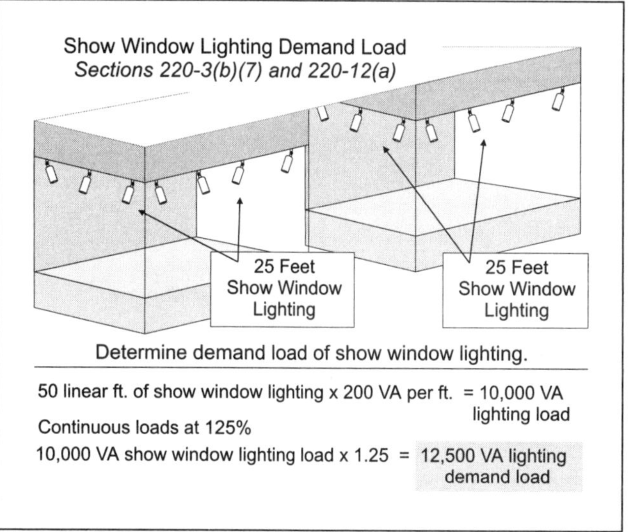

Figure 11–12
Show-Window Lighting Demand Load

❑ **Club Lighting**

What is the general lighting load for a 4,700 square-foot dance club?

(a) 4,700 VA (b) 9,400 VA (c) 11,750 VA (d) 250 kVA

• Answer: (c) 11,750 VA

4,700 square feet × 2 VA × 1.25 = 11,750 VA

❑ **School Lighting**

What is the general lighting load for a 125,000 square-foot school?

(a) 125 kVA (b) 375 kVA (c) 475 kVA (d) 550 kVA

• Answer: (c) 475 kVA

125,000 square feet × 3 VA × 1.25 = 468,750 VA

11–13 LIGHTING – MISCELLANEOUS

Show-Window Lighting [*Section 220-3(b)(7)* and *220-12(a)*]

The demand load for each linear foot of show-window lighting shall be calculated at 200 VA per foot. Show-window lighting is assumed to be a continuous load; See Example D3 in Appendix D for the requirements for show-window branch circuits.

❑ **Show-Window Load**

What is the demand load in kVA for 50 feet of show-window lighting (Figure 11–12)?

(a) 6 kVA (b) 7.5 kVA (c) 9 kVA (d) 12.5 kVA

• Answer: (d) 12.5 kVA

50 feet × 200 VA per foot = 10,000 VA × 1.25 = 12,500 VA

11–14 MULTI-OUTLET RECEPTACLE ASSEMBLY [*SECTION 220-3(b)(8)*]

Each 5 feet, or fraction of that, of multioutlet receptacle assembly shall be considered to be 180 VA for service calculations. When a multi-outlet receptacle assembly is expected to have a number of appliances used simultaneously, each foot or fraction of a foot shall be considered as 180 VA for service calculations. A multi-outlet receptacle assembly is not generally considered to be a continuous load.

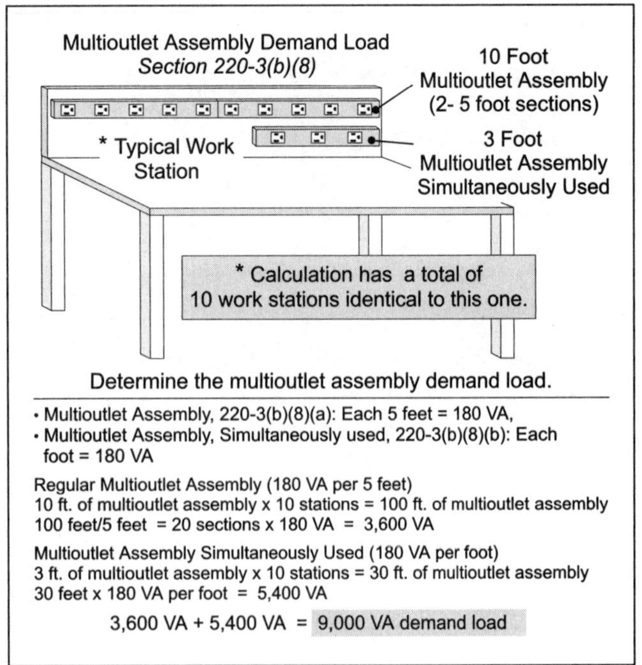

Figure 11–13
Multi-outlet Assembly Demand Load

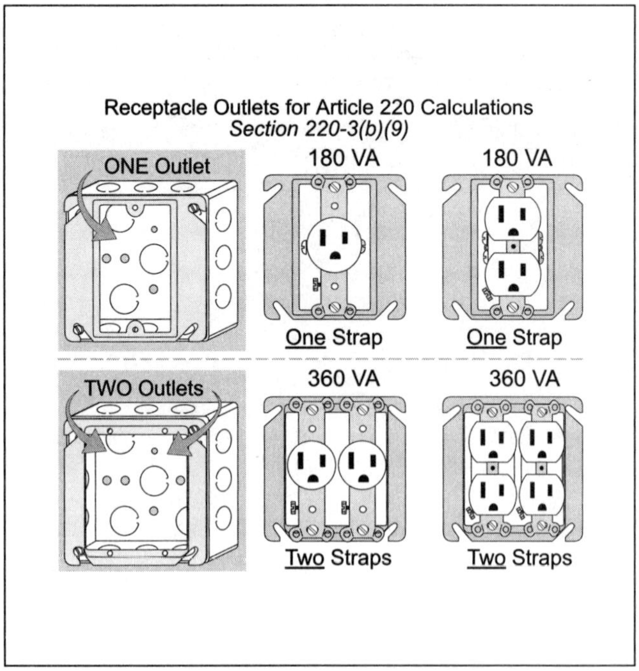

Figure 11–14
Receptacle Outlet Load Calculations

❑ **Multi-outlet Receptacle Assembly**

What is the demand load for 10 work stations that have 10 feet of multi-outlet receptacle assembly and 3 feet of multi-outlet receptacle assembly simultaneously used (Figure 11–13)?

(a) 5 kVA (b) 6 kVA (c) 7 kVA (d) 9 kVA

• Answer: (d) 9 kVA

100 feet/5 feet per section (20 sections × 180 VA) 3,600 VA

30 feet/1 foot per section (30 sections × 180 VA) <u>5,400 VA</u>

9,000 VA/1,000 = 9 kVA

11–15 RECEPTACLE VA LOAD [*SECTION 220-13*]

Receptacle VA Load

The minimum load for each commercial or industrial general use receptacle outlet shall be 180 VA per strap [*Section 220–3(b)(9)*]. Receptacles are generally not considered to be a continuous load (Figure 11–14).

Number of Receptacles Permitted on a Circuit

The maximum number of receptacle outlets permitted on a commercial or industrial circuit is dependent on the circuit ampacity. The number of receptacles per circuit is calculated by dividing the VA rating of the circuit by 180 VA for each receptacle strap.

❑ **Receptacles Per Circuit [*Section 220-3(b)(9)*]**

How many receptacle outlets are permitted on a 15-ampere, 120-volt circuit (Figure 11–15)?

(a) 10 (b) 13 (c) 15 (d) 20

• Answer: (a) 10

The total circuit VA load for a 15-ampere circuit is 120 volts × 15 ampere = 1,800 VA.

The number of receptacle outlets per circuit = 1,800 VA/180 VA = 10 receptacles.

Note: 15-ampere circuits are permitted for commercial and industrial occupancies according to the *National Electrical Code®*, but some local codes require a minimum 20-ampere rating for commercial and industrial circuits [*Section 310-5*].

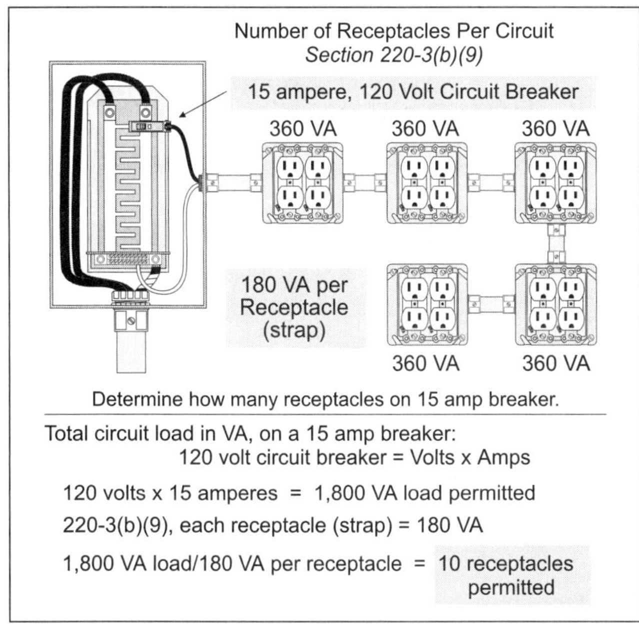

Figure 11–15
Number of Receptacles Per Circuit

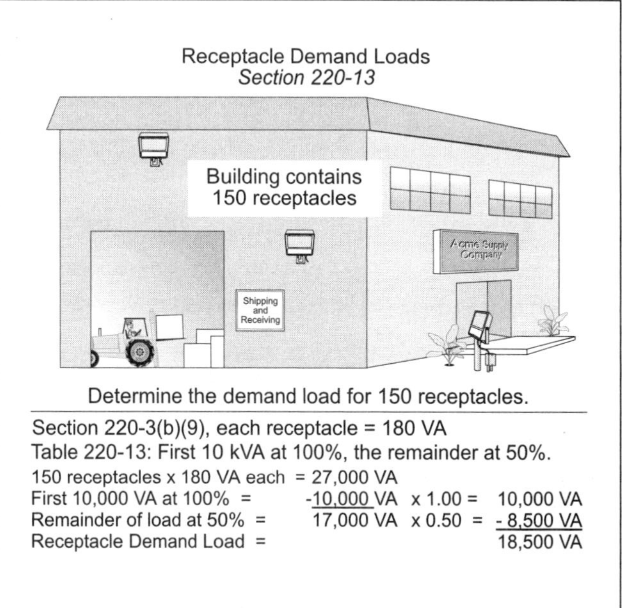

Figure 11–16
Receptacle Demand Loads

Receptacle Service Demand Load [*Section 220-13*]

The feeder and service demand load for commercial receptacles is calculated at 180 volt-ampere per receptacle strap [*Section 220-3(b)(9)*]. The demand factors of Table 220-13 can be used for that portion of the receptacle load in excess of 10 kVA, including the multi-outlet receptacle load [*Section 220–3(b)(8)*]. Receptacle loads are generally not considered to be a continuous load.

Table 220-13 Receptacle Demand Factors:

First 10 kVA at 100 percent demand factor
Remainder kVA at 50 percent demand factor

❑ **Receptacle Service Demand Load**

What is the service demand load for one hundred and fifty, 20-ampere, 120-volt general use receptacles in a commercial building (Figure 11–16)?

(a) 27 kVA (b) 14 kVA (c) 4 kVA (d) none of these

• Answer: (d) none of these

Total receptacle load (150 receptacles × 180 VA)	27,000 VA		
First 10 kVA at 100%	– 10,000 VA	× 1.00 =	10,000 VA
Remainder at 50%	17,000 VA	× 0.50 =	8,500 VA
Total receptacle demand load			18,500 VA/1,000 = 18.5 kVA

11–16 BANKS AND OFFICES GENERAL LIGHTING AND RECEPTACLES

Some testing agencies include the receptacle demand load as part of the general lighting load for banks and offices. If that is the case, the general lighting demand load for banks and offices would be calculated as 3.5 VA each square foot times 125 percent for continuous lighting load, plus the receptacle demand load after applying Table 220-13 demand factors.

Receptacle Demand [Table 220-13] The receptacle demand load is calculated at 180 volt-ampere for each receptacle outlet [*Section 220-3(b)(9)*] if the number of receptacles are known, or 1 VA for each square foot if the number of receptacles are unknown [Table 220–3 (a) Note b].

❑ **Bank General Lighting and Receptacle**

What is the general lighting demand load (including receptacles) for an 18,000 square-foot bank? The number of receptacles is unknown (Figure 11–17).

(a) 68 kVA (b) 110 kVA (c) 84 kVA (d) 93 kVA

• Answer: (d) 93 kVA

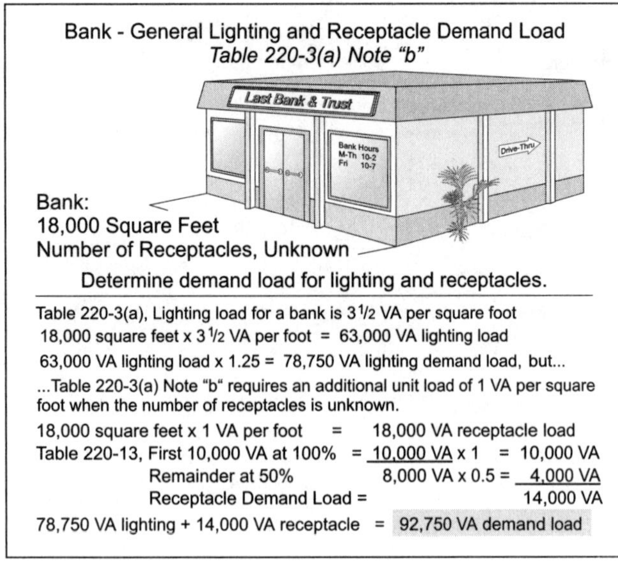

Figure 11–17
Bank – General Lighting and Receptacle Demand Load

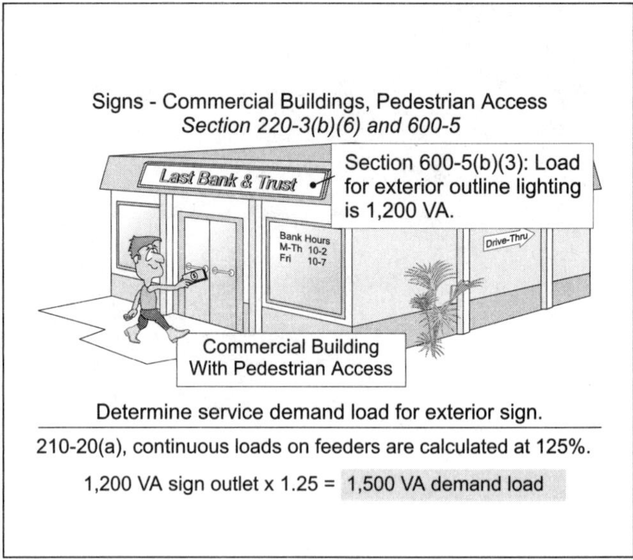

Figure 11–18
Signs – Commercial Buildings, Pedestrian Access

(a) General Lighting (18,000 square feet × 3.5 VA × 1.25)		78,750 VA
(b) Receptacle Load (18,000 square feet × 1 VA)	18,000 VA	
First 10 kVA at 100%	10,000 VA × 1.00 =	10,000 VA
Remainder at 50%	8,000 VA × 0.50 =	4,000 VA
Total demand load		92,750 VA/1,000 = 92.75 kVA

❑ **Office General Lighting and Receptacle**

What is the general lighting demand load for a 28,000 square-foot office building that has 160 receptacles?

(a) 111 kVA (b) 110 kVA (c) 128 kVA (d) 142 kVA

• Answer: (d) 142 kVA

(a) General Lighting (28,000 square feet 3.5 VA × 1.25)		122,500 VA.
(b) Receptacle Load (160 receptacles × 180 VA)	28,800 VA	
First 10,000 VA at 100%	10,000 VA × 1.00 =	10,000 VA
Remainder at 50%	18,800 VA × 0.50 =	9,400 VA
		141,900 VA/1,000 = 142 kVA

11–17 SIGNS [*SECTION 220-3(b)(6)* and *600-5*]

The *NEC®* requires each commercial occupancy that is accessible to pedestrians to be provided with at least one 20 ampere branch circuit for a sign [*Section 600-5(a)*]. The load for the required exterior signs or outline lighting shall be a minimum of 1,200 VA [*Section 220-3(b)(6)* and *600-5(b)(3)*]. A sign outlet is considered to be a continuous load, and the feeder load must be sized at 125 percent of the continuous load [*Section 210-20(a), 215-2(a), 230-42* and *384-16(d)*].

❑ **Sign Demand Load**

What is the demand load for one electric sign (Figure 11–18)?

(a) 1,200 VA (b) 1,500 VA (c) 1,920 VA (d) 2,400 VA

• Answer: (b) 1,500 VA

1,200 VA × 1.25 = 1,500 VA

11–18 NEUTRAL CALCULATIONS [*SECTION 220-22*]

The neutral load is considered the maximum unbalanced demand load between the grounded (neutral) conductor and any one ungrounded (hot) conductor as determined by the calculations in Article 220, Part B. This means that line-to-line loads are not considered when sizing the neutral conductor.

Reduction over 200 Ampere

For balanced 3-wire, single-phase and 4-wire, 3-phase wye systems, the neutral demand load can be reduced 70 percent for that portion of the unbalanced load over 200 ampere.

❑ Reduction over 200 Ampere

What is the neutral current for a balanced 400-ampere 3-wire 120/240-volt feeder?

(a) 400 ampere (b) 340 ampere (c) 300 ampere (d) none of these

• Answer: (b) 340 ampere

Total neutral load	400 ampere	
First 200 ampere at 100%	200 ampere × 1.00 =	200 ampere
Remainder at 70%	200 ampere × 0.70 =	140 ampere
		340 ampere

Reduction Not Permitted

The neutral demand load shall not be permitted to be reduced for 3-wire, single-phase 208Y/120-, or 480Y/277-volt circuits consisting of two line wires and the common conductor (neutral) of a 4-wire, 3-phase wye system. This is because the common (neutral) conductor of a 3-wire circuit connected to a 4-wire, 3-phase wye system carries approximately the same current as the phase conductors; see Section 310–15 (b)(4)(b). This can be proven with the following formula:

$I_{Neutral} = \sqrt{(L_1^2 + L_2^2) - (L_1 \times L_2)}$

$Line_1$ = Neutral Current of Line 1

$Line_2$ = Neutral Current of Line 2

❑ Three-Wire Wye Neutral Current

What is the neutral current for a balanced 300-ampere 3-wire 208Y/120-volt feeder, supplied from a 4-wire, 3-phase wye system (Figure 11–19)?

(a) 100 ampere (b) 200 ampere (c) 300 ampere (d) none of these

• Answer: (c) 300 ampere

$$I_{Neutral} = \sqrt{L_1^2 + L_2^2 - (L_1 \times L_2)} = \sqrt{(300^2 + 300^2) - (300 \times 300)} = \sqrt{180{,}000 - 90{,}000} = \sqrt{90{,}000} = 300 \text{ ampere}$$

Nonlinear Loads

The neutral demand load cannot be reduced for nonlinear loads supplied from a 4-wire, wye-connected, 3-phase system because they produce *triplen harmonic currents* that add on the neutral conductor, which can require the neutral conductor to be larger than the ungrounded conductor load; see Section 220-22 FPN No. 2.

PART C – LOAD CALCULATION EXAMPLES

MARINA [SECTION 555-6]

The *National Electrical Code*® permits a demand factor to apply to the receptacle outlets for boat slips at a marina. The demand factors of Section 555-6 are based on the number of receptacles on the feeder. The receptacles must also be balanced between the lines to determine the number of receptacles on any given line.

❑ Marina Receptacle Outlet Demand

What size 120/240-volt, single-phase service is required for a marina that has twenty 20-ampere, 120-volt receptacles and twenty 30-ampere, 240-volt receptacles (Figure 11–20)?

(a) 200 ampere (b) 400 ampere (c) 600 ampere (d) 800 ampere

• Answer: (c) 600 ampere

	Line 1	Line 2
Ten 20 ampere, 120 volt	200 ampere	200 ampere
Twenty 30 ampere, 240 volt	600 ampere	600 ampere
	800 ampere	800 ampere × 0.7 = 560 ampere per line

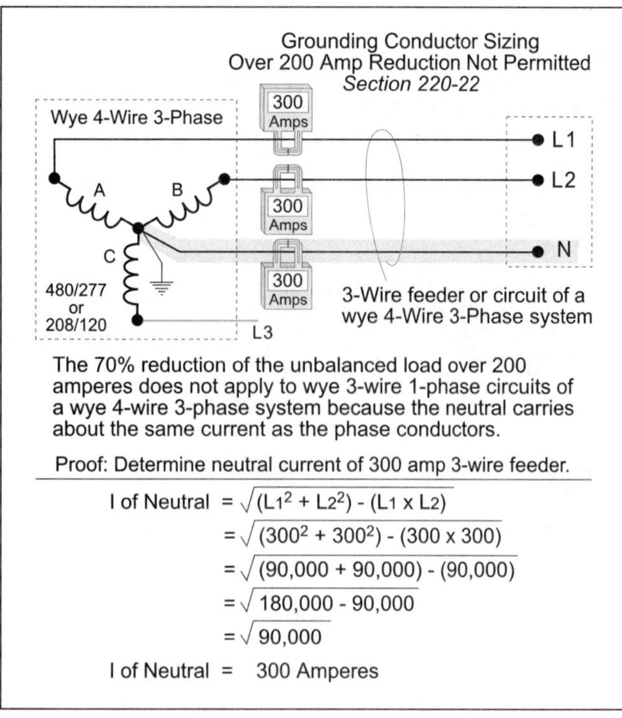

Figure 11–19
Grounding Conductor Sizing

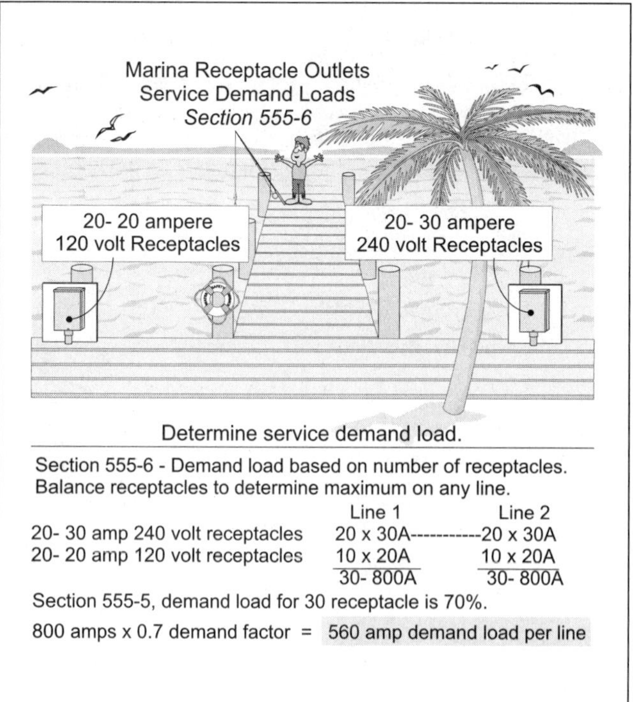

Figure 11–20
Marina Receptacle Outlet – Service Demand Loads

MOBILE/MANUFACTURED HOME PARK [SECTION 550-22]

The service demand load for a mobile/manufactured home park is sized according to the demand factors of Table 550-22 to the larger of 16,000 VA for each mobile/manufactured home lot or the calculated load for each mobile/manufactured home site according to Section 550-13.

❑ **Mobile/Manufactured Home Park**

What is the demand load for a mobile/manufactured home park that has facilities for 35 sites? The system is 120/240 volts, single-phase (Figure 11–21).

(a) 400 ampere (b) 600 ampere
(c) 800 ampere (d) 1,000 ampere

• Answer: (b) 600 ampere nearest answer

16,000 VA × 35 sites × 0.24 = 134,400 VA

I = VA/E = 134,400 VA/240 volts = 560 ampere

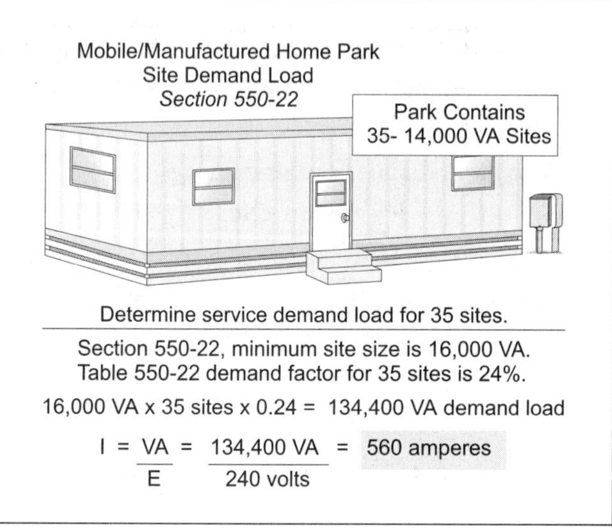

Figure 11–21
Mobile/Manufactured Home Park Site Demand Load

RECREATIONAL VEHICLE PARK [SECTION 551-73]

Recreational vehicle parks are calculated according to the demand factor of Table 551-73. The total calculated load is based on:

• 2,400 VA for each 20-ampere supply facilities site,

• 3,600 VA for each 20- and 30-ampere supply facilities site, and

• 9,600 VA for each 50-ampere, 120/240-volt supply facilities site.

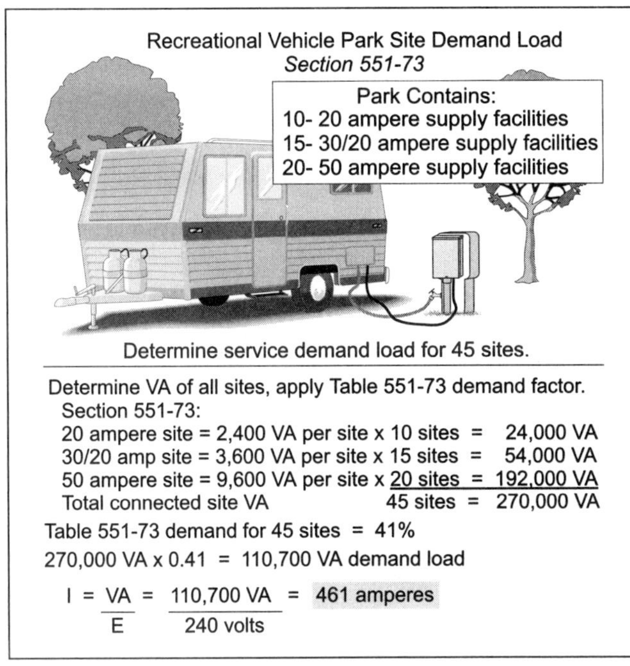

Figure 11–22
Recreational Vehicle Park Site Demand Load

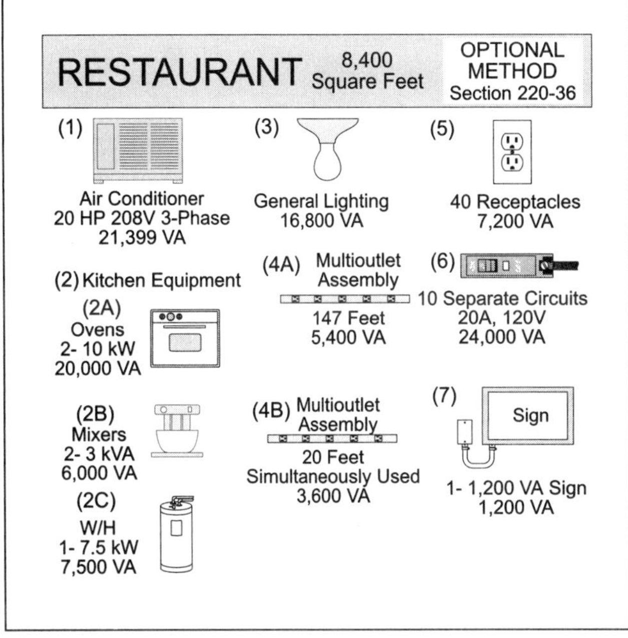

Figure 11–23
Restaurant Demand Load Optional Method

❑ Recreational Vehicle Park

What is the demand load for a recreational vehicle park that has ten 20-ampere supply facilities, fifteen 20/30-ampere supply facilities, and twenty 50-ampere supply facilities? The system is 120/240 volts, single-phase (Figure 11–22).

(a) 400 ampere (b) 500 ampere (c) 600 ampere (d) 700 ampere

• Answer: (b) 500 ampere

The service for the recreational vehicle park is sized according to the demand factors of Table 551-73

Ten 20-ampere sites (2,400 VA × 10 sites)	24,000 VA
Fifteen 20/30-ampere sites (3,600 VA × 15 sites)	54,000 VA
Twenty 50-ampere sites (9,600 VA × 20 sites)	192,000 VA
Demand factor [Table 551-73]	270,000 VA × 0.41 = 110,700 VA

I = VA/E = 110,700 VA/240 volts = 461 ampere

RESTAURANT – OPTIONAL METHOD [SECTION 220-36]

An optional method of calculating the demand service load for a restaurant is permitted. The following steps can be used to determine the service size:

Step 1: ➛ Determine the total connected load. Add the nameplate rating of all loads at 100 percent and include both the air conditioner and heat load.

Step 2: ➛ Apply the demand factors from from Table 220-36 to the total connected load (Step 1).

All-Electric Restaurant Demand Factors	
0 – 250 kVA	80%
251 – 280 kVA	70%
281 – 325 kVA	60%
Over 326 kVA	50%

Not All-Electric Restaurant Demand Factors	
0 – 250 kVA	100%
251 – 280 kVA	90%
281 – 325 kVA	80%
326 – 375 kVA	70%
376 – 800 kVA	65%
Over 800 kVA	50%

❑ Restaurant Optional Method

Using the optional method, what is the service size for the loads (Figure 11–23)?

Step 1: ➛ Determine the total connected load.

(1) Air-conditioning demand [Table 430-150 and 440-34]

VA = 208 volts × 59.4 ampere × 1.732		21,399 VA
(2) Kitchen equipment		
Ovens	10 kVA × 2 units =	20,000 VA
Mixers	3 kVA × 2 units =	6,000 VA
Water heaters	7.5 kVA × 1 unit =	7,500 VA
(3) Lighting [Table 220-3(a)] (8,400 square feet × 2 VA)		16,800 VA
(4) Multi-outlet assembly [*Section 220-3(b)(8)*]		
147 feet/5 feet = thirty 5-foot sections (180 VA × 30)		5,400 VA
Simultaneously used (180 VA × 20)		3,600 VA
(5) Receptacles [*Section 220-3(b)(9)*] (180 VA × 40 receptacles)		7,200 VA
(6) Separate circuits (120 volts × 20 ampere × 10 circuits)		24,000 VA
(7) Sign [*Section 220-3(b)(6)* and *600-5*]		1,200 VA

Totals

(1) Air-conditioning	21,399 VA
(2) Kitchen equipment	33,500 VA
(3) Lighting	16,800 VA
(4) Multi-outlet assembly	9,000 VA
(5) Receptacles	7,200 VA
(6) Separate circuits	24,000 VA
(7) Sign	1,200 VA
Total connected load	113,099 VA

Step 2: ➛ Apply the demand factors from Table 220-36 to the total connected load.

All-Electric If the restaurant is all electric, a demand factor of 80 percent is permitted to apply to the first 250 kVA.

113,099 VA × 0.8 = 90,479 VA

I = VA/(E × $\sqrt{3}$) = 90,479 VA/(208 volts × 1.732) = 251 ampere

Not All-Electric If the restaurant is not all electric, then a demand factor of 100 percent applies to the first 250 kVA.

113,099 VA at 100% = 113,099 VA

I = VA/(E × $\sqrt{3}$) = 113,099 VA/(208 volts × 1.732) = 314 ampere

SCHOOL – OPTIONAL METHOD [SECTION 220-34]

An optional method of calculating the demand service load for a school is permitted. The following steps can be used to determine the service size:

Step 1: ➛ Determine the total connected load. Add the nameplate rating of all loads at 100 percent and select the larger of the air-conditioning versus heat load.

Step 2: ➛ (a) Determine the average VA each square foot by dividing the total connected load (Step 1) by the square feet of the building.

(b) Determine the demand VA each square foot by applying the demand factors of Table 220-34 to the average VA each square foot (Step 2).

(c) Determine the school net VA by multiplying the demand VA each square foot (Step 3) by the square footage of the school building.

❑ School Optional Method

What is the service size for the following loads? The system voltage is 208Y/120 volt, 3-phase (Figure 11–24).

(1) Air-conditioning	50,000 VA
(2) Cooking equipment	40,000 VA
(3) Lighting	100,000 VA
(4) Multi-outlet assembly	10,000 VA
(5) Receptacles	40,000 VA
(6) Separate circuits	40,000 VA
	280,000 VA

(a) 300 ampere (b) 400 ampere
(c) 500 ampere (d) 600 ampere

- Answer: (c) 500 ampere

Determine the average VA per square foot. Total connected load/square feet area:
280,000 VA/10,000 square feet =
28 VA per square foot.

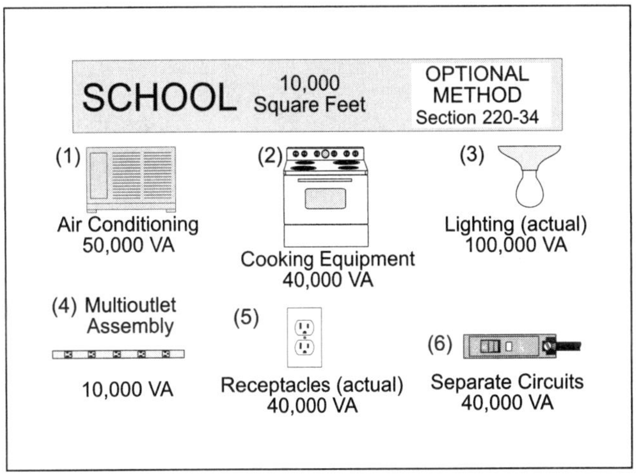

Figure 11–24
School – Optional Service Demand Load

Apply the demand factors from Table 220-34 to the VA per square foot

Average VA per square foot	28 VA	
First 3 VA at 100%	– 3 VA × 1.00 =	3.00 VA
	25 VA	
Next 17 VA at 75%	– 17 VA × 0.75 =	12.75 VA
Remainder at 25%	8 VA × 0.25 =	2.00 VA
		17.75 VA

Net VA per square foot = 17.75 VA × 10,000 square feet = 177,500 VA
Service Size – I = VA/(E × $\sqrt{3}$) = 177,500 VA/(208 volts × 1.732) = 493 ampere

Service Demand Load Using the Standard Method

For exam preparation purposes, you are not expected to calculate the total demand load for a commercial building. However, you are expected to know how to determine the demand load for the individual loads (Steps 1 through 10 below). For your own personal knowledge, you can use the following steps to determine the total demand load for a commercial building for the purpose of sizing the service.

Part A – Determine the Demand Load for Each Type of Load

Step 1: ➻ Determine the general lighting load: Table 220-3(a) and Table 220-11.
Step 2: ➻ Determine the receptacle demand load, [*Section 220-13*].
Step 3: ➻ Determine the appliance demand load at 100 percent.
Step 4: ➻ Determine the demand load for show windows at 125 percent, [*Section 220-12*].
Step 5: ➻ Determine the demand load for multi-outlet assembly, [*Section 220-3(b)(8)*].
Step 6: ➻ Determine the larger of:
Air-conditioning, largest air-conditioning load at 125 percent plus all other loads at 100 percent, [*Section 440-33*]
Heat at 100 percent, [*Section 220-15*].
Step 7: ➻ Determine the demand load for educational cooking equipment, [*Section 220-19* Note 5].
Step 8: ➻ Determine the kitchen equipment demand load, [Table 220-20].
Step 9: ➻ All other noncontinuous loads at 100 percent, [*Section 220-10(b)*].
Step 10: ➻ All other continuous loads at 125 percent, [*Section 215-2, 215-3, 230-42*].

Part B – Determine the Total Demand Load

Add up the individual demand loads of Steps 1 through 10.

Part C – Determine the Service Size

Divide the total demand load (Part B) by the system voltage.
Single-Phase = I = VA/E Three-Phase = I = VA/(E × 1.732)

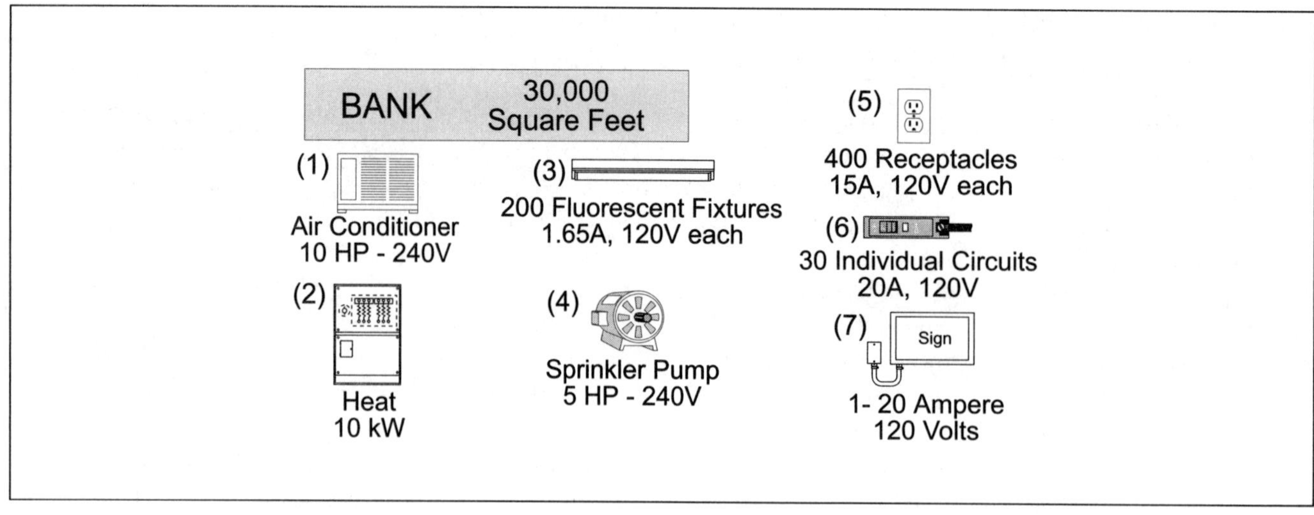

Figure 11–25
Bank

PART D – LOAD CALCULATION EXAMPLES

BANK (120/240 Volt, Single-Phase) (Figure 11–25)

(1) Air-conditioning [Article 440, 430-24, and Table 430-148]
10 horsepower single-phase (230 volts × 50 ampere × 1.25) 14,375 VA

(2) Heat [*Section 220-15* and *220-21*] 10 kW omit

(3) Lighting [Table 220-12(b)]
General Lighting (30,000 square feet × 3.5 VA = 105,000 VA* × 1.25) 131,250 VA
Actual Lighting (200 units × 1.65 ampere × 120 volts × 1.25 = 49,500 VA, omit)

(4) Motor [Table 430-148] 5 horsepower single-phase (230 volts × 28 ampere) 6,440 VA
(Some exams use 240 volts × 28 ampere = 6,720 VA)

(5) Receptacles [*Section 220-13*] (400 receptacles × 180 VA) 72,000 VA
–10,000 VA × 1.00 = 10,000 VA*
62,000 VA × 0.50 = 31,000 VA*

(6) Separate Circuits (30 circuits × 20 ampere × 120 volts) 72,000 VA*

(7) Sign [*Section 220-3(b)(6),* and *600-5(b)(3)*] (1,200 VA* × 1.25) 1,500 VA

*Neutral load

SUMMARY	Overcurrent Protection	Conductor	Neutral
(1) Air conditioner	14,375 VA	14,375 VA *1	0 VA
(3) Lighting	131,250 VA	131,250 VA	105,000 VA *2
(4) Motor	16,100 VA *3	6,440 VA	0 VA
(5) Receptacles	41,000 VA	41,000 VA	41,000 VA
(6) Separate Circuits	72,000 VA	72,000 VA	72,000 VA
(7) Sign	+1,500 VA	+1,500 VA	+1,200 VA *2
	276,225 VA	266,565 VA	219,200 VA

*1 Includes 25 percent for the largest motors (14,375 × 1.25)

*2 The neutral reflects the feeder conductor load at 100 percent, not 125 percent [*Section 220-22*]!

*3 Feeder protection is sized according to the largest motor short-circuit ground-fault protection device [*Section 430-63*]. 28 ampere × 2.5 = 70 ampere, 70 ampere × 230 volts = 16,100 VA

Feeder/service protection device I = VA/E = (276,225 VA /240 volts) = 1,151 ampere

Feeder and service conductor I = VA/E = 266,565 VA/240 volts = 1,111 ampere

Neutral Ampere The neutral conductor is permitted to be reduced according to the requirements of Section 220-22 for that portion of the unbalanced load that exceed 200 ampere. There shall be no reduction of the neutral capacity for that portion of the load that consists of nonlinear loads supplied from a 4-wire, wye-connected, 3-phase system. Since this system is a 120/240-volt, single-phase system, we are permitted to reduce the neutral by 70 percent for that portion that exceeds 200 ampere.

I = VA/E I = 219,200/240 volts	913 ampere	
First 200 ampere	– 200 ampere × 1.00 =	200 ampere
Remainder	713 ampere × 0.7 =	+ 499 ampere
		699 ampere

❑ **Summary Questions**

The service overcurrent protection device must be sized no less than 1,151 ampere. The service conductors must have an ampacity no less than the overcurrent protection device rating [*Section 240-3(c)*]. The grounded (neutral) conductor must not be less than 699 ampere.

➸ **Overcurrent Protection: What size service overcurrent protection device is required?**

(a) 800 ampere (b) 1,000 ampere (c) 1,200 ampere (d) 1,600 ampere

• Answer: (c) 1,200 ampere [*Section 240-6(a)*].

➸ **Service Conductor Size: What size service THHN conductors are required in each raceway if the service is parallel in four raceways?**

(a) 250 kcmil (b) 300 kcmil (c) 350 kcmil (d) 400 kcmil

• Answer: (c) 350 kcmil

1,200 ampere/4 raceways = 300 ampere per raceway, sized based on 75°C terminal rating [*Section 110-14(c)(2)*]

350 kcmil rated 310 ampere × 4 = 1,240 ampere [*Section 240-3(c)* and Table 310-16]

Note: 300 kcmil THHN has an ampacity of 320 ampere at 90°C, but we must size the conductors at 75°C, not 90°C.

➸ **Neutral Size: What size service grounded (neutral) conductor is required in each of the four raceways?**

(a) No. 1/0 (b) No. 2/0 (c) 250 kcmil (d) 350 kcmil

• Answer: (b) No. 2/0

The grounded (neutral) conductor must be sized no less than:

(1) $12^1/_2$ percent area of the line conductor, 350,000 × 4 = 1,400,000 × 0.125 = 175,000 [*Section 250-24(b)*]
175,000/4 = 43,750 kcmil, No. 3 per raceway [Chapter 9, Table 8].

(2) When paralleling conductors, no grounded (neutral) conductor can be smaller than No. 1/0 [*Section 310-4*].

(3) The service neutral conductor must have an ampacity of at least 699 ampere.
699 ampere/4 = 175 ampere, Table 310-16, No. 2/0 is required at 75°C [*Section 110-14(c)*]

➸ **Grounding Electrode [*Section 250-66(b)*]: What size grounding electrode conductor is required to a concrete encased electrode?**

(a) No. 4 (b) No. 1/0 (c) No. 2/0 (d) No. 3/0

• Answer: (a) No. 4

OFFICE BUILDING (480Y/277 Volt, Three-Phase) (Figure 11–26)

(1) Air-conditioning (Table 430-150 and 440-34)
460 volts × 7.6 ampere × 1.732 = (6,055 VA × 15 units) + (6,055 VA × 0.25) 92,339 VA

(2) Lighting (*NEC®*) [Table 220-3(a) and 220-10(b)]
28,000 square feet × 3.5 VA × 1.25 = 122,500 VA (omit)
Lighting Electric Discharge (Actual) [*Section 215-2* and *230-42*]
500 lights × 0.75 ampere × 277 volts = 103,875 VA* × 1.25 129,844 VA

(3) Lighting, Track (Actual) [*Section 220-12(b)*] 150 VA per 2 feet
200 feet/2 feet = 100 sections × 150 VA = 15,000 VA* × 1.25 18,750 VA

(4) Receptacles [*Section 220-13*](Actual) (200 receptacles × 180 VA) 36,000 VA
−10,000 VA × 100% = 10,000 VA*
26,000 VA × 50% = 13,000 VA*

(5) Multi-outlet Assembly [*Section 220-3(b)(8)*]
147 feet/5 feet = thirty 5-foot sections (30 sections × 180 VA) 5,400 VA*
Simultaneously used (20 feet × 180 VA) 3,600 VA*

(6) Separate Circuits (noncontinuous loads) (120 volts × 20 ampere × 30 circuits) 72,000 VA*

(7) Sign [*Section 220-3(b)(6)* and *600-5 (b)(3)*] (1,200 VA* × 1.25) 1,500 VA

*Neutral load

SUMMARY	Overcurrent Protection	Conductor	Neutral
(1) Air-conditioning	92,339 VA	92,339 VA	0 VA
(2) Lighting (actual)	129,844 VA	129,844 VA	103,875 VA *1
(3) Lighting (track)	18,750 VA	18,750 VA	15,000 VA *1
(4) Receptacles (actual)	23,000 VA	23,000 VA	23,000 VA
(5) Multi-outlet Assembly	9,000 VA	9,000 VA	9,000 VA
(6) Separate Circuits	72,000 VA	72,000 VA	72,000 VA
(7) Sign	1,500 VA	1,500 VA	1,200 VA *1
	346,433 VA	346,433 VA	224,075 VA

*1 The neutral reflects the feeder load at 100 percent, not 125 percent.

Feeder/service protection device I = VA/(E × $\sqrt{3}$), I = 346,433 VA/(480 × 1.732) = 416 ampere

Feeder and service conductor ampere I = VA/(E × $\sqrt{3}$), I = 346,433 VA/(480 × 1.732) = 416 ampere

Neutral conductor I = VA/(E × $\sqrt{3}$), I = 224,075 VA/(480 volts × 1.732) = 270 ampere*

❑ Summary

The service overcurrent protection device and conductor must be sized no less than 416 ampere, and the grounded (neutral) conductor must not be less than 283 ampere.

➾ Overcurrent Protection: What size service overcurrent protection device is required?

(a) 300 ampere (b) 350 ampere
(c) 450 ampere (d) 500 ampere

• Answer: (c) 450 ampere [*Section 240-6(a)*]

Since the total demand load for the overcurrent device is 416 ampere, the minimum service permitted is 450 ampere.

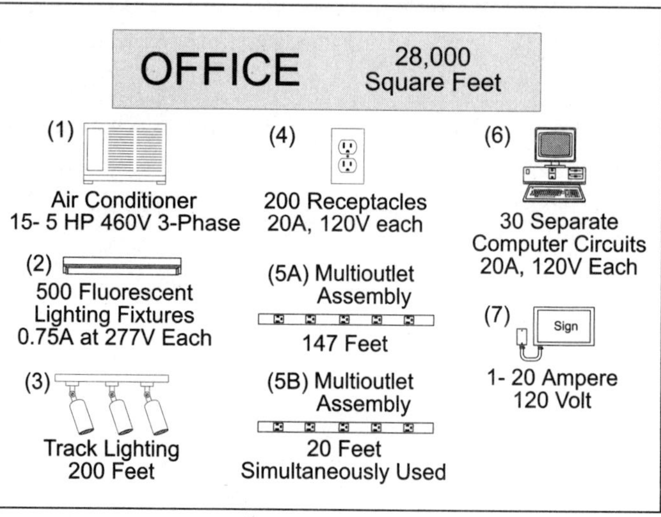

Figure 11–26
Office

➻ **Service Conductor Size: What size service conductors are required if total calculated demand load is 416 ampere?**

(a) 300 kcmil (b) 400 kcmil (c) 500 kcmil (d) 600 kcmil

• Answer: (d) 600 kcmil rated 420 ampere [*Section 240-3(b)* and *Table 310-16*]

The service conductors must have an ampacity of at least 416 ampere, protected by a 450 ampere overcurrent protection device [*Section 240-3(b)*], and the conductors must be selected based on 75°C insulation rating [*Section 110-14(c)(2)*].

➻ **Neutral Size: What size service grounded (neutral) conductor is required for this service?**

(a) No. 1/0 (b) 3/0 (c) 250 kcmil (d) 300 kcmil

• Answer: (d) 300 kcmil

The grounded (neutral) conductor must be sized no less than:

(1) The required grounding electrode conductor [*Section 250-24(b)* and *Table 250-66*], No. 2.

(2) An ampacity of at least 270 ampere, Table 310-16, 300 kcmil is rated 280 ampere at 75°C [*Section 110-14(c)*].

* Note: Because there is 125 ampere of electric discharge lighting [103,875 VA/(480 × 1.732)], the neutral conductor is not permitted to be reduced for that portion of the load in excess of 200 ampere [*Section 220-22*].

RESTAURANT – Standard Load Calculation (208Y/120 Volt, Three-Phase) (Figure 11–27)

(1) Air-conditioning [Table 430-150, and 440-34] (208 volts × 59.4 ampere × 1.732 × 1.25) = 26,749 VA

(2) Kitchen equipment [*Section 220-20*]

Ovens	(10 kW × 2 units)	20.0 kW
Mixer	(3 kW × 2 units)	6.0 kW
Water heaters	(7.5 kW × 1 unit)	7.5 kW
Total connected		33.5 kW × 0.7 × 1,000 = 23,450 kW

(3) Lighting *NEC*® [Table 220-3(a) and 220-10(b)]
8,400 square feet 3 2 VA 3 1.25 = 21,000 VA, omit

(4) Lighting, Track 150 VA per two feet [*Section 220-12(b)*]

50 feet/2 = 25 sections 3 150 VA 3 1.25	4,688 VA
100 lights 3 1.65 ampere 3 120 volts 3 1.25	24,750 VA

(5) Multi-outlet Assembly [*Section 220-3(b)(8)*]

50/5 = ten 5-foot sections (10 sections 3 180 VA)	1,800 VA
Simultaneous used (20 feet 3 180 VA)	3,600 VA
(6) Receptacles (Actual) [*Section 220-13*] (40 receptacles 3 180 VA)	7,200 VA
(7) Separate Circuits [*Section 215-2(a)* and *230–42*](120 volts × 20 ampere × 10 circuits)	24,000 VA
(8) Sign [*Section 220-3(b)(6)* and *600-5 (b)(3)*] (1,200 VA × 1.25)	1,500 VA

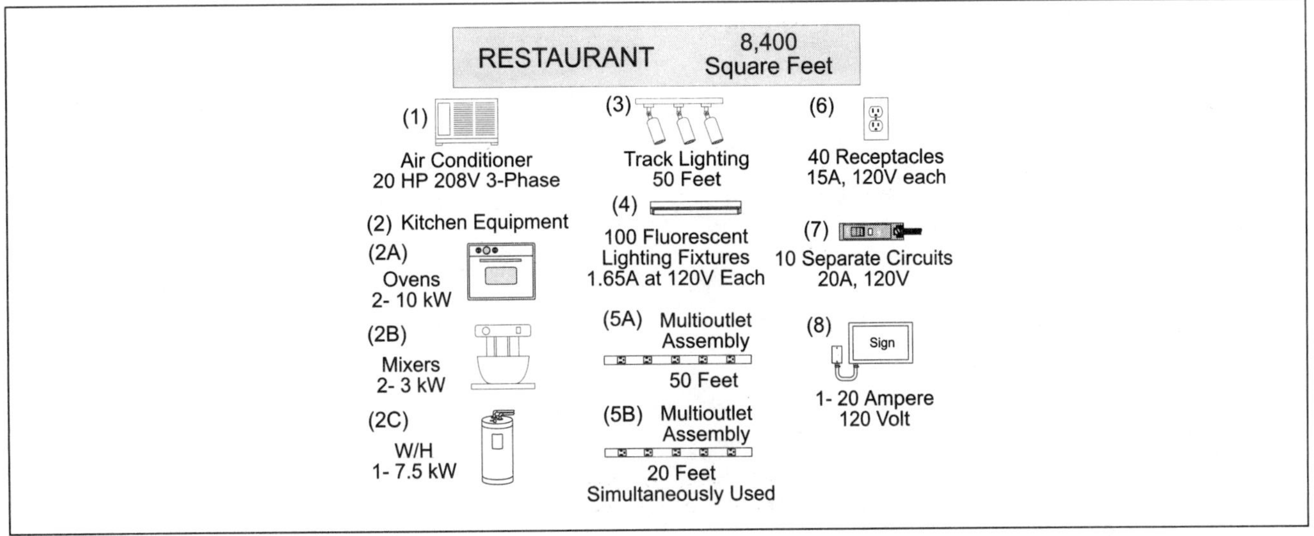

Figure 11–27 Restaurant

SUMMARY	Overcurrent Protection	Conductor
(1) Air-conditioning	26,749 VA	26,749 VA
(2) Kitchen equipment	23,450 VA	23,450 VA
(3) Lighting (track)	4,688 VA	4,688 VA
(4) Lighting (actual)	24,750 VA	24,750 VA
(5) Multi-outlet Assembly	5,400 VA	5,400 VA
(6) Receptacles (actual)	7,200 VA	7,200 VA
(7) Separate Circuits	24,000 VA	24,000 VA
(8) Sign	1,500 VA	1,500 VA
	117,737 VA	117,737 VA

Feeder and service protection device $I = VA/(E \times \sqrt{3}) = 117{,}737\ VA/(208 \times 1.732) = 327$ ampere
Feeder and service conductor $I = VA/(E \times \sqrt{3}) = 117{,}737\ VA/(208 \times 1.732) = 327$ ampere

❑ **Summary Questions**

The service overcurrent protection device must be sized no less than 327 ampere. The service conductors must be no less than 327 ampere.

➛ **Overcurrent Protection: What size service overcurrent protection device is required?**

(a) 300 ampere (b) 350 ampere (c) 450 ampere (d) 500 ampere

• Answer: (b) 350 ampere [*Section 240-6(a)*]

Since the total demand load for the overcurrent device is 327 ampere, the minimum service permitted is 350 ampere [*Section 240-6(a)*].

➛ **Service Conductors Size: What size service conductors are required?**

(a) No. 4/0 (b) 250 kcmil (c) 300 kcmil (d) 400 kcmil

• Answer: (d) 400 kcmil

The service conductors must have an ampacity of at least 327 ampere and be protected by a 350-ampere protection device [*Section 240-3(b)*] next size up is permitted, and the conductors must be selected according to Table 310-16, 400 kcmil based on 75°C insulation rating [*Section 110-14(c)(2)*] is rated 335 ampere.

Unit 11 – Commercial Load Calculations Summary Questions

Part A – General

11–2 Conductor Ampacity [Article 100]

1. The _____ of a conductor is the rating in ampere a conductor can carry continuously without exceeding its insulation temperature rating. The allowable ampacities as listed in Table 310-16 are affected by ambient temperature, current flow, conductor insulation, and conductor bundling [*Section 310-10*].
   (a) load rating (b) ampacity (c) demand load (d) continuous factor

11–3 Conductor Overcurrent Protection [*Section 240-3*]

2. The purpose of conductor _____ is to protect the conductors against excessive or dangerous temperatures. If the ampacity of a conductor does not correspond with the standard ampere rating of a fuse or circuit breaker, the next size up protection device is permitted. This applies only if the conductors do not supply multi-outlet receptacles and if the next size overcurrent protection device does not exceed 800 ampere.
   (a) short-circuit protection (b) ground-fault protection
   (c) overload protection (d) overcurrent protection

3. The following is a list of some of the standard ampere ratings for overcurrent protection devices (fuses and inverse time circuit breakers): 15, 25, 35, 45, 80, 90, 110, 175, 250, and 350.
   (a) True (b) False

11–4 Voltages [*Section 220-2(a)*]

4. Unless other voltages are specified, branch-circuit, feeder, and service loads shall be computed at a nominal system voltage of _____ .
   (a) 600Y/347 (b) 240/120 (c) 208Y/120 (d) any of these

11–5 Rounding an Ampere [*Section 220-2(b)*]

5. There are no *NEC*® rules for rounding when a calculation results in a fraction of an ampere, but it does contain the following note: "except where the computations result in a _____ of an ampere or larger, such fractions may be dropped."
   (a) 0.05 (b) 0.5 (c) 0.49 (d) 0.51

Part B – Loads

11–6 Air-Conditioning

6. What is the feeder or service demand load required for the air-conditioning of a 6-unit office building? Each unit contains one air conditioner rated 3 horsepower 230 volts.
   (a) 13 kVA (b) 24 kVA (c) 45 kVA (d) 51 kVA

11–7 Dryers

7. What size branch circuit conductor and overcurrent protection is required for a 6 kW dryer rated 240 volts, located in the laundry room of a multifamily dwelling?
(a) No. 12 with a 20-ampere OCPD
(b) No. 10 with a 20-ampere OCPD
(c) No. 12 with a 30-ampere OCPD
(d) No. 10 with a 30-ampere OCPD

8. What is the feeder and service demand for eight 6.75 kW dryers?
(a) 70 kW (b) 54 kW (c) 35 kW (d) 27 kW

11–8 Electric Heat

9. • What size conductor and protection is required for a single-phase, 15 kW, 480-volt heat strip that has a 1.6-ampere blower motor?
(a) No. 10 THHN with 30-ampere protection
(b) No. 6 THHN with 45-ampere protection
(c) No. 10 THHN with 40-ampere protection
(d) No. 4 THHN with 70-ampere protection

10. What is the feeder and service demand load in VA for a building that has four 20 kW, 230-volt, single-phase heat strips that have a 5.4-ampere blower motor for each unit?
(a) 100 kVA (b) 50 kVA (c) 125 kVA (d) 85 kVA

11–9 Kitchen Equipment

11. What is the branch circuit demand load in ampere for one 11.4 kW oven rated 240 volts?
(a) 60 ampere (b) 27 ampere (c) 48 ampere (d) 33 ampere

12. • What is the feeder and service demand load for the following?

Water heater	9 kW
Booster heater	12 kW
Mixer	3 kW
Oven	2 kW
Dishwasher	1.5kW
Waste disposal	1 kW

(a) 19 kW (b) 21 kW (c) 29 kW (d) 12 kW

13. What is the feeder and service demand load for the following?

Water heater	14 kW
Booster heater	11 kW
Mixer	7 kW
Oven	9 kW
Dishwasher	1.5 kW
Waste disposal	3 kW

(a) 25 kW (b) 30 kW (c) 20 kW (d) 45 kW

11–10 Laundry Equipment

14. What is the feeder and service demand load for 7 washing machines at 1,500 VA each?
(a) 10,500 VA (b) 15,000 VA (c) 1,125 VA (d) none of these

11–11 Lighting – Demand Factors [Table 220-3(a) and 220-11]

15. What is the general lighting feeder or service demand load for a 24-room motel, 685 square feet each?
(a) 10 kVA (b) 15 kVA (c) 20 kVA (d) 25 kVA

16. What is the general lighting and receptacle feeder or service demand load for a 250,000 square-foot storage warehouse that has 200 receptacles?
(a) 25 kVA (b) 55 kVA (c) 95 kVA (d) 105 kVA

11–12 Lighting without Demand Factors [Table 220-3(a), 215-2, and 230-42]

17. What is the feeder and service (overcurrent protection device) general lighting demand load for a 3,200 square-foot dance club?
(a) 3,200 VA (b) 6,400 VA (c) 8,000 VA (d) 12,000 VA

18. What is the feeder and service general lighting load for a 90,000 square-foot school?
(a) 238 kVA (b) 338 kVA (c) 90 kVA (d) 270 kVA

11–13 Lighting – Miscellaneous

19. How many 2 × 4 fluorescent fixtures, each rated 277 volts, 0.8 ampere, can be connected to a 20-ampere circuit? The four lamps are rated 40 watts each and the fixture is to be on for more than 3 hours.
(a) 5 (b) 7 (c) 9 (d) 20

20. How many 360-watt, 120-volt incandescent fixtures can be connected to a 20-ampere, 120-volt circuit continuously?
(a) 5 (b) 7 (c) 9 (d) 12

21. • What is the VA demand feeder and service load for 130 feet of show-window lighting?
(a) 26 kVA (b) 33 kVA (c) 9 kVA (d) 11 kVA

11–14 Multi-outlet Receptacle Assembly [*Section 220-3(b)(8)*]

22. What is the feeder demand load for 50 feet of multi-outlet assembly and 10 feet of multi-outlet assembly simultaneously used?
(a) 3,600 VA (b) 12,000 VA (c) 7,200 VA (d) 5,500 VA

11–15 Receptacle VA Load [*Section 220-13*]

23. How many receptacle outlets are permitted on a 20-ampere, 120-volt circuit?
(a) 10 (b) 13 (c) 15 (d) 20

24. What is the service demand load for 110 receptacles (15 or 20 ampere, 125 volts) in a commercial building?
(a) 5 kVA (b) 10 kVA (c) 15 kVA (d) 20 kVA

25. What is the service demand load for the general lighting and receptacles for a 30,000 square-foot bank?
(a) 111 kVA (b) 152 kVA (c) 123 kVA (d) 175 kVA

26. • What is the service general lighting demand and receptacle load for a 10,000 square-foot office building with 75 receptacles?
(a) 25 kVA (b) 35 kVA (c) 45 kVA (d) 55 kVA

11–17 Signs [*Section 220-3(b)(b) and 600-5*]

27. What is the feeder demand load for sizing the overcurrent protection device for one electric sign?
(a) 1,400 VA (b) 1,500 VA (c) 1,920 VA (d) 2,400 VA

11–18 Neutral Calculations [*Section 220-22*]

28. What is the neutral current for a balanced 150-ampere, 3-wire, 208Y/120-volt feeder?
(a) 0 ampere (b) 150 ampere (c) 250 ampere (d) 300 ampere

Part C – Load Calculations

Marina [*Section 555-6*]

29. A marina has the following: 24 slips (20-ampere, 240-volt receptacles) and 30 slips designed for boats over 20 feet (30-ampere, 240-volt receptacles). What size service is required for the marina?
(a) 400 ampere (b) 550 ampere (c) 900 ampere (d) 1,000 ampere

30. What size conductor is required for the service if the total demand load per phase is 570 ampere and the service is paralleled in two raceways?
(a) 4/0 (b) 250 kcmil (c) 300 kcmil (d) 350 kcmil

Mobile/Manufactured Home Park [*Section 550-22*]

31. What is the feeder and service load for a mobile home park that has the facilities for 42 sites? The system is 120/240 volt, single-phase.
(a) 644 ampere (b) 512 ampere (c) 732 ampere (d) 452 ampere

Recreational Vehicle Park [*Section 551-73*]

32. A recreational vehicle park has 17 sites with only 20-ampere, 240-volt receptacles, 35 sites with 20- and 30-ampere, 240-volt receptacles, and 10 sites with 50-ampere, 240-volt receptacles. The feeder and service demand load is approximately _____ .
(a) 355 ampere (b) 450 ampere (c) 789 ampere (d) 1,114 ampere

Restaurant – Optional Method [*Section 220-36*]

33. A restaurant has a total connected load of 400 kVA and is all-electric. What is the demand load for the service?
(a) 200 kVA (b) 325 kVA (c) 300 kVA (d) 275 kVA

34. A restaurant has a total connected load of 400 kVA and is not all-electric. What is the demand load for the service?
(a) 260 kVA (b) 325 kVA (c) 300 kVA (d) 275 kVA

School – Optional Method [*Section 220-34*]

35. A 28,000 square-foot school has a total connected load of 590,000 VA. What is the demand load?
(a) 350 kVA (b) 400 kVA (c) 450 kVA (d) 500 kVA

★ Challenge Questions

General Commercial Calculations

36. • If the service high-leg conductor only supplies 3-phase loads, what size service conductor is required for the high-leg? Three-phase loads: A/C (10 horsepower 230 volt, 3-phase) and Heat (10 kW 230 volt, 3-phase).
(a) No. 10 TW (b) No. 12 THW (c) No. 10 THHN (d) No. 8 THW

37. • A commercial office building has forty-six 250-watt, 120-volt lights installed on a 208Y/120-volt, 3-phase system. If the fixtures are on continuously, how many 20-ampere, 120-volt circuits would be required?
(a) 9 circuits (b) 7 circuits (c) 5 circuits (d) 4 circuits

Conductor Sizing

38. • Three parallel service raceways are installed. Each raceway contains three 500 kcmil THHN conductors. What size bonding jumper is required for each service raceway?
(a) 250,000 (b) No. 2/0 (c) No. 1/0 (d) No. 4/0

39. • If each service raceway contains 300 kcmil conductors, what size bonding jumper is required for the raceway?
(a) No. 1 (b) No. 2 (c) No. 1/0 (d) No. 4

Kitchen Equipment

40. What is the kitchen equipment demand load for: one 14 kW range, one 1.5 kW water heater, one $^3/_4$ kW mixer, one 2 kW dishwasher, one 3 kW booster heater, and one 3 kW coffee machine?
(a) 17 kW (b) 14 kW (c) 15 kW (d) 24 kW

11–11 Lighting – Demand Factors [Table 220-3(a) and 220-11]

41. • Each unit of a 100-unit hotel is 12 × 15 feet. In addition, there is an office that is 60 × 20 feet, and there are hallways of 120 square feet. What is the general lighting demand load?
(a) 29 kVA (b) 37 kVA (c) 23 kVA (d) 25 kVA

11–15 Receptacle VA Load [*Section 220-13*]

42. If, in the hall and other areas of a motel, there are 100 receptacle outlets (not in the motel rooms), the demand load added to the service for these receptacles would be _____ .
(a) 0 VA (b) 10,000 VA (c) 14,000 VA (d) 20,000 VA

43. What is the receptacle demand load for a 20,000 square-foot office building?
(a) 10,000 VA (b) 40,000 VA (c) 30,000 VA (d) 15,000 VA

11–16 Banks and Offices General Lighting and Receptacle

44. What is the general lighting and general use receptacle load for a 30,000 square-foot bank?
(a) 162 kVA (b) 123 kVA (c) 173 kVA (d) 151 kVA

11–18 Neutral Calculations [*Section 220-22*]

A 208Y/120-volt, 3-phase service has a total connected load of 1,450 ampere:
600 ampere of these are phase-to-phase loads
300 ampere are balanced 120-volt fluorescent lighting
550 ampere of other 120-volt loads

45. • The neutral demand for this service is _____ .
(a) 1,150 ampere (b) 650 ampere (c) 745 ampere (d) 420 ampere

Part C – Load Calculations

Marina Calculations [*Section 555*]

46. A marina shore power facility has twenty 20-ampere, 240-volt receptacles, seventeen 30-ampere, 240-volt receptacles, and seven 50-ampere, 240-volt receptacles. After applying demand factors, the service demand load for the shore power boxes is _____ .
(a) 1,260 ampere (b) 630 ampere (c) 1,160 ampere (d) 625 ampere

Mobile/Manufactured Home Parks [*Section 550*]

47. A 75-site mobile home park is designed for mobile homes that have a 14,000 VA load per site. The service demand load for the park is _____ .
(a) 1,200 kVA (b) 264 kVA (c) 222 kVA (d) 201 kVA

Motel

48. A 40-unit motel (300 square feet in each unit) has 3 kVA of air-conditioning and 4 kW of heat in each unit, and every two units share one 1.5 kW water heater. What is the demand load for the motel?
(a) 281 kVA (b) 202 kVA (c) 226 kVA (d) 251 kVA

Recreational Vehicle Park [Article 551]

49. A recreational vehicle park has 42 sites, 3 sites rated 50 ampere, 240/120 volts, 30 sites rated 30/20 ampere, and 9 sites rated 20 ampere. The minimum feeder demand load for these sites would be _____ .
(a) 158 kVA (b) 101 kVA (c) 139 kVA (d) 65 kVA

Restaurant Calculations

50. • The branch circuit rating required for a 12 kW range in a restaurant would be _____ .
(a) 50 ampere (b) 45 ampere (c) 35 ampere (d) 30 ampere

51. • A new restaurant has a total connected lighting load of 30 kVA. The kitchen equipment includes two gas stoves, one gas grill, three gas ovens, one 75-gallon gas water heater, one 5 kW dishwasher, two 2 kW coffee makers, five 2 kW kitchen appliances on their own circuit, and ten 1.5 kVA small appliance circuits. Using the optional method, the service demand load would be closest to _____ .
(a) 64 kVA (b) 50 kVA (c) 45 kVA (d) 38 kVA

School – Optional Method [*Section 220-34*]

52. • Using the optional method, what is the demand load (VA per square foot) for a 10,000 square foot school that has a total connected load of 320 kVA?
(a) 15.75 VA per square foot (b) 18.75 VA per square foot
(c) 29.00 VA per square foot (d) 12.75 VA per square foot

53. Using the optional method, what is the demand load (kVA) for a 20,000 square-foot school that has a total connected load of 160 kVA?
(a) 135 kVA (b) 106 kVA (c) 120 kVA (d) 112 kVA

Unit 12

Delta/Delta and Delta/Wye Transformer Calculations

OBJECTIVES

After reading this unit, the student should be able to briefly explain the following concepts:

Part A – Delta/Delta Transformers
- Current flow
- Delta transformer voltage
- Delta high-leg
- Delta primary and secondary line currents
- Delta primary or secondary phase currents
- Delta phase versus line
- Delta current triangle
- Delta transformer balancing
- Delta transformer sizing
- Delta panel schedule in kVA
- Delta panelboard and conductor sizing
- Delta neutral current
- Delta maximum unbalanced load
- Delta/delta example

Part B – Delta/Wye Transformers
- Wye transformer voltage
- Wye voltage triangle
- Wye transformer current line current
- Phase current
- Wye phase versus line wye transformer loading and balancing
- Wye transformer sizing
- Wye panel schedule in kVA
- Wye panelboard and conductor sizing
- Wye neutral current
- Wye maximum unbalanced load
- Delta/wye example
- Delta versus wye

After reading this unit, the student should be able to briefly explain the following terms:

- Delta connected
- kVA rating
- Line
- Line current
- Line voltage
- Phase
- Phase current
- Phase load – delta
- Phase load – wye
- Phase voltage
- Ratio
- Balanced load
- Unbalanced load
- Delta secondary
- Wye secondary
- Winding
- Wye connected
- Primary/secondary voltage – ampere
- Three-phase load
- Single-phase load

INTRODUCTION

This unit deals with 3-phase delta/delta and delta/wye transformer sizing and balancing. The system voltages used in this Unit were selected because of their common industry configuration. Delta/delta and delta/wye concepts are opposite of each other, and it is almost impossible to remember the differences between the systems. But only a few of the concepts contained in this Unit are ever on an electrician's exam.

DEFINITIONS

Delta Connected

Delta connected means the windings of three single-phase transformers are connected in series to form a closed circuit. A line can be traced from one point of the delta system, through all transformers, then back to the original starting point (series). Many call it a *delta high-leg* system because the voltage from Line 2 to to ground is 208 volts (Figure 12–1).

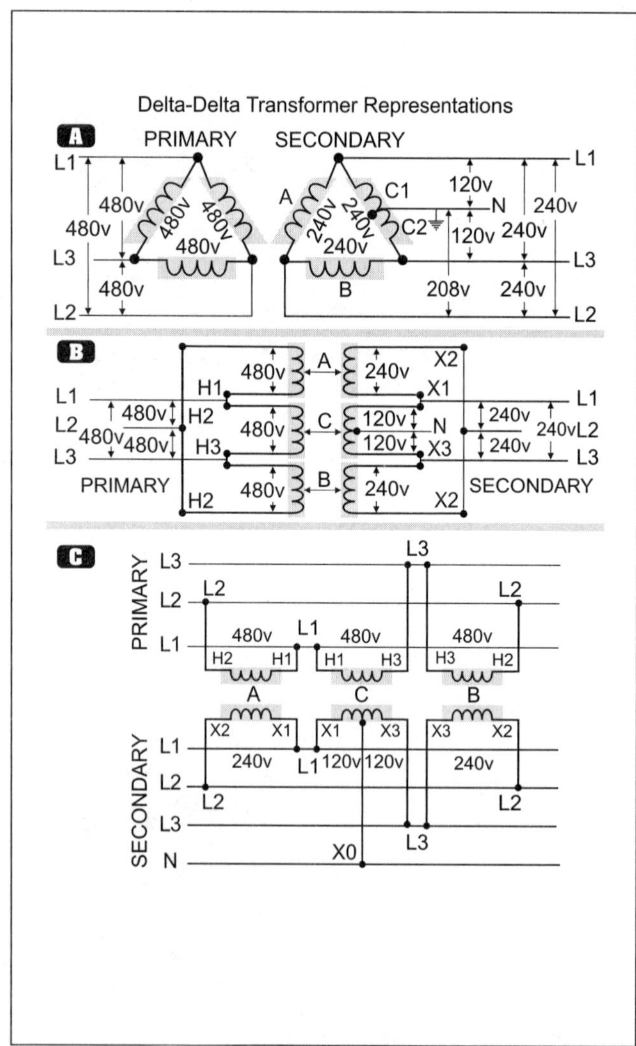

Figure 12–1
Delta-Delta Transformer Representations

Figure 12–2
Transformer Definitions

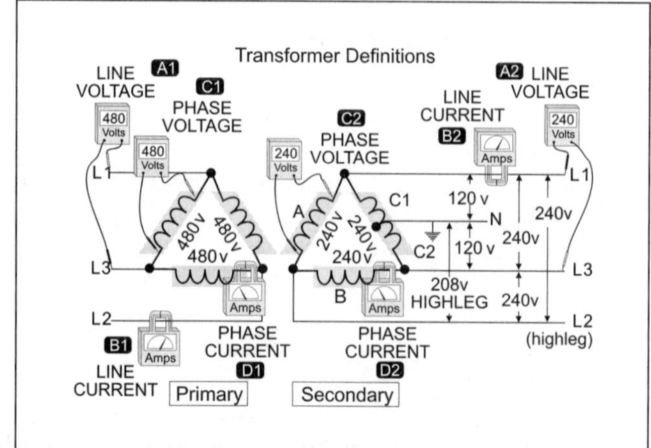

Figure 12–3
Transformer Definitions

kVA Rating

Transformers are rated in kilovolt-ampere (kVA) and are sized according to the VA rating of the loads they supply.

Line

The line is considered the electrical system supply (hot, ungrounded) conductors (Figure 12–2 and Figure 12– 3).

Line Current

The line current for both delta and wye systems is the current on the line (hot, ungrounded) conductors, calculated according to the following formulas (Figure 12–3, B1 and B2):

$$\textbf{Line Current (single-phase)} = \frac{\textbf{Line Power}}{\textbf{Line Voltage, } \mathbf{I_{Line}}} = \frac{\mathbf{VA_{Line}}}{\mathbf{E_{Line}}}$$

$$\textbf{Line Current (three-phase)} = \frac{\textbf{Line Power}}{(\textbf{Line Voltage} \times \sqrt{3}), \mathbf{I_{Line}}} = \frac{\mathbf{VA_{Line}}}{(\mathbf{E_{Line}} \times \sqrt{3})}$$

Line Voltage

The line voltage is the voltage that is measured between any two-line (hot ungrounded) conductors. It is also called line-to-line voltage. Line voltage is greater than phase voltage for wye systems, and line voltage is the same as *phase voltage* for delta systems (Figure 12–3, A1 and A2).

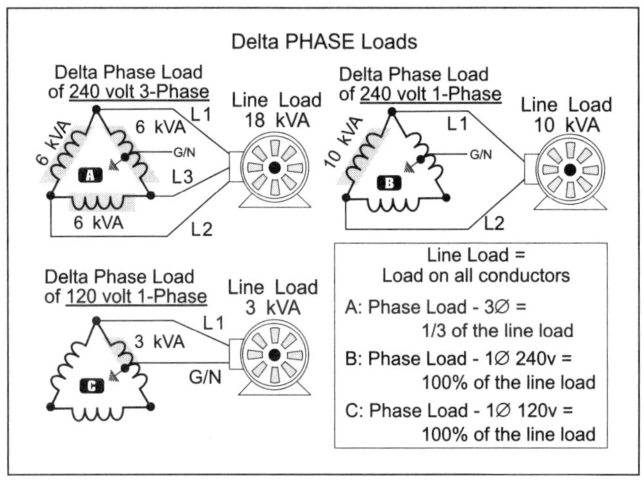

Figure 12–4
Delta Phase Loads

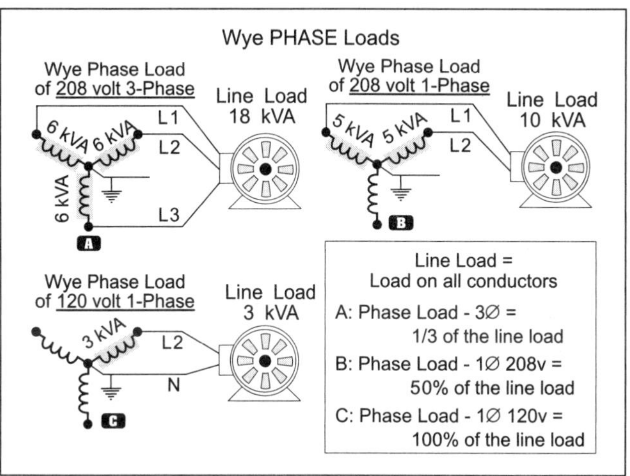

Figure 12–5
Wye Phase Loads

Phase

The *phase* winding is the coil shaped conductors that serve as the primary or secondary of the transformer (Figure 12–3, C1 and C2).

Phase Current

The phase current is the current of the transformer winding. For delta systems, the phase current is less than the line current. For wye systems, the phase current is the same as the line current (Figure 12–3, D1 and D2).

Phase Load Delta

The phase load is the load on the transformer winding (Figure 12–4).

The phase load of a three-phase, 240-volt load = 1/3 of the line load.

The phase load of a single-phase, 240-volt load = line load.

The phase load of a single-phase, 120-volt load = line load.

Phase Load Wye

The phase load is the load on the transformer winding (Figure 12–5).

The phase load of a three-phase, 208-volt load = $^1/_3$ of the line load.

The phase load of a single-phase, 208-volt load = $^1/_2$ of the line load.

The phase load of a single-phase, 120-volt load = line load.

Phase Voltage

The phase voltage is the internal transformer voltage generated across any one winding of a transformer. For a delta secondary, the phase voltage is equal to the line voltage. For wye secondaries, the phase voltage is less than the line voltage (Figure 12–6).

Ratio (Voltage)

The ratio is the relationship between the number of primary winding turns as compared to the number of secondary winding turns. The ratio is a comparison between the primary phase voltage to the secondary phase voltage. For typical delta/delta systems, the ratio is 2:1. For typical wye systems, the ratio is 4:1 (Figure 12–7).

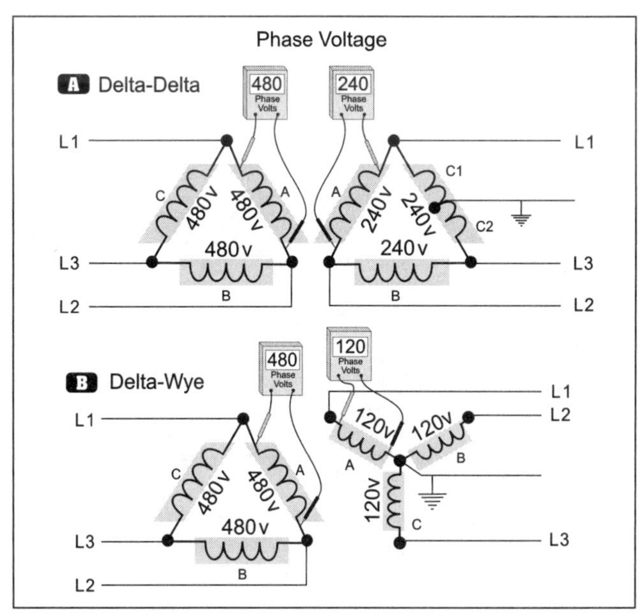

Figure 12–6
Phase Voltage

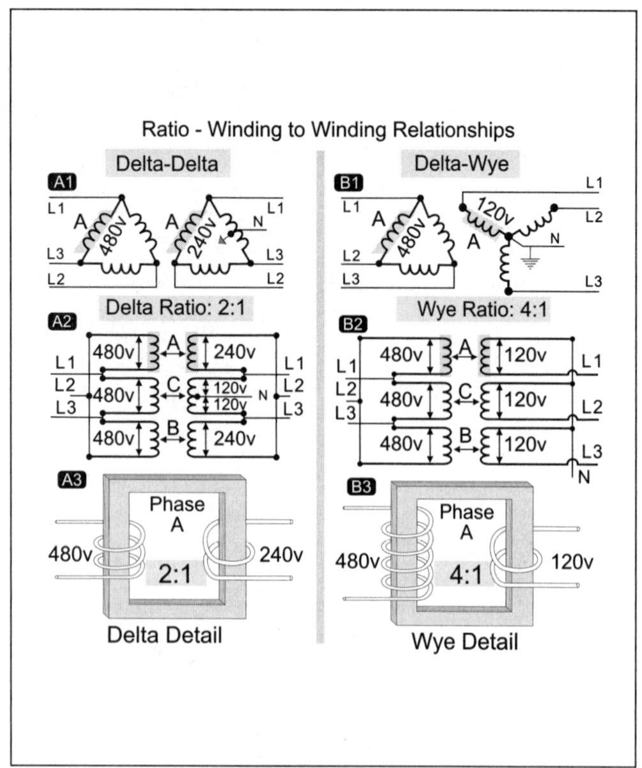

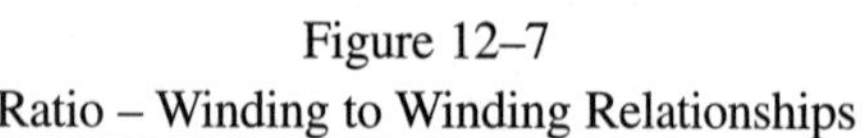
Figure 12–7
Ratio – Winding to Winding Relationships

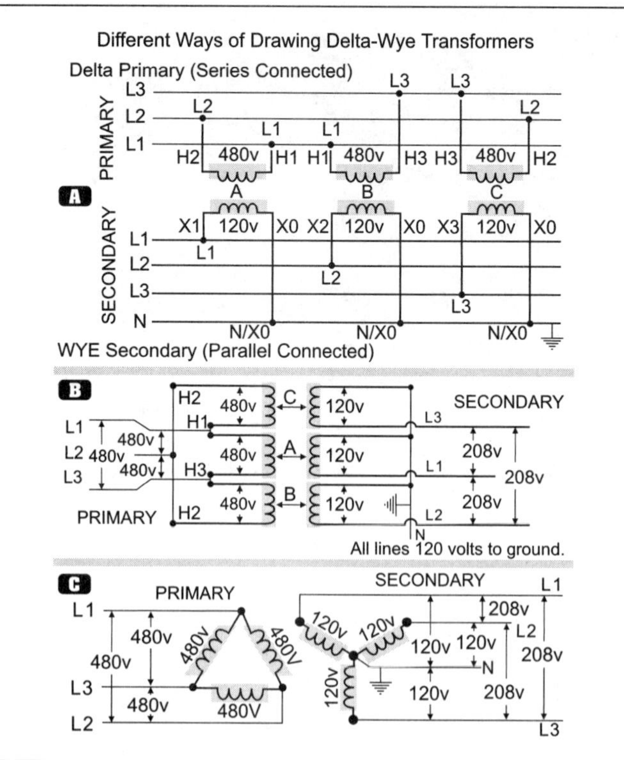

Figure 12–8
Different Ways of Drawing Delta-Wye Transformers

Unbalanced Load (Neutral Current)

The unbalanced load is the load on the secondary grounded (neutral) conductors.

Delta Secondary – Unbalanced Current

The unbalanced load is calculated using:

Line 1 Neutral Current less Line 3 Neutral Current, $I_{Neutral} = I_{Line1} - I_{Line3}$

Note: Line 2 (high-leg) and its voltage to ground is approximately 208 volts, therefore no neutral loads are connected to this line.

Wye Secondary – Unbalanced Current

The unbalanced load is calculated using:

$$I_{Neutral} = \sqrt{(L_1^2 + L_2^2 + L_3^2) - (L_1 \times L_2 + L_2 \times L_3 + L_1 \times L_3)}$$

Winding

The *primary winding* is the winding(s) on the input side of a transformer and the *secondary winding(s)* is the output side of a transformer.

Wye Connected

Wye connected means a connection of three single-phase transformers of the same rated voltage to a common point (neutral) and the other ends are connected to the line conductors (Figure 12–8).

12–1 CURRENT FLOW

When a load is connected to the secondary of a transformer, current will flow through the secondary conductor windings. The current flow in the secondary creates an electromagnetic field that opposes the primary electromagnetic field. The secondary flux lines effectively reduce the strength of the primary flux lines. As a result, less *counter-electromotive force* is generated in the primary winding conductors. With less CEMF to oppose the primary applied voltage, the primary current automatically increases in direct proportion to the secondary current (Figure 12–9).

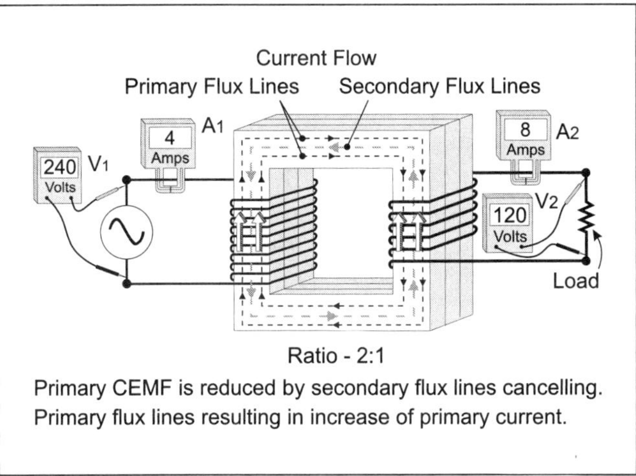

Figure 12–9
Current Flow

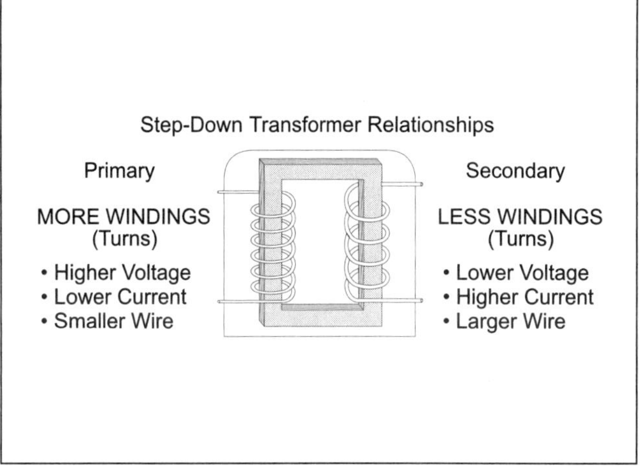

Figure 12–10
Step-Down Transformer Relationships

Note: The primary and secondary line currents are inversely proportional to the voltage ratio of the transformer. This means that the winding with the most number of turns will have a higher voltage and lower current as compared to the winding with the least number of turns, which will have a lower voltage and higher current (Figure 12–10).

The following Tables show the current relationship between kVA and voltage for common size transformers:

Single-Phase Transformers I = VA/E			
kVA Rating	**Current at 208 Volts**	**Current at 240 Volts**	**Current at 480 Volts**
7.5	36 ampere	31 ampere	16 ampere
10	48 ampere	42 ampere	21 ampere
15	72 ampere	63 ampere	31 ampere
25	120 ampere	104 ampere	52 ampere
37.5	180 ampere	156 ampere	78 ampere

Three-Phase Transformers I = VA/(E × $\sqrt{3}$)			
kVA Rating	**Current at 208 Volts**	**Current at 240 Volts**	**Current at 480 Volts**
15	42 ampere	36 ampere	18 ampere
22.5	63 ampere	54 ampere	27 ampere
30	83 ampere	72 ampere	36 ampere
37.5	104 ampere	90 ampere	45 ampere
45	125 ampere	108 ampere	54 ampere
50	139 ampere	120 ampere	60 ampere
75	208 ampere	180 ampere	90 ampere
112.5	313 ampere	271 ampere	135 ampere

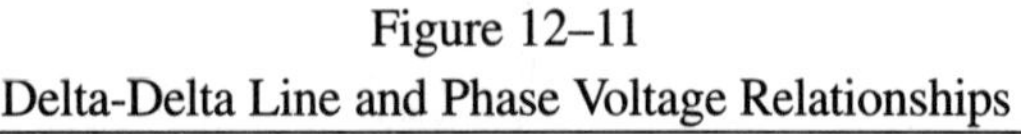

Figure 12–11
Delta-Delta Line and Phase Voltage Relationships

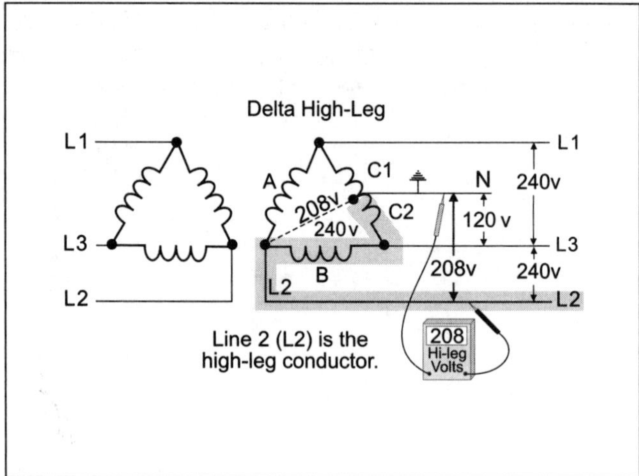

Figure 12–12
Delta High-Leg

PART A – DELTA/DELTA TRANSFORMERS

12–2 DELTA TRANSFORMER VOLTAGE

In a delta configured transformer, the line voltage equals the phase voltage ($E_{Line} = E_{Phase}$) (Figure 12–11).

Primary Delta Voltage

LINE Voltage	PHASE Voltage
L_1 to L_2 = 480 volts	Phase winding A = 480 volts
L_2 to L_3 = 480 volts	Phase winding B = 480 volts
L_3 to L_1 = 480 volts	Phase winding C = 480 volts

Secondary Delta Voltage

LINE Voltage	PHASE Voltage	NEUTRAL Voltage
L_1 to L_2 = 240 volts	Phase winding A = 240 volts	Neutral to L_1 = 120 volts
L_2 to L_3 = 240 volts	Phase winding B = 240 volts	Neutral to L_2 = 208 volts
L_3 to L_1 = 240 volts	Phase winding C = 240 volts	Neutral to L_3 = 120 volts

12–3 DELTA HIGH-LEG

The term high-leg (bastard leg, identified with an orange cover [*Seciton 384-3(e)*]) is used to identify the conductor that has a higher voltage to ground. The high-leg voltage is the vector sum of the voltage of transformer "A" and "C_1", or transformers "B" and "C_2," which equals 120 volts × 1.732 = 208 volts for a 120/240 volt secondary.

Note: The actual voltage is often less than the nominal system voltage because of voltage drop (Figure 12–12).

❑ **High-Leg Voltage**

What is the high-leg voltage if the secondary voltage is 230/115 volts?

(a) 115 volts **(b) 230 volts** **(c) 199 volts** **(d) none of these**

• Answer: (c) 199 volts

115 volts × 1.732 = 199.18 volts

12–4 DELTA PRIMARY AND SECONDARY LINE CURRENTS

In a delta configured transformer, the line current does not equal the phase current.

The primary or secondary line current of a transformer can be calculated by the formula:

$$I_{Line} = \frac{VA_{Line}}{E_{Line} \times \sqrt{3}}$$

❑ **Primary Line Current**

What is the primary line current for a 150 kVA, 480- to 240/120-volt, 3-phase transformer (Figure 12–13, Part A)?

(a) 416 ampere (b) 360 ampere
(c) 180 ampere (d) 144 ampere

- Answer: (c) 180 ampere

$I_{Line} = VA_{Line}/(E_{Line} \times \sqrt{3})$

I_{Line} = 150,000 VA/(480 volts × 1.732) = 180 ampere

❑ **Secondary Line Current**

What is the secondary line current for a 150 kVA, 480- to 240-volt, 3-phase transformer (Figure 12–13, Part B)?

(a) 416 ampere (b) 360 ampere (c) 180 ampere (d) 144 ampere

- Answer: (b) 360 ampere

$I_{Line} = VA_{Line}/(E_{Line} \times \sqrt{3})$

I_{Line} = 150,000 VA/(240 volts × 1.732) = 360 ampere

12–5 DELTA PRIMARY OR SECONDARY PHASE CURRENTS

The phase current of a transformer winding is calculated by dividing the phase VA by the phase volts:

$$I_{Phase} = VA_{Phase}/E_{Phase}$$

The phase load of a three-phase, 240-volt load = line load/3.
The phase load of a single-phase, 240-volt load = line load.
The phase load of a single-phase, 120-volt load = line load.

❑ **Primary Phase Current**

What is the primary phase current for a 150 kVA, 480 to 240/120 volt, 3-phase transformer (Figure 12–14, Part A)?

(a) 416 ampere (b) 360 ampere (c) 180 ampere (d) 104 ampere

- Answer: (d) 104 ampere

$$VA_{Phase} = \frac{150{,}000 \text{ VA}}{3 \text{ Phases}} = 50{,}000 \text{ VA}$$

$$I_{Phase} = \frac{VA_{Phase}}{E_{Phase}}$$

I_{Phase} = 50,000 VA/480 volts = 104 ampere

❑ **Secondary Phase Current**

What is the secondary phase current for a 150 kVA, 480- to 240/120-volt, 3-phase transformer (Figure 12–14, Part B)?

(a) 416 ampere (b) 360 ampere
(c) 208 ampere (d) 104 ampere

- Answer: (c) 208 ampere

Phase power = 150,000 VA/3 = 50,000 VA

$$I_{Phase} = \frac{VA_{Phase}}{E_{Phase}} = 50{,}000 \text{ VA/240 volts}$$

I_{Phase} = 208 ampere

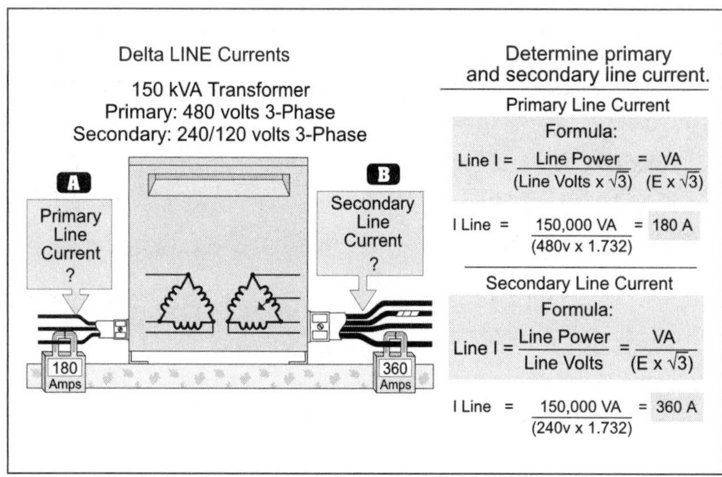

Figure 12–13
Delta Line Currents

12–6 DELTA PHASE VERSUS LINE

Since each line conductor from a delta transformer is actually connected to two transformer windings (phases), the effects of loading on the line (conductors) can be different than on the phase (winding).

❑ **Phase VA (three-phase)**

A 36 kVA, 240-volt, 3-phase load has the following effect on a delta system (Figure 12–15):

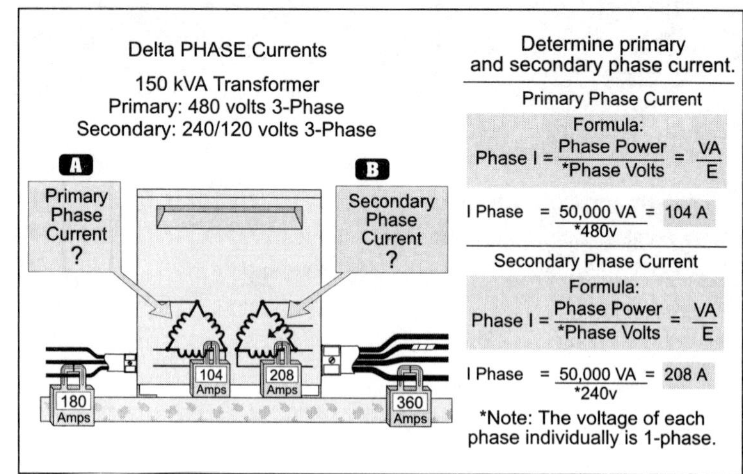

Figure 12–14
Delta Phase Currents

LINE: Total line power = 36 kVA

$I_{Line} = VA_{Line}/(E_{Line} \times \sqrt{3})$

$I_{Line} = 36{,}000\ VA/(240\ volts \times \sqrt{3})$

$I_{Line} = 87$ ampere, or

$I_{Line} = I_{Phase} \times \sqrt{3}$

$I_{Line} = 50\ ampere \times 1.732 = 87$ ampere

PHASE: Phase power = 12 kVA (winding)

$I_{Phase} = VA_{Phase}/E_{Phase}$

$I_{Phase} = 12{,}000\ VA/240\ volts = 50$ ampere, or

$I_{Phase} = I_{Line}/\sqrt{3}$

$I_{Phase} = 87\ ampere/1.732 = 50$ ampere

❑ **Phase VA (single-phase) 240 Volt**

A 10 kVA, 240-volt, single-phase load has the following effect on a delta/delta system (Figure 12–16):

LINE: Total line power = 10 kVA

$I_{Line} = VA_{Line}/E_{Line}$

$I_{Line} = 10{,}000\ VA/240\ volts = 42$ ampere

PHASE: Phase power = 10 kVA (winding)

$I_{Phase} = VA_{Phase}/E_{Phase}$

$I_{Phase} = 10{,}000\ VA/240\ volts = 42$ ampere

❑ **Phase VA (single-phase) 120 Volt**

A 3 kVA, 120-volt, single-phase load has the following effect on the system (Figure 12–17):

LINE: Line power = 3 kVA

$I_{Line} = VA_{Line}/E_{Line}$

$I_{Line} = 3{,}000\ VA/120\ volts = 25$ ampere

PHASE: Phase power = 3 kVA (C_1 or C_2 winding)

$I_{Phase} = VA_{Phase}/E_{Phase}$

$I_{Phase} = 3{,}000\ VA/120\ volts = 25$ ampere

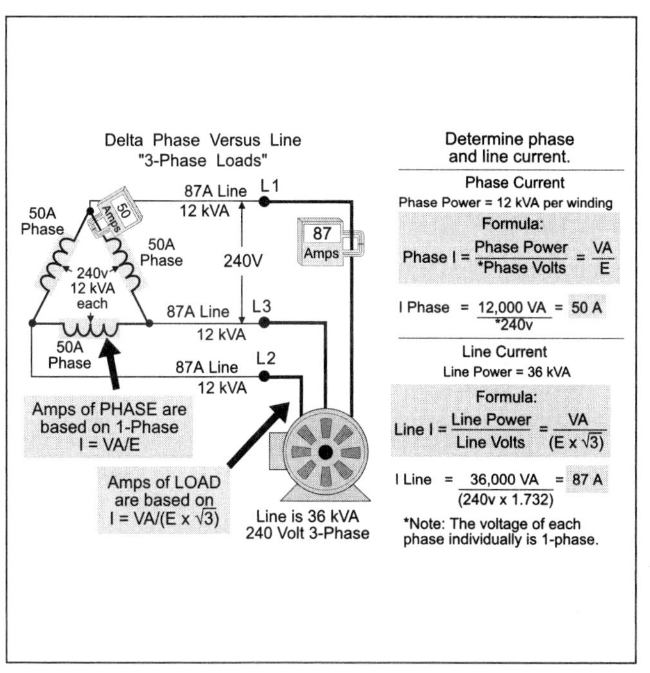

Figure 12–15
Delta Phase versus Line Loads (3-Phase)

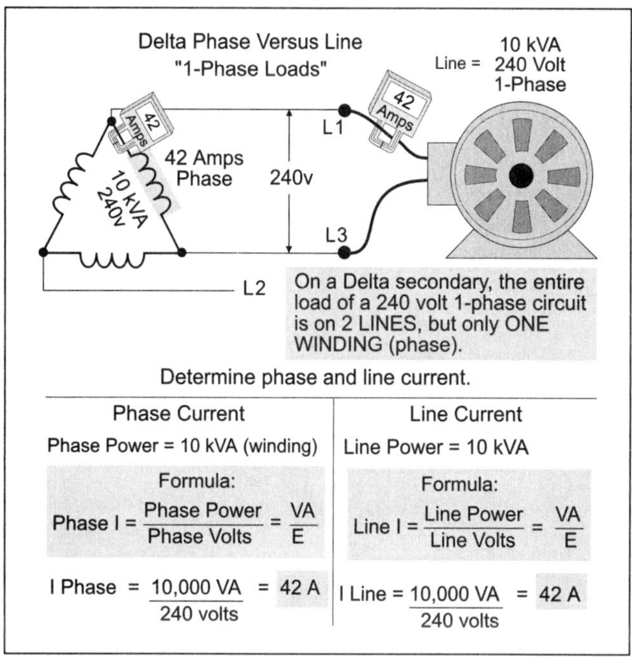

Figure 12–16
Delta Phase versus Line Load (1-Phase)

12–7 DELTA CURRENT TRIANGLE

The 3-phase line and phase current of a delta system are not equal; the difference is the square root of 3 ($\sqrt{3}$).

$$\mathbf{I_{Line} = I_{Phase} \times \sqrt{3} \qquad I_{Phase} = I_{Line} / \sqrt{3}}$$

The delta triangle can be used to calculate delta 3-phase line and phase currents. Place your finger over the desired item and the remaining items show the formula to use (Figure 12–18).

12–8 DELTA TRANSFORMER BALANCING

To properly size a delta/delta transformer, the transformer phases (windings) must be balanced. The following steps are used to balance the transformer:

Step 1: ➛ Determine the VA rating of all loads.

Step 2: ➛ Balance 3-phase loads: $^1/_3$ on Phase A, $^1/_3$ on Phase B, and $^1/_3$ on Phase C.

Step 3: ➛ Balance single-phase, 240-volt loads (largest to smallest): 100 percent on Phase A or 100 percent on Phase B. It is permissible to place some 240-volt, single-phase load on Phase C when necessary for balance.

Step 4: ➛ Balance the 120-volt loads (largest to smallest): 100 percent on C1 or 100 percent on C2.

❑ Delta Transformer Balancing

Balance and size a 480 to 240/120 volt 3-phase transformer for the following loads: One 36 kVA 3-phase heat strip, two 10 kVA single-phase 240-volt loads, and three 3 kVA 120-volt loads (Figure 12–19).

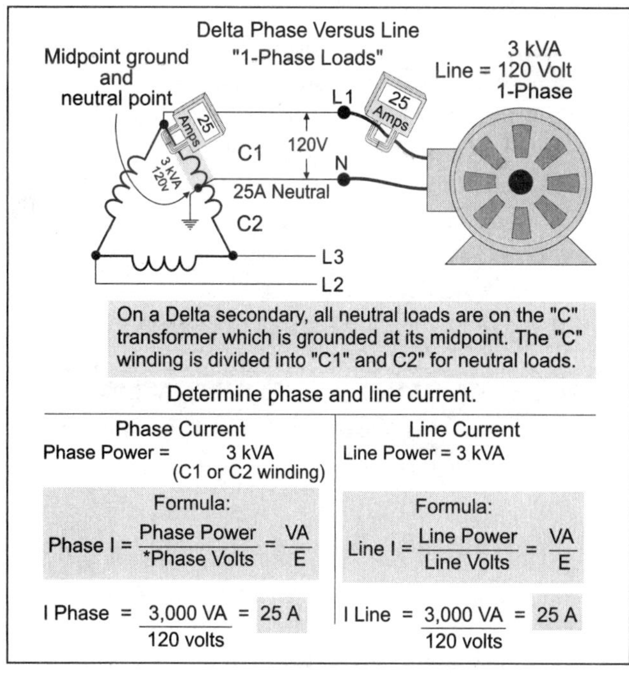

Figure 12–17
Delta Phase versus Line Load (1-Phase)

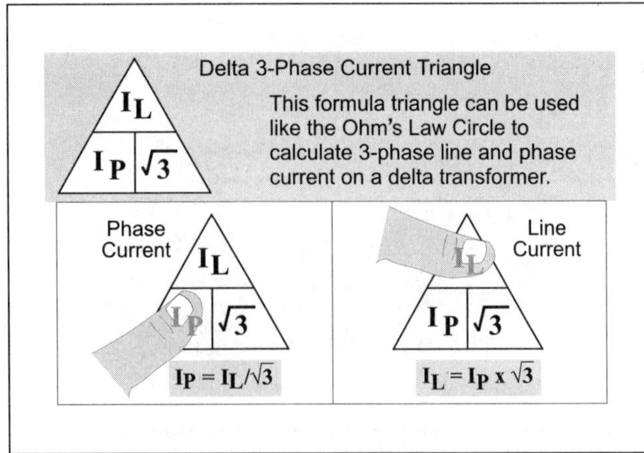

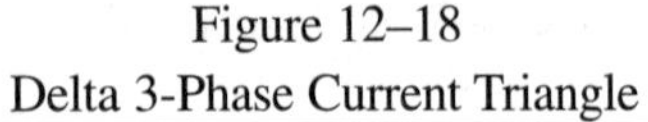

Figure 12–18
Delta 3-Phase Current Triangle

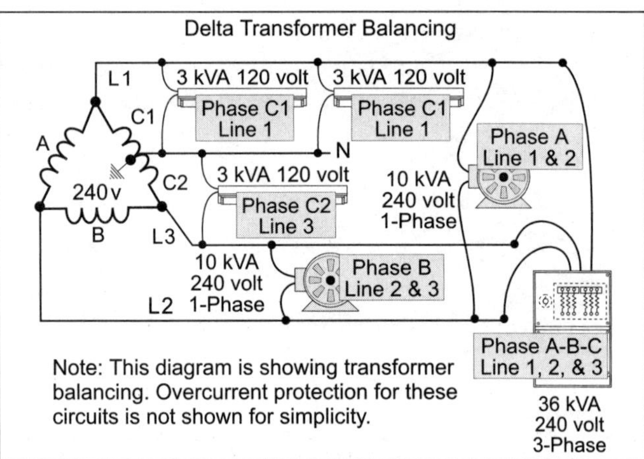

Figure 12–19
Delta Transformer Balancing

Note: For convenience of formatting, the symbol "Ø" is used to represent the word "phase."

	Phase A (L_1 and L_2)	Phase B (L_2 and L_3)	C_1 (L_1)	C_2 (L_3)	Line Total
One 36 kVA, 240 volt, 3Ø	12 kVA	12 kVA	6 kVA	6 kVA	36,000 VA
Two 10 kVA, 240 volt, 1Ø	10 kVA	10 kVA			20,000 VA
Three 3 kVA, 120 volt, 1Ø			6 kVA *	3 kVA *	9,000 VA
	22 kVA	22 kVA	12 kVA	9 kVA	65,000 VA

*Indicates neutral (120-volt) loads.

12–9 DELTA TRANSFORMER SIZING

Once you balance the transformer, size the transformer according to the load of each phase. The "C" transformer must be sized using two times the highest of "C_1 " or "C_2." The "C" transformer is actually a single unit. If one side has a larger load, that side determines the transformer size.

❑ Delta Transformer Sizing

What size 480- to 240/120-volt, 3-phase transformer is required for the following loads: one 36 kVA 3-phase heat strip, two 10 kVA single-phase, 240-volt loads, and three 3 kVA 120-volt loads?

(a) three single-phase 25 kVA transformers
(b) one 3-phase 75 kVA transformer
(c) a or b
(d) none of these

• Answer: (c) a or b

Phase winding A = 22 kVA
Phase winding B = 22 kVA
Phase winding C = (12 kVA of $C_1 \times 2$) = 24 kVA

12–10 DELTA PANEL SCHEDULE IN kVA

When balancing a panelboard in kVA, be sure that 3-phase loads are balanced $^1/_3$ on each line and 240-volt, single-phase loads are balanced on each line.

❑ **Delta Panel Schedule in kVA**

Balance a 240/120-volt, 3-phase panelboard for the following loads in kVA: one 36 kVA 3-phase heat strip, two 10 kVA single-phase, 240-volt loads, and three 3 kVA 120-volt loads.

	Line 1	Line 2	Line 3	Line Total
36 kVA, 240 volt, 3Ø	12 kVA	12 kVA	12 kVA	36,000 VA
10 kVA, 240 volt, 1Ø	5 kVA	5 kVA		10,000 VA
10 kVA, 240 volt, 1Ø		5 kVA	5 kVA	10,000 VA
Three 3 kVA, 120 volt, 1Ø	6 kVA *		3 kVA *	9,000 VA
	23 kVA	22 kVA	20 kVA	65,000 VA

*Indicates neutral (120-volt) loads.

12–11 DELTA PANELBOARD AND CONDUCTOR SIZING

When selecting the sizing of *panelboards and conductors*, you must balance the line loads in ampere.

❑ **Delta Conductors Sizing**

Balance a 240/120-volt, 3-phase panelboard for the following loads in ampere: one 36 kVA 3-phase heat strip, two 10 kVA single-phase, 240-volt loads, and three 3 kVA 120-volt loads.

	Line 1	Line 2	Line 3	Line Ampere
36 kVA, 240 volt, 3Ø	87 Amperes	87 Amperes	87 Amperes	36,000 VA/(240 volts × 1.732)
10 kVA, 240 volt, 1Ø	42 Amperes	42 Amperes		10,000 VA/240 volts
10 kVA, 240 volt, 1Ø		42 Amperes	42 Amperes	10,000 VA/240 volts
Three 3 kVA, 120 volt, 1Ø	50 Amperes *		25 Amperes *	3,000 VA/120 volts
	179 Amperes	171 Amperes	154 Amperes	

*Indicates neutral (120-volt) loads.

Why balance the panel in ampere? Why not take the VA per phase and divide by phase voltage?

- Answer: Line current of a 3-phase load is calculated by the formula

$I_L = VA/(E_{Line} \times \sqrt{3})$

I_L = 36,000 VA/(240 volts × 1.732) = 87 ampere per line.

Note: If you took the per line power of 12,000 VA and divided by one line voltage of 120 volts, you come up with an incorrect line current of 12,000 VA/120 volts = 100 ampere.

12–12 DELTA NEUTRAL CURRENT

The neutral current for a delta secondary is calculated as $Line_1$ neutral current less $Line_3$ neutral current.

❑ **Delta Neutral Current**

What is the neutral current for the following loads: one 36 kVA 3-phase heat strip, two 10 kVA, single-phase 240-volt loads, and three 3 kVA 120-volt loads?

In our continuous example, Line 1 neutral current = 50 ampere and Line 3 neutral current = 25 ampere.

(a) 0 ampere (b) 25 ampere (c) 50 ampere (d) none of these

- Answer: (b) 25 ampere

Neutral current = 50 ampere less 25 ampere

Neutral current = 25 ampere

	Line 1	Line 2	Line 3	Ampere Calculation
36 kVA, 240 volts, 3Ø	87 Amperes	87 Amperes	87 Amperes	36,000 VA/(240 volts × 1.732)
10 kVA, 240 volts, 1Ø	42 Amperes	42 Amperes		10,000 VA/240 volts
10 kVA, 240 volts, 1Ø		42 Amperes	42 Amperes	10,000 VA/240 volts
Three 3 kVA, 120 volts, 1Ø	50 Amperes *		25 Amperes *	3,000 VA/120 volts

*Indicates neutral (120-volt) loads.

12–13 DELTA MAXIMUM UNBALANCED LOAD

The *maximum unbalanced load* (neutral) is the largest phase 120-volt neutral current. The neutral current is the actual current on the neutral conductor.

❑ Maximum Unbalanced Load

What is the maximum unbalanced load for the following loads: one 36 kVA 3-phase heat strip, two 10 kVA single-phase, 240-volt loads, and three 3 kVA 120-volt loads?

(a) 0 ampere (b) 25 ampere (c) 50 ampere (d) none of these

- Answer: (c) 50 ampere

Maximum unbalanced current equals the line with the largest neutral current.

	Line 1	Line 2	Line 3	Ampere Calculation
36 kVA 240 volt, 3Ø	87 Amperes	87 Amperes	87 Amperes	36,000 VA/(240 volts × 1.732)
10 kVA 240 volt, 1Ø	42 Amperes	42 Amperes		10,000 VA/240 volts
10 kVA 240 volt, 1Ø		42 Amperes	42 Amperes	10,000 VA/240 volts
Three 3 kVA 120 volt, 1Ø	50 Amperes*		25 Amperes*	3,000 VA/120 volts

*Indicates neutral (120-volt) loads.

12–14 DELTA/DELTA EXAMPLE

One – Dishwasher 4.5 kW, 120v
One – A/C 10 horsepower 240 volt, 3Ø
One – Heaters 18 kW 240 volt 3Ø
Two – Ranges 14 kW, 240 volt 1Ø
Two – 3 horsepower motors, 240 volt, 1Ø
One – Water heater 10 kW, 240 volt, 1Ø
Eight – Lighting circuit 1.5 kW 120 volt

Motor VA

VA (single-phase) = Table Volts × Table Ampere, Table 430-148
3 horsepower VA = 230 volts × 17 ampere = 3,910 VA, Table 430-150
VA (3-phase) = Table Volts × Table Ampere × $\sqrt{3}$, Table 430-150
10 horsepower VA = 230 volts × 28 ampere × 1.732 = 11,154

Note: Some exam testing agencies use 240 volts instead of the table volts.

	Phase A (L_1 and L_2)	Phase B (L_2 and L_3)	C_1 (L_1)	C_2 (L_3)	Line Total
Heat 18 kW (omit A/C)	6,000 VA	6,000 VA	3,000 VA	3,000 VA	18,000 VA
Ranges 14 kW, 240 volt, 1Ø	14,000 VA	14,000 VA			28,000 VA
Water heater, 10 kW 240 volt, 1Ø	10,000 VA				10,000 VA
3 horsepower 240 volt, 1Ø		3,910 VA			3,910 VA
3 horsepower 240 volt, 1Ø		3,910 VA			3,910 VA
Dishwasher 4.5 kW, 120 volt			4,500 VA *		4,500 VA
Lighting (8 – 1.5 kW), 120 volt	______	______	4,500 VA *	7,500 VA*	12,000 VA
	30,000 VA	27,820 VA	12,000 VA	10,500 VA	80,320 VA

*Indicates neutral (120-volt) loads.

Note: Phase totals (30,000 VA, 27,820 VA, 22,500 VA) should add up to the Line total (80,320 VA). This is done as a check to make sure all items have been accounted for and added correctly.

10 horsepower VA = Table Volts × Table Ampere × 1.732
VA = 230 volts × 28 ampere × 1.732 = 11,154 VA (omit)

➛ Transformer Size

What size transformers are required?

(a) three 30 kVA single-phase transformers
(b) one 90 kVA 3-phase transformer
(c) a or b
(d) none of these

- Answer: (c) a or b

Phase A = 30 kVA
Phase B = 28 kVA
Phase C = 24 kVA (12 kVA × 2)

➵ High-Leg Voltage

What is the high-leg voltage to the ground?

(a) 120 volts (b) 208 volts (c) 240 volts (d) none of these

- Answer: (b) 208 volts

The vector sum of "A" and "C1" or transformers "B" and "C2," = 120 volts × 1.732 = 208 volts.

➵ Neutral kVA

What is the maximum kVA on the neutral?

(a) 3 kVA (b) 6 kVA (c) 7.5 kVA (d) 9 kVA

- Answer: (d) 9 kVA

If Phase C2 loads are not on, then phase C1 neutral loads would impose a 9 kVA load to neutral. (4.5 kW + 4.5 kW). This does not include the 6 kVA on transformer "C" from the heat load since there are no neutrals involved.

➵ Maximum Unbalanced Current

What is the maximum unbalanced load on the neutral?

(a) 25 ampere (b) 50 ampere (c) 75 ampere (d) 100 ampere

- Answer: (c) 75 ampere

I = P/E
I = 9,000 VA/120 volts = 75 ampere
Note: The neutral must be sized to the carry maximum unbalanced neutral current, which in this case is 75 ampere.

➵ Neutral Current

What is the current on the neutral?

(a) 0 ampere (b) 13 ampere (c) 25 ampere (d) none of these

- Answer: (b) 13 ampere

C1 = 9,000 VA/120 volts = 75 ampere
C2 = 7,500 VA/120 volts = 62.5 ampere
Neutral current = 75 ampere less 62.5 ampere = 12.5 ampere

➵ Phase VA Load (single-phase)

What is the phase load of each 3 horsepower 240-volt, single-phase motor?

(a) 1,855 VA (b) 3,910 VA (c) 978 VA (d) none of these

- Answer: (b) 3,910 VA

VA = 230 volts × 28 ampere = 3,910 VA
On a delta system, a 240-volt, single-phase load is 100 percent on one transformer.

➵ Phase VA Load (three-phase)

What is the phase load for the 3-phase, 10 horsepower motor?

(a) 11,154 VA (b) 3,718 VA (c) 5,577 VA (d) none of these

- Answer: (b) 3,718 VA

VA = Volts × Ampere × 1.732
VA = 230 volts × 28 ampere × 1.732, VA = 11,154 VA
11,154 VA/3 phases = 3,718 VA per phase

➵ High-Leg Conductor Size

If only the 3-phase 18 kW load is on the high-leg, what is the current on the high-leg conductor?

(a) 25 ampere (b) 43 ampere (c) 73 ampere (d) 97 ampere

- Answer: (b) 43 ampere

To calculate current on any conductor:
$I = P/(E \times \sqrt{3})$
I = 18,000 VA/(240 volts × 1.732) = 43 ampere

➵ **Voltage Ratio**

What is the phase voltage ratio of the transformer?

(a) 4:1 (b) 1:4 (c) 1:2 (d) 2:1

• Answer: (d) 2:1

480 primary phase volts to 240 secondary phase volts

❑ **Panel Loading – VA**

Balance the loads on the panelboard in VA.

	Line 1	Line 2	Line 3	Line Total
Heat 18 kW 240 volts, 3Ø	6,000 VA	6,000 VA	6,000 VA	18,000 VA
Ranges 14 kW 240 volts, 1Ø	7,000 VA	7,000 VA		14,000 VA
Ranges 14 kW 240 volts, 1Ø		7,000 VA	7,000 VA	14,000 VA
Water heater 240 volts, 1Ø	5,000 VA	5,000 VA		10,000 VA
3 horsepower 240 volts, 1Ø Motor		1,955 VA	1,955 VA	3,910 VA
3 horsepower 240 volts, 1Ø Motor		1,955 VA	1,955 VA	3,910 VA
Dishwasher 1Ø 120V	4,500 VA*			4,500 VA
Lighting (8 circuits) 1Ø 120V	4,500 VA*	______	7,500 VA *	4,500 VA
	27,000 VA	28,910 VA	24,410 VA	80,320 VA

* Indicates neutral (120-volt) loads.

❑ **Panel Sizing – Ampere**

Balance the loads from the previous example on the panelboard in ampere.

	Line 1	Line 2	Line 3	Ampere Calculation
Heat 18 kW 240 volts, 3Ø	43 amp	43 amp	43 amp	18,000 VA/(240 volts × 1.732)
Ranges 14 kW 240 volts, 1Ø	58 amp	58 amp		14,000 VA/240 volts
Ranges 14 kW 240 volts, 1Ø		58 amp	58 amp	14,000 VA/240 volts
Water heater 240 volts, 1Ø	42 amp	42 amp		10,000 VA/240 volts
3 horsepower 240 volts, 1Ø Motor		17 amp	17 amp	FLC
3 horsepower 240 volts, 1Ø Motor		17 amp	17 amp	FLC
Dishwasher 1Ø 120 volts	38 amp*			4,500 VA/120 volts
Lighting (8 circuits) 1Ø 120 volts	38 amp*	______	63 amp*	7,500 VA/120 volts
	219 amp	235 ampere	198 amp	

*Indicates neutral (120-volt) loads.

PART B – DELTA/WYE TRANSFORMERS

12–15 WYE TRANSFORMER VOLTAGE

In a wye configured transformer, the line voltage does not equal the phase voltage (Figure 12–20).

Reminder: In a delta configured transformer, line current does not equal phase current.

Primary Voltage (Delta)

LINE Voltage	PHASE Voltage
L_1 to L_2 = 480 volts	Phase Winding A = 480 volts
L_2 to L_3 = 480 volts	Phase Winding B = 480 volts
L_3 to L_1 = 480 volts	Phase Winding C = 480 volts

Secondary Voltage (Wye)

LINE Voltage	PHASE Voltage	NEUTRAL Voltage
L_1 to L_2 = 208 volts	Phase Winding A = 120 volts	Neutral to L_1 = 120 volts
L_2 to L_3 = 208 volts	Phase Winding B = 120 volts	Neutral to L_2 = 120 volts
L_3 to L_1 = 208 volts	Phase Winding C = 120 volts	Neutral to L_3 = 120 volts

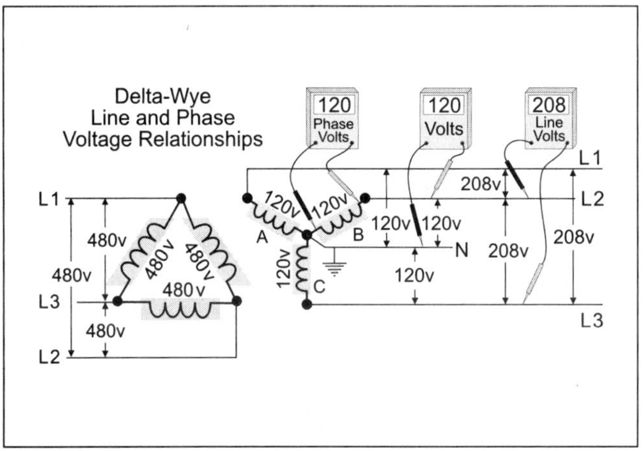

Figure 12–20
Delta-Wye Line and Phase Voltage Relationships

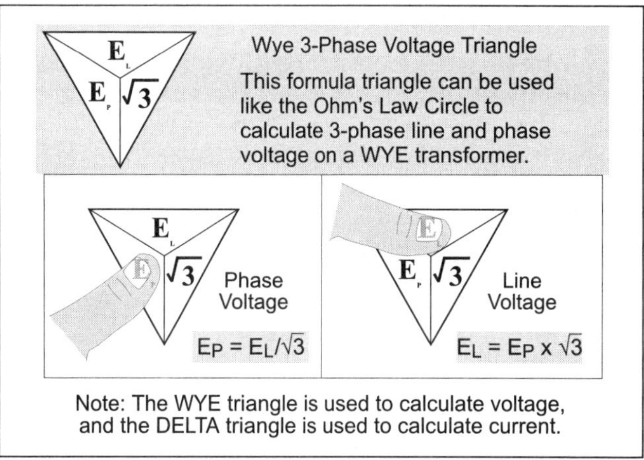

Figure 12–21
Wye 3-Phase Voltage Triangle

12–16 WYE LINE AND PHASE VOLTAGE TRIANGLE

The line voltage and phase voltage of a wye system are not the same. They differ by a factor of $\sqrt{3}$ (1.732).

Line Voltage $I_{Line} = E_{Phase} \times \sqrt{3}$ **Phase Voltage $I_{Phase} = E_{Line}/\sqrt{3}$**

The wye voltage triangle (Figure 12–21) can be used to calculate wye 3-phase line and phase voltage. Place your finger over the desired item, and the remaining items show the formula to use. These formulas can be used when either one of the phase or line voltages is known.

12–17 WYE TRANSFORMERS CURRENT

In a wye configured transformer, the 3-phase and single-phase 120-volt line current equals the phase current ($I_{Phase} = I_{Line}$) (Figure 12–22).

12–18 WYE LINE CURRENT

The line current of both delta and wye transformers can be calculated by the following formula:

$I_{Line} = VA_{Line}/(E_{Line} \times \sqrt{3})$

❑ **Primary Line Current Question**

What is the primary line current for a 150 kVA, 480- to 208Y/120-volt, 3-phase transformer (Figure 12–23, Part A)?

(a) 416 ampere (b) 360 ampere
(c) 180 ampere (d) 144 ampere

• Answer: (c) 180 ampere

$I_{Line} = VA_{Line}/(E_{Line} \times \sqrt{3})$ = 150,000 VA/(480 volts × 1.732) = 180 ampere

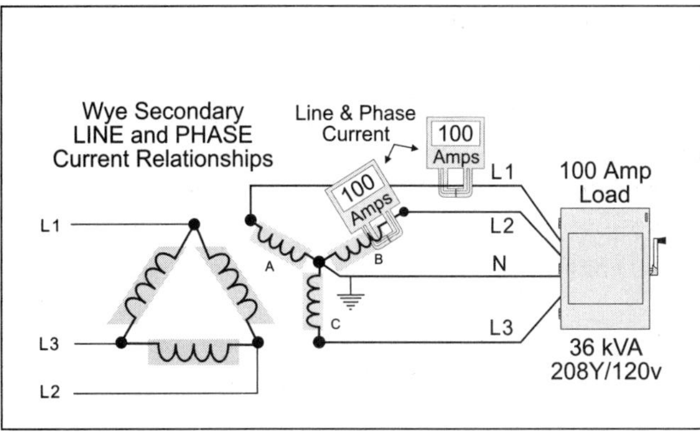

Figure 12–22
Wye Secondary Line and Phase Current Relationships

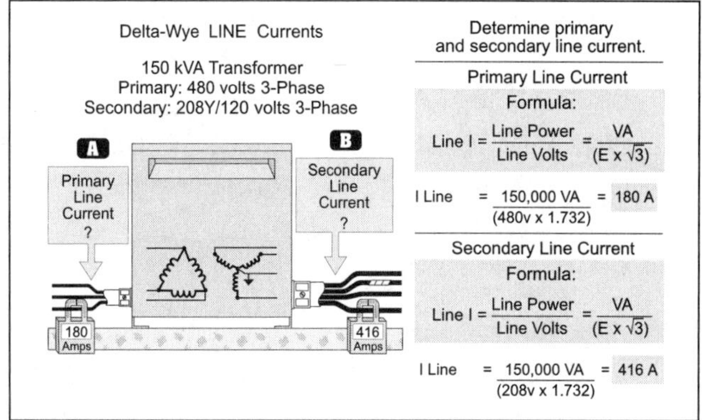

Figure 12–23
Delta-Wye Line Currents

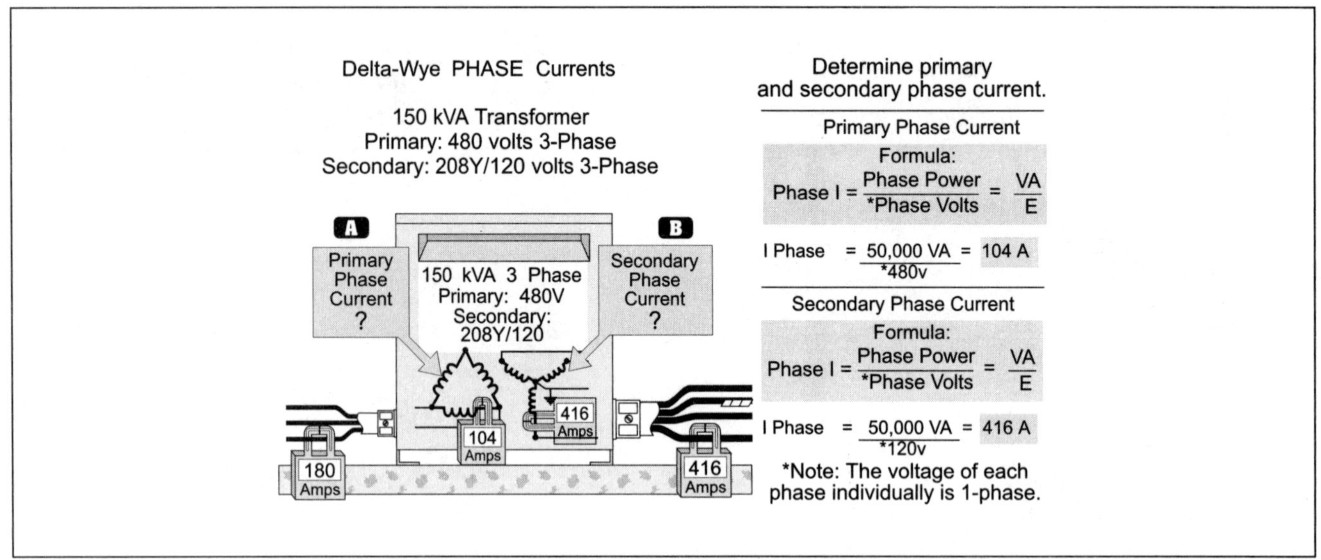

Figure 12–24
Delta-Wye Phase Currents

❑ **Secondary Line Current Question**

What is the secondary line current for a 150 kVA, 480- to 208Y/120-volt, 3-phase transformer (Figure 12–23, Part B)?

(a) 416 ampere (b) 360 ampere (c) 180 ampere (d) 144 ampere

- Answer: (a) 416 ampere

$I_{Line} = VA_{Line}/(E_{Line} \times \sqrt{3})$ = 150,000 VA/(208 volts × 1.732) = 416 ampere

12–19 WYE PHASE CURRENT

The phase current of a wye configured transformer winding is the same as the line current. The phase current of a delta configured transformer winding is less than the line current by $\sqrt{3}$ (1.732) .

❑ **Primary Phase Current Question**

What is the primary phase current for a 150 kVA, 480- to 208Y/120-volt, 3-phase transformer (Figure 12–24, Part A)?

(a) 416 ampere (b) 360 ampere (c) 180 ampere (d) 104 ampere

- Answer: (d) 104 ampere

$I_{Line} = \frac{VA_{Line}}{E_{Line} \times \sqrt{3}}$ = 150,000 VA/(480 volts × 1.732) = 180 ampere

$I_{Phase} = \frac{VA_{Phase}}{E_{Phase}}$ = 50,000 VA/480 volts = 104 ampere, or $I_{Phase} = I_{Line}/\sqrt{3}$ = 180 ampere/1.732 = 104 ampere

❑ **Secondary Phase Current Question**

What is the secondary phase current for a 150 kVA, 480- to 208Y/120-volt, 3-phase transformer (Figure 12-24, Part B)?

(a) 416 ampere (b) 360 ampere (c) 208 ampere (d) 104 ampere

- Answer: (a) 416 ampere

$I_{Line} = \frac{VA_{Line}}{E_{Line} \times \sqrt{3}}$ = 150,000 VA/(208 volts × 1.732) = 416 ampere, or

$I_{Phase} = \frac{VA_{Phase}}{E_{Phase}}$ = 50,000 VA/120 volts = 416 ampere

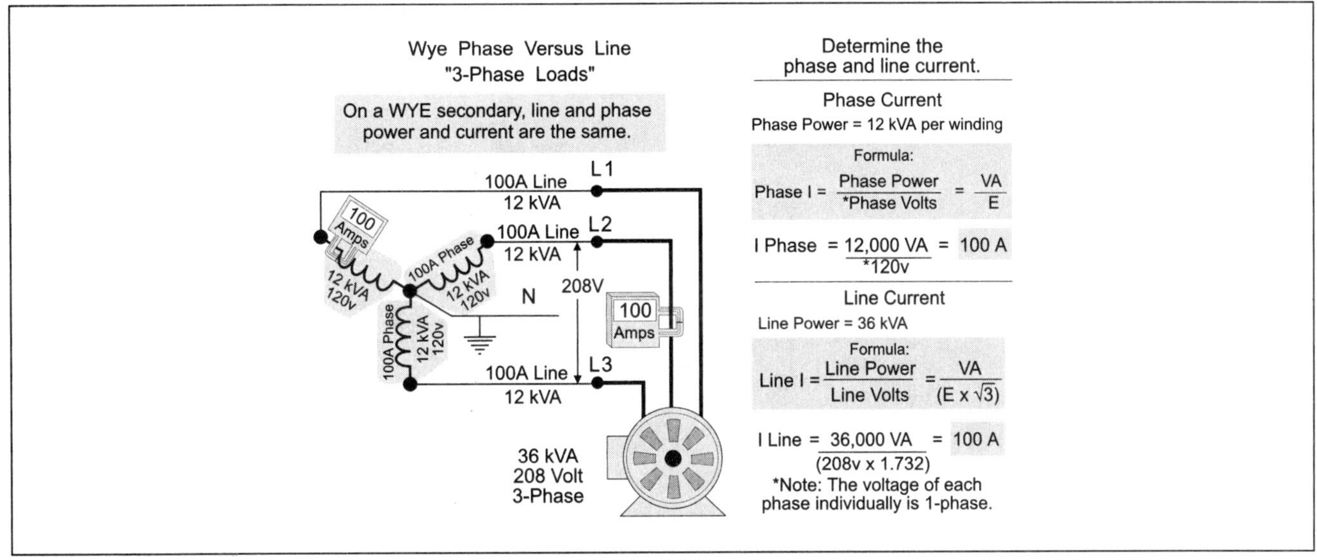

Figure 12–25
Wye Phase versus Line Load (3-Phase)

12–20 WYE PHASE VERSUS LINE CURRENT

Since each line conductor from a wye transformer is connected to a different transformer winding (phase), the effects of 3-phase loading on the line are the same as the phase (Figure 12–25).

❑ **Phase VA Load (three-phase)**

A 36 kVA, 208-volt, 3-phase load has the following effect on the system:

LINE: Line power = 36 kVA

$I_{Line} = VA_{Line}/(E_{Line} \times \sqrt{3})$

$I_{Line} = 36{,}000 \text{ VA}/(208 \text{ volts} \times \sqrt{3}) = 100 \text{ ampere}$

PHASE: Phase power = 12 kVA (any winding)

$I_{Phase} = VA_{Phase}/E_{Phase}$

$I_{Phase} = 12{,}000 \text{ VA}/120 \text{ volts} = 100 \text{ ampere}$

❑ **Phase VA Load (single-phase) 208 Volt**

A 10 kVA, 208-volt, single-phase load has the following effect on the system (Figure 12–26):

LINE: Line power = 10 kVA

$I_{Line} = VA_{Line}/E_{Line}$

$I_{Line} = 10{,}000 \text{ VA}/208 \text{ volts} = 48 \text{ ampere}$

PHASE: Phase power = 10 kVA (winding)

$I_{Phase} = VA_{Phase}/E_{Phase}$

$I_{Phase} = 5{,}000 \text{ VA}/120 \text{ volts} = 42 \text{ ampere}$

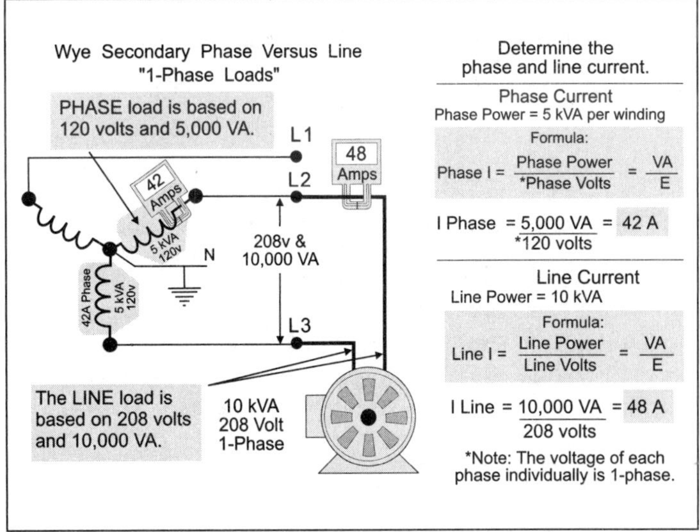

Figure 12–26
Wye Phase versus Line Load (1-Phase)

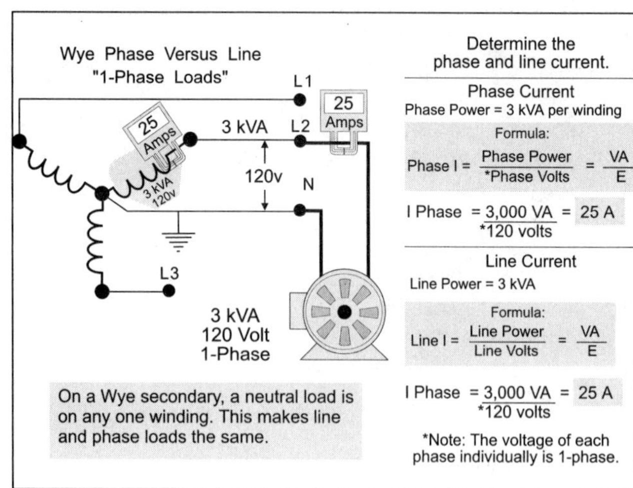

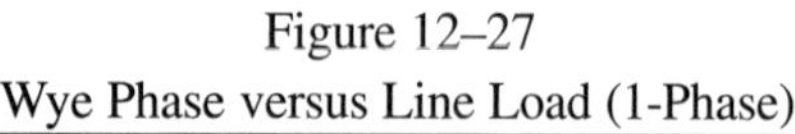

Figure 12–27
Wye Phase versus Line Load (1-Phase)

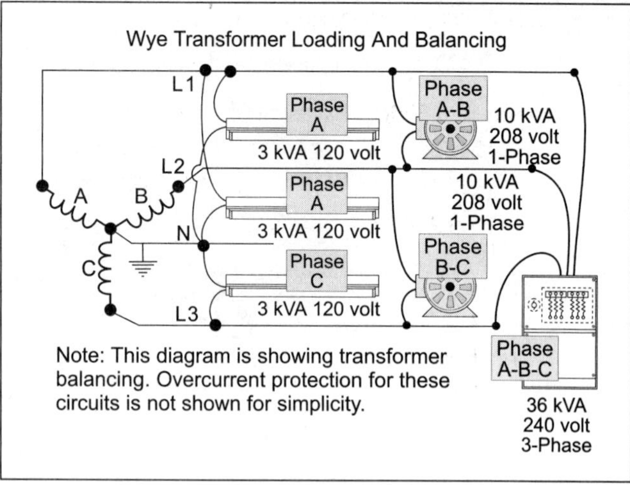

Figure 12–28
Wye Transformer Loading and Balancing

❑ Phase VA Load (single-phase) 120 Volt

A 3 kVA, 120-volt, single-phase load has the following effect on the system (Figure 12–27):

LINE: Line power = 3 kVA

$I_{Line} = VA_{Line}/E_{Line}$

I_{Line} = 3,000 VA/120 volts = 25 ampere

PHASE: Phase power = 3 kVA (any winding)

$I_{Phase} = VA_{Phase}/E_{Phase}$

I_{Phase} = 3,000 VA/120 volts = 25 ampere

12–21 WYE TRANSFORMER LOADING AND BALANCING

To properly size a delta/wye transformer, the secondary transformer phases (windings) or the line conductors must be balanced. Balancing the panel (line conductors) is identical to balancing the transformer for wye configured transformers. The following steps should be helpful to balance the transformer:

Step 1: ➛ Determine the VA rating of all loads.

Step 2: ➛ Balance 3-phase loads: $^1/_3$ on Phase A, $^1/_3$ on Phase B, and $^1/_3$ on Phase C.

Step 3: ➛ Balance single-phase, 208-volt loads (largest to smallest): 50 percent on each phase (A to B, B to C, or A to C).

Step 4: ➛ Balance the 120-volt loads (largest to smallest): 100 percent on any phase.

❑ Transformer Loading and Balancing

Balance and size a 480- to 208Y/120-volt, 3-phase phase transformer for the following loads: One 36 kVA, 3-phase heat strip, two 10 kVA, single-phase, 208-volt loads, and three 3 kVA, 120-volt loads (Figure 12–28).

	Phase A (L_1)	Phase B (L_2)	Phase C (L_3)	Line Total
36 kVA, 208 volt, 3Ø	12 kVA	12 kVA	12 kVA	36 kVA
10 kVA, 208 volt, 1Ø	5 kVA	5 kVA		10 kVA
10 kVA, 208 volt, 1Ø		5 kVA	5 kVA	10 kVA
Three 3 kVA, 120 volt, 1Ø	6 kVA *		3 kVA *	9 kVA
	23 kVA	22 kVA	20 kVA	65 kVA

* Indicates neutral (120-volt) loads.

12–22 WYE TRANSFORMER SIZING

Once you balance the transformer, you can size the transformer according to the load of each phase.

❑ Transformer Sizing Example

What size 480- to 208Y/120-volt, 3-phase, phase transformer is required for the following loads: one 36 kVA, 3-phase heat strip, two 10 kVA single-phase, 208-volt loads, and three 3 kVA 120-volt loads?

(a) three single-phase 25 kVA transformers
(b) one 3-phase 75 kVA transformer
(c) a or b
(d) none of these

• Answer: (c) a or b
Phase A = 23 kVA
Phase B = 22 kVA
Phase C = 20 kVA

	Phase A (L_1)	Phase B (L_2)	Phase C (L_3)	Line Total
36 kVA, 208 volt, 3Ø	12 kVA	12 kVA	12 kVA	36 kVA
10 kVA, 208 volt, 1Ø	5 kVA	5 kVA		10 kVA
10 kVA, 208 volt, 1Ø		5 kVA	5 kVA	10 kVA
Three 3 kVA, 120 volt, 1Ø	6 kVA *		3 kVA *	9 kVA
	23 kVA	22 kVA	20 kVA	65 kVA

* Indicates neutral (120-volt) loads.

12–23 WYE PANEL SCHEDULE IN kVA

When balancing a panelboard in kVA, be sure that 3-phase loads are balanced $^1/_3$ on each line and 208-volt single-phase loads are balanced $^1/_2$ on each line.

❑ Panel Schedule, kVA

Balance and size a 208Y/120-volt, 3-phase phase panelboard in kVA for the following loads: one 36 kVA, 3-phase heat strip, two 10 kVA, single-phase, 208-volt loads, and three 3 kVA, 120-volt loads.

	Line 1	Line 2	Line 3	Line Total
36 kVA, 208 volt, 3Ø	12 kVA	12 kVA	12 kVA	36,000 VA
10 kVA, 208 volt, 1Ø	5 kVA	5 kVA		10,000 VA
10 kVA, 208 volt, 1Ø		5 kVA	5 kVA	10,000 VA
Two 3 kVA, 120 volt, 1Ø	6 kVA *			6,000 VA
One 3 kVA, 120 volt, 1Ø			3 kVA *	3,000 VA
	23 kVA	22 kVA	20 kVA	65,000 VA

* Indicates neutral (120-volt) loads.

12–24 WYE PANELBOARD AND CONDUCTOR SIZING

When selecting and sizing the panelboard and conductors, we must balance the line loads in ampere.

❑ Panelboard Conductors Sizing

Balance and size a 208Y/120-volt, 3-phase panelboard in ampere for the following loads: one 36 kVA, 3-phase heat strip, two 10 kVA, single-phase, 208-volt loads, and three 3 kVA, 120-volt loads.

	Line 1	Line 2	Line 3	Ampere Calculation
36 kVA, 208 volt, 3Ø	100 amp	100 amp	100 amp	36,000/(208 volts × 1.732)
10 kVA, 208 volt, 1Ø	48 amp	48 amp		10,000 VA/208 volts
10 kVA, 208 volt, 1Ø		48 amp	48 amp	10,000 VA/208 volts
Two 3 kVA, 120 volt, 1Ø	50 amp*			6,000 VA/120 volts
One 3 kVA, 120 volt, 1Ø	______	______	25 amp*	3,000 VA/120 volts
	198 amp	196 amp	173 amp	

*Indicates neutral (120-volt) loads.

12–25 WYE NEUTRAL CURRENT

To determine the neutral current of a wye system, we use the following formula:

$$I_{Neutral} = \sqrt{(L_1^2 + L_2^2 + L_3^2) - (L_1 \times L_2 + L_2 \times L_3 + L_1 \times L_3)}$$

❑ Neutral Current

Balance and size the neutral current for the following loads: one 36 kVA, 3-phase heat strip, two 10 kVA, single-phase, 208-volt loads, and three 3 kVA, 120-volt loads. L_1 = 50 ampere, L_2 = 0 ampere, L_3 = 25 ampere.

(a) 0 ampere (b) 25 ampere (c) 43 ampere (d) 50 ampere

• Answer: (c) 43 ampere

Based on the previous example.

$$I_{Neutral} = \sqrt{(L_1^2 + L_2^2 + L_3^2) - (L_1 \times L_2 + L_2 \times L_3 + L_1 \times L_3)}$$

$$I_{Neutral} = \sqrt{(50^2 + 0^2 + 25^2) - [(50 \times 0) + 0 \times 25)\ (50 \times 25)]} = \sqrt{(2{,}500 + 625) - 1{,}250} = \sqrt{1{,}875} = 43.3 \text{ ampere}$$

12–26 WYE MAXIMUM UNBALANCED LOAD

The maximum unbalanced load (neutral) is the largest phase 120-volt neutral current with other lines off.

❑ Maximum Unbalanced Load

Balance and size the maximum unbalanced load for the following loads: one 36 kVA, 3-phase heat strip, two 10 kVA, single-phase, 208-volt loads, and three 3 kVA, 120-volt loads. L1 = 50 ampere, L2 = 0 ampere, L3 = 25 ampere.

(a) 0 ampere (b) 25 ampere (c) 50 ampere (d) none of these

• Answer: (c) 50 ampere

The maximum unbalanced current equals the largest line neutral current.

	Line 1	Line 2	Line 3	Ampere Calculation
36 kVA, 208 volt, 3Ø	100 amp	100 amp	100 amp	36,000/(208 volts × 1.732)
10 kVA, 208 volt, 1Ø	48 amp	48 amp		10,000 VA/208 volts
10 kVA, 208 volt, 1Ø		48 amp	48 amp	10,000 VA/208 volts
Two 3 kVA, 120 volt, 1Ø	50 amp*			6,000 VA/120 volts
One 3 kVA, 120 volt, 1Ø			25 amp*	3,000 VA/120 volts
	198 amp	196 amp	173 amp	

* Indicates neutral (120-volt) loads.

12–27 DELTA/WYE EXAMPLE

Dishwasher (4.5 kW) 120 volts
A/C motor (10 horsepower) 208 volts, 3Ø
Heat (18 kW) 208 volts, 3Ø Eight – Lighting circuits (1.5 kW), 120 volts
Two – Ranges (14 kW) 208 volts, 1Ø
Two – 3 horsepower motors, 208 volts, 1Ø
Water heater (10 kW) 208 volts, 1Ø

	Phase A (L_1)	Phase B (L_2)	Phase C (L_3)	Line Total
Heat (A/C omitted *)	6,000 VA	6,000 VA	6,000 VA	18.0 kVA
Ranges 14 kW 208 volt, 1Ø	7,000 VA	7,000 VA		14.0 kVA
Ranges 14 kW 208 volt, 1Ø		7,000 VA	7,000 VA	14.0 kVA
Water Heater 10 kW 208 volt, 1Ø	5,000 VA		5,000 VA	10.0 kVA
3 horsepower 208 volt, 1Ø Motor	1,945 VA	1,945 VA		3.89 kVA
3 horsepower 208 volt, 1Ø Motor		1,945 VA	1,945 VA	3.89 kVA
Dishwasher 4.5 kW 120 volt			4,500 VA *	4.5 kVA
Lighting (8 – 1.5 kW circuits)	6,000 VA *	3,000 VA *	3,000 VA *	12.0 kVA
	25,945 VA	26,890 VA	27,445 VA	80.28 kVA

Note: The phase totals (26 kVA, 27 kVA and 27.5 kVA) should add up to the line total (80.5 kVA). This method is used as a check to make sure all items have been accounted for and added correctly.

Motor VA

VA Single-phase = Table Volts × Table Ampere
3 horsepower VA = E × I = 208 volts × 18.7 ampere = 3,890 VA
VA Per Phase 3,890/2 = 1,945 VA per phase

VA Three-phase = Table Volts × Table Ampere × $\sqrt{3}$
10 horsepower A/C, VA = Table Volts × Table Ampere × 1.732
10 horsepower A/C, VA = 208 volts × 30.8 ampere × 1.732 = 11,096 VA (omit, smaller than 18 kW heat)

➛ Transformer Sizing

What size transformers are required?

(a) three 30 kVA, single-phase transformers
(b) one 90 kVA, 3-phase transformer
(c) a or b
(d) none of these

• Answer: (c) a or b
Phase A = 26 kVA
Phase B = 27 kVA
Phase C = 28 kVA

➛ Maximum kVA on Neutral

What is the maximum kVA on the neutral?

(a) 3 kVA (b) 6 kVA (c) 7.5 kVA (d) 9 kVA

• Answer: (c) 7.5 kVA
120-volt loads only, do not count phase to phase (208-volt) loads.

➛ Maximum Neutral Current

What is the maximum current on the neutral?

(a) 50 ampere (b) 63 ampere (c) 75 ampere (d) 100 ampere

• Answer: (b) 63 ampere
I = P/E = 7,500 VA/120 volts = 63 ampere.
The neutral must be sized to carry the maximum unbalanced neutral current, which in this case is 63 ampere.

➛ Neutral Current Question

What is the current on the neutral?

(a) 25 ampere (b) 33 ampere (c) 50 ampere (d) 63 ampere

• Answer: (b) 33 ampere

$I_{Neutral} = \sqrt{L_1^2 + L_2^2 + L_3^2 - (L_1 \times L_2 + L_2 \times L_3 + L_1 \times L_3)}$

L_1 = 6,000 VA/120 volts = 50 ampere
L_2 = 3,000 VA/120 volts = 25 ampere
L_3 = 7,500 VA/120 volts = 63 ampere

$I_{Neutral} = \sqrt{(50^2 + 25^2 + 63^2) - [(50 \times 25) + (25 \times 63) + 50 \times 63)}$

$I_{Neutral} = \sqrt{(2{,}500 + 625 + 3{,}969) - (1{,}250 + 1{,}575 + 3{,}150)} = \sqrt{7{,}049 - 5{,}975} = \sqrt{1{,}119} = 33$ ampere

➛ Phase VA (three-phase)

What is the per phase load of each 3 horsepower 208-volt, single-phase motor?

(a) 1,945 VA (b) 3,910 VA (c) 978 VA (d) none of these

• Answer: (a) 1,945 VA
VA (single-phase) = Table Volts × Table Ampere
3 horsepower VA = 208 volts × 18.7 ampere = 3,890 VA
208-volt, single-phase load is on two transformer phases = 3,890 VA/2 phases = 1,945 VA

➛ Phase VA (single-phase)

What is the per phase load for the 3-phase, 10 horsepower A/C motor?

(a) 11,154 VA (b) 3,698 VA (c) 5,577 VA (d) none of these

• Answer: (b) 3,698 VA

10 horsepower VA = Table volts × Table ampere × 1.732
VA = 208 volts × 30.8 ampere × 1.732 = 11,095 VA
Three-phase load is on three phases = 11,095 VA/3 phases = 3,698 VA per phase

➵ Voltage Ratio

What is the phase voltage ratio of the transformer?

(a) 4:1 (b) 1:4 (c) 1:2 (d) 2:1

- Answer: (a) 4:1

480 primary phase volts to 120 secondary phase volts

❑ Balance Panel – VA

Balance the loads on the panelboard in VA.

	Line 1	Line 2	Line 3	Line Total
Heat 18 kW(A/C 11 kW/ omitted)	6,000 VA	6,000 VA	6,000 VA	18.0 kVA
Ranges 14 kW, 208 volts, 1Ø	7,000 VA	7,000 VA		14.0 kVA
Ranges 14 kW, 208 volts, 1Ø		7,000 VA	7,000 VA	14.0 kVA
Water heater 10 kW, 208 volts, 1Ø	5,000 VA		5,000 VA	10.0 kVA
3 horsepower 208 volts, 1Ø	1,945 VA	1,945 VA		3.9 kVA
3 horsepower 208 volts, 1Ø		1,945 VA	`,945 VA	3.9 kVA
Dishwasher 4.5 kW, 120 volt			4,500 VA *	4.5 kVA
Lighting (8 – 1.5 kW circuits)	6,000 VA *	3,000 VA *	3,000 VA *	12.0 kVA
	25,945 VA	26,890 VA	27,445 VA	80.3 kVA

*Indicates neutral (120-volt) loads.

❑ Panelboard Balancing and Sizing

Balance the previous example loads on the panelboard in ampere.

	Line 1	Line 2	Line 3	Ampere Calculations
Heat 18 kW Heat 3Ø	50.0 amp	50.0 amp	50.0 amp	18,000 VA/(208 volts × 1.732)
Range 14 kW, 208 volts, 1Ø	67.0 amp	67.0 amp		14,000 VA/208 volts
Range 14 kW, 208 volts, 1Ø		67.0 amp	67.0 amp	14,000 VA/208 volts
Water heater 10 kW, 208 volt,1Ø	48.0 amp		48.0 amp	10,000 VA/208 volts
3 horsepower 208 volts, 1Ø Motor	18.7 amp	18.7 amp		Motor FLC
3 horsepower 208 volts, 1Ø Motor		18.7 amp	18.7 amp	Motor FLC
Dishwasher 1Ø 120 volt			38.0 amp	4,500 VA/120 volts
Lighting (8 – 1.5 kW circuits)	50.0 amp	25.0 amp	25.0 amp	1,500 VA/120 volts
	233.7 amp	246.4 amp	246.7 amp	

12–28 DELTA VERSUS WYE

Wye

Phase CURRENT is the SAME as line current.

$I_{Phase} = I_{Line}$ or $I_{Line} = I_{Phase}$

Wye Phase VOLTAGE is DIFFERENT than line voltage.
Line Voltage is greater than Phase Voltage by the square root of 3 ($\sqrt{3}$).
$E_{Line} = E_{Phas} \times \sqrt{3}$ or $E_{Phase} = E_{Line} / \sqrt{3}$

Delta

Delta phase VOLTAGE is the SAME as the line voltage.
$E_{Phase} = E_{Line}$ or $E_{Line} = E_{Phase}$

Delta phase CURRENT is DIFFERENT than the line current.
Line current is greater than Phase Current by the square root of 3 (1.732).
$I_{Line} = I_{Phase} \times \sqrt{3}$ or $I_{Phase} = I_{Line} / \sqrt{3}$

Unit 12 – Delta/Delta and Delta/Wye Transformers Summary Questions

Definitions

1. Delta connected means the windings of three single-phase transformers (same rated voltage) are connected in _____ . A delta-connected transformer is represented by the greek letter delta Δ.
(a) series (b) parallel (c) series-parallel (d) a and b

2. Which system is called the high-leg system because the voltage from one conductor to ground is between 190 and 208 volts-to-ground?
(a) delta (b) wye

3. The _____ is the electrical system ungrounded conductors.
(a) load (b) line (c) phase (d) system

4. The_____ voltage is the voltage measured between any ungrounded conductors.
(a) load (b) line (c) phase (d) system

5. • Line voltage of a _____ system is greater than phase voltage and line voltage of a _____ system is the same as phase voltage.
(a) delta, delta (b) wye, wye (c) delta, wye (d) wye, delta

6. The_____ is the coil shape conductor that serve as the primary or secondary of the transformer.
(a) line (b) load (c) phase (d) none of these

7. The phase current is the current of the transformer winding. For _____ systems, the phase current is less than the line current. For _____ systems, the phase current is the same as the line current.
(a) wye, wye (b) delta, delta (c) wye, delta (d) delta, wye

8. The phase load is the load on the transformer winding. For delta systems, the phase load is _____ .
(a) three-phase, 240-volt load = line load/3
(b) single-phase, 240-volt load = line load
(c) Single-phase, 120-volt load = line load
(d) all of these

9. The phase load is the load on the transformer winding. For wye systems, the phase load is _____ .
(a) three-phase, 208-volt load = line load/3
(b) single-phase, 208-volt load = line load/2
(c) single-phase, 120-volt load = line load
(d) all of these

10. The phase voltage is the internal transformer voltage generated across any one "winding" of a transformer. For _____ systems, the phase voltage is equal to the line voltage. For _____ systems, the phase voltage is less than the line voltage.
(a) wye, wye (b) delta, delta (c) wye, delta (d) delta, wye

11. The ratio is the relationship between the number of primary winding turns compared to the number of secondary winding turns. The ratio is a comparison between the primary phase voltage and the secondary phase voltage. For typical delta/delta systems, the ratio is ____, and for typical wye systems, the ratio is _____ .
(a) 1:2, 1:4 (b) 2:1, 4:1 (c) 4:1, 2:1 (d) none of these

12. What is the turns ratio of a 480- to 208Y/120-volt transformer?
(a) 4:1 (b) 1:4 (c) 1:2 (d) 2:1

13. • The unbalanced load is the load on the secondary grounded (neutral) conductors. For _____ systems, the formula is: $I_{Neutral} = \sqrt{(L_1^2 + L_2^2 + L_3^2) - (L_1 \times L_2 + L_2 \times L_3 + L_1 \times L_3)}$
(a) delta (b) wye (c) a or b (d) none of these

14. The unbalanced (neutral) current for a _____ system can be calculated by: $I_{Neutral} = I_{Line\ 1} - I_{Line\ 3}$.
(a) delta (b) wye (c) a or b (d) none of these

15. _____ connected means a connection of three single-phase transformers to a common point (neutral), and the other end is connected to the line conductors.
(a) Delta (b) Wye (c) a or b (d) none of these

12–1 Current Flow

16. When a load is connected to the secondary of a transformer, current flows through the secondary conductor winding. The current flow in the secondary creates an electromagnetic field that opposes the primary electromagnetic field.
(a) True (b) False

17. The primary and secondary line currents are directly proportional to the ratio of the transformer.
(a) True (b) False

Part A – Delta/Delta Transformer

12–2 Delta Transformer Voltage

18. In a delta configured transformer, the line voltage equals the phase voltage.
(a) True (b) False

12–3 Delta High-Leg

19. If the secondary voltage of a delta/delta transformer is 220/110 volts, the high-leg voltage to ground (or neutral) is _____ volts.
(a) 190 (b) 196 (c) 202 (d) 208

12–4 Delta Primary and Secondary Line Currents

20. In a delta configured transformer, the line current equals the phase current.
(a) True (b) False

21. What is the primary line current for a 45 kVA, 480- to 240/120-volt, 3-phase transformer?
(a) 124 ampere (b) 108 ampere (c) 54 ampere (d) 43 ampere

22. What is the secondary line current for a 45 kVA, 480- to 240-volt, 3-phase transformer?
(a) 124 ampere (b) 108 ampere (c) 54 ampere (d) 43 ampere

12–5 Delta Primary or Secondary Phase Currents

23. The phase current of a transformer winding is calculated by dividing the phase load by the phase volts: I Phase = VAPhase/EPhase.
(a) True (b) False

24. What is the primary phase current for a 15 kVA, 480- to 240/120-volt, single-phase transformer?
(a) 124 ampere (b) 62 ampere (c) 45 ampere (d) 31 ampere

25. What is the secondary phase current for a 15 kVA, 480- to 240/120-volt, single-phase transformer?
(a) 124 ampere (b) 62 ampere (c) 45 ampere (d) 31 ampere

12–6 Delta Phase versus Line

26. A 15 kVA, 240-volt, 3-phase load has the following effect on a delta system:
    (a) I_{Line} = 15,000 VA/(240 volts × $\sqrt{3}$) = 36 ampere
    (b) I_{Phase} = 5,000 VA/240 volts = 21 ampere
    (c) $I_{Line} = I_{Phase} \times \sqrt{3}$ = 21ampere × 1.732 = 36 ampere
    (d) all of the above

27. • A 5 kVA, 240-volt, single-phase load has the following effect on a delta/delta system:
    (a) Line Power = 5 kVA (b) Phase Power = 5 kVA
    (c) I_{Phase} = 5,000 VA/240 volts = 21 ampere (d) all of the above

28. • A 2 kVA, 120-volt, single-phase load has the following effect on a delta system:
    (a) Line Power = 2 kVA (b) I_{Line} = 2,000 VA/120 volts = 17 ampere
    (c) Phase Power = 3 kVA (C1 or C2 winding) (d) a and b

12–7 Delta Current Triangle

29. The line and phase current of a delta system are not equal. The difference between line and phase current can be described by which of the following equations?
    (a) $I_{Line} = I_{Phase} \times \sqrt{3}$ (b) $I_{Phase} = I_{Line}/\sqrt{3}$
    (c) a and b (d) none of these

12–8 and 12–9 Delta Transformer Balancing and Sizing

30. To properly size a delta/delta transformer, the transformer must be balanced in kVA. Which of the following steps should be used to balance the transformer?
    (a) Balance 3-phase loads, 1/3 on Phase A, 1/3 on Phase B, and 1/3 on Phase C.
    (b) Balance single-phase, 240-volt loads, 100 percent on Phase A or Phase B.
    (c) Balance the 120-volt loads, 100 percent on Phase C1 or C2.
    (d) all of these

31. • Balance and size a 480- to 240/120-volt, 3-phase delta transformer for the following loads: One 18 kVA 3-phase heat strip, two 5 kVA, single-phase, 240-volt loads, and three 2 kVA, 120-volt loads. What size transformer is required?
    (a) three single-phase 12.5 kVA transformers
    (b) one 3-phase 37.5 kVA transformer
    (c) a or b
    (d) none of these

12–10 Delta Panel Schedule in kVA

32. When balancing a panelboard in kVA be sure that 3-phase loads are balanced 1/3 on each line and 240-volt, single-phase loads are balanced on each line.
    (a) True (b) False

33. • Balance a 120/240-volt delta 3-phase panelboard in kVA for the following loads: one 18 kVA 3-phase heat strip, two 5-kVA, single-phase, 240-volt loads, and three 2-kVA, 120-volt loads.
    (a) The largest line load is 13 kVA. (b) The smallest line load is 10 kVA.
    (c) Total line load is equal to 34 kVA. (d) all of these

12–11 Delta Panelboard and Conductor Sizing

34. When selecting and sizing panelboards and conductors, we must balance the line loads in ampere. Balance a 120/240-volt, 3-phase panelboard in ampere for the following loads: one 18 kVA, 3-phase heat strip, two 5 kVA, single-phase 240-volt loads, and three 2 kVA, 120-volt loads.
    (a) The largest line is 98 ampere. (b) The smallest line is 81 ampere.
    (c) Each line equals to 82 ampere. (d) a and b

12–12 Delta Neutral Current

35. What is the neutral current for: one 18 kVA, 3-phase heat strip, two 5 kVA, single-phase, 240-volt loads, and three 2 kVA, 120-volt loads?
(a) 17 ampere (b) 34 ampere (c) 0 ampere (d) a and b

12–13 Delta Maximum Unbalanced Load

36. • What is the maximum unbalanced load in ampere for: One 18 kVA, 3-phase heat strip, two 5 kVA, single-phase, 240-volt loads, and three 2 kVA, 120-volt loads?
(a) 25 ampere (b) 34 ampere (c) 0 ampere (d) none of these

37. What is the phase load for a 18 kVA, 3-phase heat strip?
(a) 18 kVA (b) 9 kVA (c) 6 kVA (d) 3 kVA

38. • What is the transformer winding phase load for a 5 kVA, single-phase, 240-volt load?
(a) 5 kVA (b) 2.5 kVA (c) 1.25 kVA (d) none of these

39. If only a 3-phase, 12 kW load is on the high-leg, what is the current on the high-leg conductor?
(a) 29 ampere (b) 43 ampere (c) 73 ampere (d) 97 ampere

40. What is the ratio of a 480- to 240/120-volt transformer?
(a) 4:1 (b) 1:4 (c) 1:2 (d) 2:1

Part B Delta/Wye Transformers

12–15 Wye Transformer Voltage

41. In a wye configured transformer, the line voltage equals the phase voltage.
(a) True (b) False

12–17 Wye Transformers Current

42. In a wye configured transformer, the line current equals the phase current.
(a) True (b) False

12–18 Wye Line Current

43. The line current of both delta and wye 3-phase transformers can be calculated by the formula: $I_{Line} = VA_{Line}/(E_{Line} \times \sqrt{3})$.
(a) True (b) False

44. What is the primary line current for a 22 kVA, 480- to 208Y/120-volt, 3-phase transformer?
(a) 61 ampere (b) 54 ampere (c) 26 ampere (d) 22 ampere

45. What is the secondary line current for a 22 kVA, 480- to 208Y/120-volt, 3-phase transformer?
(a) 61 ampere (b) 54 ampere (c) 27 ampere (d) 22 ampere

12–19 Wye Phase Current

46. The phase current of a wye configured transformer winding is the same as the line current. The phase current of a delta configured transformer winding is less than the line current by $\sqrt{3}$.
(a) True (b) False

47. • What is the primary phase current for a 22 kVA, 480- to 208Y/120-volt, 3-phase transformer?
(a) 15 ampere (b) 22 ampere (c) 36 ampere (d) 61 ampere

48. • What is the secondary phase current for a 22 kVA, 480- to 208Y/120-volt, 3-phase transformer?
(a) 15 ampere (b) 22 ampere (c) 36 ampere (d) 61 ampere

12–20 Wye Phase versus Line Current

49. Since each line conductor from a wye transformer is connected to a different transformer winding (phase), the effects of loading on the line are the same as the phase. A 12 kVA, 208-volt, 3-phase load has the following effect on the system:
(a) $I_{Line} = VA_{Line}/(E_{Line} \times \sqrt{3})$ = 12,000 VA/(208 volts × $\sqrt{3}$) = 33 ampere
(b) Phase Power = 4 kVA
(c) $I_{Phase} = VA_{Phase}/E_{Phase}$ = 4,000 VA/120 volts = 30 ampere
(d) all of these

50. A 5 kVA, 208-volt, single-phase load has the following effect on the system:
(a) Line Power = 5 kVA
(b) I_{Line} = VA/E = 5,000 VA/208 volts = 24 ampere
(c) $I_{Phase} = VA_{Phase}/E_{Phase}$ = 2,500 VA/120 volts = 21 ampere
(d) all of these

51. A 2 kVA, 120-volt, single-phase load has the following effect on the system:
(a) Line Power = 2 kVA
(b) I_{Line} = VA/E = 2,000 VA/120 volts = 17 ampere
(c) $I_{Phase} = VA_{Phase}/E_{Phase}$ = 2,000 VA/120 volts = 17 ampere
(d) all of these

52. • What is the phase load of a 3 horsepower 208-volt, single-phase motor?
(a) 1,945 VA (b) 3,890 VA (c) 978 VA (d) none of these

53. • What is the per phase load for a 3-phase, 7.5 horsepower 208-volt motor?
(a) 2,906 VA (b) 8,718 VA (c) 4.359 VA (d) none of these

12–21 and 12–22 Wye Transformer Balancing and Sizing

54. To properly size a delta/wye transformer, the secondary transformer phases (windings) or the line conductors must be balanced. Balancing the panel (line conductors) is identical to balancing the transformer for wye configured transformers. Which of the following steps should be used to balance the transformer?
(a) balance 3-phase loads, 1/3 on each phase
(b) balance single-phase, 208-volt loads 50 percent on each phase
(c) balance the 120-volt loads 100 percent on any phase
(d) all of these

55. • Balance and size a 480- to 208Y/120-volt, 3-phase transformer for the following loads: one 18 kVA, 3-phase heat strip, two 5 kVA, single-phase, 208-volt loads, and three 2 kVA, 120-volt loads. What size transformer is required?
(a) three single-phase 12.5 kVA transformers (b) one 3-phase 37.5 kVA transformer
(c) a or b (d) none of these

12–23 Wye Panel Schedule in kVA

56. When balancing a panelboard in kVA be sure that 3-phase loads are balanced $^1/_3$ on each line and 208-volt, single-phase loads are balanced $^1/_2$ on each line. Balance a 208Y/120-volt, 3-phase panelboard in kVA for the following loads: one 18 kVA, 3-phase heat strip, two 5 kVA, single-phase, 208-volt loads, and three 2 kVA, 120-volt loads.
(a) the largest line is 12.5 kVA (b) the smallest line is 10.5 kVA
(c) a and b (d) none of these

12–24 Wye Panelboard and Conductor Sizing

57. • When selecting and sizing panelboards and conductors, we must balance the line loads in ampere. Balance a 208Y/120-volt, 3-phase panelboard in ampere for the following loads: one 18 kVA, 208-volt, 3-phase heat strip, two 5 kVA, single-phase, 208-volt loads, and three 2 kVA, 120-volt loads.
(a) The largest line current is 108 ampere. (b) The smallest line is 91 ampere.
(c) a and b (d) none of these

12–25 Wye Neutral Current

58. To determine the neutral current for a 3-wire wye system we must use the following formula:
$I_{Neutral} = \sqrt{(L_1^2 + L_2^2 + L_3^2) - (L_1 \times L_2 + L_2 \times L_3 + L_1 \times L_3)}$
(a) True (b) False

59. • What is the neutral current for: one 18 kVA, 3-phase heat strip, two 5 kVA, single-phase, 208-volt loads and three 2 kVA, 120-volt loads?
(a) 47 ampere (b) 29 ampere (c) 34 ampere (d) a and b

12–26 Wye Maximum Unbalanced Load

60. • What is the maximum unbalanced load in ampere for: one 18 kVA, 3-phase heat strip, two 5 kVA, single-phase, 208-volt loads, and three 2 kVA, 120-volt loads.
(a) 17 ampere (b) 34 ampere (c) 0 ampere (d) a and b

12–27 Delta versus Wye

61. Which of the following statements are true for wye systems?
(a) Phase current is the same as line current: $I_{Phase} = I_{Line}$.
(b) Phase voltage is the same as the line voltage.
(c) $E_{Line} = E_{Phase} \times \sqrt{3}$
(d) a and c

62. Which of the following statements are true for delta systems?
(a) Phase voltage is the same as line voltage: $E_{Phase} = E_{Line}$.
(b) Phase current is the same as the line current.
(c) Line current is greater than phase current: $I_{Line} = I_{Phase} \times \sqrt{3}$.
(d) a and c

✯ Challenge Questions

Part A – Delta/Delta Transformers

The following statement applies to the next seven questions:

The following loads are to be connected to a 3-phase delta/delta, 480-volt to 240-volt transformer.
Heat 18 kW, 240-volt, 3-phase heat
A/C 25 horsepower 240-volt, 3-phase synchronous motor
Two motors 1 horsepower 240-volt, single-phase motors
Two 2 horsepower 120-volt motors
Three 1,900 watt, 120-volt lighting loads

Balance the above loads as closely as possible, then answer the next seven questions.

63. • Three single-phase transformers are connected in series. What size transformers are required for the above loads?
(a) 10/10/10 kVA (b) 15/15/15 kVA (c) 10/10/20 kVA (d) 15/15/10 kVA

64. • The line VA load of the 25 horsepower 3-phase synchronous motor would be _____ VA.
(a) 9,700 (b) 21,113 (c) 32,031 (d) 28,090

65. • After balancing the loads, the maximum unbalanced neutral current is _____ ampere.
(a) 27 (b) 1.5 (c) 48 (d) 148

66. • The transformer phase load of one 1 horsepower motor would be _____ VA.
(a) 1,380 (b) 2,760 (c) 2,300 (d) 1,150

67. • The line load of one of the 1 horsepower motors would be _____ VA.
(a) 1,380 (b) 2,760 (c) 2,300 (d) 1,150

68. • The transformer phase VA load of the 25 horsepower 3-phase synchronous motor is closest to _____ VA.
(a) 3,520 (b) 7,038 (c) 21,100 (d) 32,100

69. • If the only load on the high-leg is the 3-phase 25 horsepower motor, the high-leg conductor would have to be sized a minimum of _____ ampere.
(a) 13 (b) 66 (c) 18 (d) 22

The following statement applies to the next two questions:
A 3-phase 37.5 kVA delta/delta transformer has a primary of 480 volts and a secondary of 240/120 volts.

70. • The primary line current is closest to _____ ampere.
(a) 45 (b) 90 (c) 50 (d) 65

71. • The maximum overcurrent device size for the 37.5 kVA delta/delta transformer is _____ ampere.
(a) 40 (b) 45 (c) 50 (d) 60

The following statement applies to the next question:
Seven heaters 1,500 watt, 120 volt
Five 3 kW, 240-volt, single-phase heaters
Equipment circuits 30 kW, 120 volt
Lighting circuits 40 kW, 120

72. • If the only loads on the center tapped transformer (C1 and C2) are the 120-volt loads, what is the total kW load on this transformer?
(a) 30–40 kW (b) 50–60 kW (c) 80–90 kW (d) over 100 kW

73. • The total load on the center tapped 240/120-volt transformer is 100 kW, which consists of 80 kW of 120 volt balanced loads, and 20 kW of 240-volt loads. What is the maximum unbalanced neutral current?
(a) 167 ampere (b) 0 ampere (c) 192 ampere (d) 333 ampere

74. • On a delta/delta 3-phase transformer, a 30,000 VA, 240-volt, 3-phase load would have a phase current of _____ ampere on the secondary winding.
(a) 125 (b) 72 (c) 42 (d) none of these

75. • What is the secondary line current of a 3-phase, 45 kVA, delta/delta transformer that has a turn ratio of 2:1 and a primary voltage of 460?
(a) 113 ampere (b) 55 ampere (c) 36 ampere (d) cannot be determined

76. • A 2,200/220-volt, 3-phase transformer is connected delta/delta. The secondary phase current is 300 ampere and the secondary line current is 240 volts. What is the primary line current?
(a) 40 ampere (b) 20 ampere (c) 50 ampere (d) 30 ampere

77. • For a given 3-phase transformer, the connection giving the highest secondary voltage is _____ .
(a) wye primary, delta secondary (b) delta primary, delta secondary
(c) wye primary, wye secondary (d) delta primary, wye secondary

Part B – Delta/Wye Transformers

78. • A delta/wye, 480/208- 120-volt, 3-phase transformer has a secondary line current of 200 ampere. What is the primary line current?
(a) 31 ampere (b) 72 ampere (c) 87 ampere (d) 53 ampere

79. • If the secondary phase voltage of a delta/wye connected transformer is 277 volts, and a 10 kVA, 3-phase load was connected to that secondary, the line current would be _____ ampere.
(a) 7 (b) 12 (c) 32 (d) none of these

The following statement applies to the next two questions:

Motor 15 horsepower 208-volt, 3-phase
Four lighting circuits 3 kW, 120 volt
Five heaters 2.5 kW, 208 volt
Two dryers 1.5 kW, 208-volt dryers
Cooktop 8 kW, 208 volt

80. • Each 2.5 kW heat strip has a line load of _____ .
(a) 1,250 watts (b) 12,500 watts (c) 2,500 watts (d) no load

81. • The load per line of the 3-phase motor can be determined by which of these formulas?
(a) $\sqrt{3}$ × Volts × Ampere (b) (Volts × Ampere)/3
(c) Volts × Ampere (d) none of these

82. • If a 3-phase motor has a total load of 12,000 VA, and is connected to a delta/wye transformer, the load on one phase of the secondary will be _____ VA.
(a) 12,000 (b) 4,000 (c) 3,000 (d) 1,500

83. • A 3-phase 480Y/208-120-volt transformer provides power to nine 2,000 VA, 120-volt lights and nine 2,000 watt, 208-volt heaters. The maximum neutral current is _____ ampere.
(a) 50 (b) 75 (c) 87 (d) 95

84. • There are five 2,500 watt, 208-volt heat strips on the secondary of a 3-phase 208Y/120-volt transformer. What is the load per phase for each heat strip?
(a) 1,403 watts (b) 1,250 watts (c) 2,500 watts (d) 12,500 watts

85. • The phase load of the 10 horsepower 3-phase 208Y/120-volt motor is closest to _____ VA.
(a) 3,700 (b) 5,500 (c) 4,300 (d) 11,100

The following statement applies to the next question:

The system is 3-phase 208Y/120-volt and the loads are:
Nine – 2 kVA, 120-volt lighting loads
Five – 3 kVA, 120-volt loads

86. • The neutral conductor must be size to carry _____ ampere.
(a) 10 (b) 25 (c) 90 (d) 100

Index

I

J

K

L

M

N

O

P

R

S

T

V

W